Current Trends in
Chemical Engineering

Current Trends in Chemical Engineering

J.M.P.Q. Delgado
Building Physics Laboratory - LFC
Department of Civil Engineering
Faculty of Engineering of University of Porto
4200 465 Porto, Portugal

2010

STUDIUM PRESS LLC
P.O. Box 722200, Houston, Texas 77072, USA
Tel. (P): (713) 541-9400, Fax: (713) 541-9401
e-mail: studiumpress@studiumpress.com
Website: http://www.studiumpress.in

Current Trends in Chemical Engineering

© 2010

This book contains information obtained from authentic and highly regarded sources. Reprinted material is quoted with one acknowledgement, and sources are indicated. A wide variety of references are listed. Reasonable efforts have been made to publish reliable data and information, but the editor and the publisher cannot assume responsibility for the validity of all materials or for the consequences of their use.

All rights are reserved under International and Pan-American Copyright Conventions. Apart from any fair dealing for the purpose of private study, research, criticism or review, as permitted under the Copyright Act, 1956, no part of this publication may be reproduced, stored in a retrieval system or transmitted, in any form or by any means—electronic, electrical, chemical, mechanical, optical, photocopying, recording or otherwise—without the prior permission of the copyright owner.

ISBN: 1-933699-75-2

Published by:
STUDIUM PRESS, LLC
P.O. Box-722200, Houston, Texas-77072, USA
Tel. 713-541-9400; Fax: 713-541-9401
E-mail: studiumpress@studiumpress.com

Printed at:
Thomson Press (India) Ltd.

Preface

Chemical engineering deals with the application of physical science (in particular chemistry and physics), life sciences (in particular biology, microbiology and biochemistry) and mathematics for the processing of raw materials or chemicals into more useful or valuable forms. Besides this, modern chemical engineering is also concerned with pioneering processing techniques such as nanotechnology, fuel cells and biomedical engineering. This is an important form of research and development with direct applications in biochemical processing, biomedical devices, electronic and optical material and devices, advanced materials, safety and environmental protection.

The purpose of this book is to provide recent information about important research in chemical engineering. The book is divided in eleven chapters that intend to be a short monograph in which the authors summarize the current state of knowledge for benefit of professional colleagues.

Chapter 1: Fluid dynamics is governed by a set of well established differential equations that are based on conservation laws of physics. Advances in computational physics allow simulation of two and three-dimensional complex flows through complex geometries. These simulations, judiciously employed, can offer much insight into flows through porous media. In particular, Lattice Boltzmann equation method (LBE) is very attractive to investigate flows in complex geometries such as porous media. In this chapter LBE is used to simulate flow in porous media. It is aimed, (i) to investigate the range of validity of Darcy-Forchheimer equation, (ii) to examine the macroscopic transport properties (*i.e.*, the permeability and the inertial parameter), and (iii) to understand the effect of the continuum and the slip-flow regimes on the permeability of porous media.

Chapter 2: In this chapter two approaches have been adopted to adequately simulate the transport phenomena occurred in a Solid Oxide Fuel Cell (SOFC). The first one is a one dimensional, relatively simple model, providing a rather simple and stable solution by using analytical manipulations and combined numerical techniques. The other is a three-dimensional approach including the coupled solution for flow, heat transfer, mass transport and electrochemistry.

It is found that higher fuel consumption corresponds to higher current density values in an equivalent manner. Furthermore, it is clearly depicted that the overpotential increases along the anode, resulted by the consequent increment of the fuel consumption with the distance. Finally, the overpotential increases with current density, which is in accordance with previously presented experimental and theoretical results.

Chapter 3: This chapter is a brief review on the state of affairs in studies of structure and properties of grain boundaries (GBs) in metals. It is demonstrated that GBs structure and properties can be understood only based on simulation combined with a number of various experimental techniques, such as high-resolution electron microscopy (HREM), diffusion investigations, emission Mossbauer spectroscopy, etc.

Models of structure and energy of equilibrium and non-equilibrium grain boundaries are considered, specific features of various point defects formed in GBs are analyzed, diffusion properties of GBs in coarse-grained, submicrocrystalline and nanostructured materials are described.

A new method of investigation of grain boundaries based on the use of the emission Mossbauer spectroscopy is described in detail, and its capabilities are analyzed. It is demonstrated that the results of the Mossbauer studies do not completely agree with Fisher's model of grain boundary diffusion, and a new recently suggested model which may be considered as the Fisher's model modification is described. This model is in agreement both with the results of traditional diffusion experiments and with the NGR data. The results of the Mossbauer studies of GBs corresponding to determination of mechanisms of GB diffusion, evaluation of segregation factors, estimation of an extent of non-equilibrium state of GBs, etc. are discussed.

Further goals and nearest projects in studies of GBs are considered.

Chapter 4: In this review chapter, we introduce the recently developed Lattice Monte Carlo method for addressing and solving phenomenologically-based thermal diffusion problems especially for composite and porous materials. We describe in detail the application of this method to calculate effective thermal diffusivities and to determine temperature profiles including conditions of temperature-dependent material properties and materials exhibiting a phase change. Where possible, results of the method are compared with results of exact or finite element methods. Excellent agreement is demonstrated.

Chapter 5: This chapter investigates the thermal properties of a new type of hollow sphere structures. For this new type, the sphere shell is perforated by several holes in order to open the inner sphere volume and surface. The effective thermal conductivity of perforated sphere structures in several kinds of arrangements is numerically evaluated for different geometrical parameters such as: hole diameter of the perforated hollow spheres, hollow spheres' wall thickness and joining element dimension. The results are compared to classical configurations without perforation. In addition the influence of different joining techniques, *i.e.* sintering and adhesively bonding, on the thermal conductivity has been compared. Three dimensional finite element analysis was used in order to investigate the heat conductivity of simple cubic, body-centered cubic, face-centered cubic and hexagonal unit cell models. A linear dependency behavior was found for the heat conductivity of different hole diameters for several kinds of arrangements when the results were plotted over the average density for homogeneous models.

Chapter 6: Dehydration constitutes a common and high energy demanding operation on food processing. Reduction of water concentration contributes to preserve the food, preventing the degradation during storage and transportation. In addition, dehydration reduces the transport cost and is a common previous stage in extraction processes, avoiding the interference of water and decreasing the need for solvent. Dehydration by using hot air has been and still is the most common way at industrial scale. This work reviews the main features of this drying method, as well as it also presents new trends and modern technologies.

The high importance of the relationship between water activity and moisture content is shown, not only to identify optimal storage conditions or final moisture contents on the dried product but also as a tool to be used on modeling. The identification of significant resistances to mass transfer is mandatory in order to design adequately the drying process and address accurately modeling. Drying modeling from mechanistic theories involves considering some assumptions, which will affect not only to the accuracy of the model but also to the solution method in terms of complexity. Modeling results may be used in optimization. Actually, drying optimization objectives are based on quality as well as energy aspects, although always must be kept in mind that the main goal of drying is preservation. Despite drying by forced air has been widely addressed in literature, it still present some limitations, specially a low drying rate affecting the energy efficiency. The application of new technologies as additional energy sources may be considered, in this sense, a new drying technology is introduced, the ultrasonic assisted drying.

Chapter 7: The phenomenon of dispersion, transverse or radial and longitudinal or axial, in porous media is reviewed for a great deal of information from the literature. Dispersion plays an important part, for example, in contaminant transport in ground water flows, in miscible displacement of oil and gas and in reactant and product transport in packed bed reactors.

There are several variables that must be considered, in the analysis of dispersion in porous media, like the viscosity and density of the fluid, particle size distribution, particle shape, effect of fluid velocity and effect of temperature or Schmidt number, etc.

Empirical correlations are presented for the prediction of the dispersion coefficients over the entire range of practical values of Schmidt number, Sc, and Peclet number, Pe_m.

Chapter 8: This chapter reviews synthesis methods, properties and applications of nanostructured materials. Mechanical alloying, sol-gel and CVD as some of the dominant and promising synthesis methods along with sonochemistry as a relatively novel one are presented through a concise description of the subject and illustrated by the results obtained from some of the research projects performed in our Advanced Materials and Nanotechnology Research Laboratory in K.N.T University of Technology. Phase Transients, Diffusion and mechanical properties of nanostructured materials including strength, hardness, ductility and superplasticity are discussed with the main focus laid on the structure-properties relationship. Although it was not possible to summarize all the applications of nanostructured materials, some inspiring examples of biomedical, electronic and catalytic applications are selectively described, in order to give the reader an insight into the vast known and potential applications of nanostructured materials.

Chapter 9: In this chapter a computational procedure for determining the convective coefficient of heat exchange between the porous substrate and the working fluid for a porous medium was detailed.

Macroscopically uniform laminar and turbulent flows through a periodic cell formed by square and elliptic rods were computed. Quantitative agreement was obtained when comparing laminar results herein with simulations of other authors. For turbulent flows, Low and High Reynolds turbulence models were employed in order to obtain the interfacial heat transfer coefficient. Correlations for determining the heat transfer coefficient were compared.

Chapter 10: A supercritical fluid (SCF) is any substance at a temperature and pressure above its thermodynamic critical point. It can diffuse through solids like a gas, and dissolve materials like a liquid. Additionally, close to the critical point, small changes in pressure or temperature result in great changes in density, allowing many properties to be manipulated. Supercritical fluids are suitable as a substitute for organic solvents in a variety of industrial and laboratory processes, like supercritical fluid extraction, dry cleaning, supercritical fluid chromatography, chemical reactions, nano and micro particle formation, and many others. Carbon dioxide and water are the most commonly used supercritical fluids.

Chapter 11: This chapter presents the current state of development concerning different routes for biodiesel production, focusing on their chemical and technological aspects. Their relative advantages and disadvantages are presented and discussed in detail, with a focus on their industrial implementation and exploration. The full biodiesel production chain is taken into account, with a brief description of the potential feedstocks that can be used. Not only is the conventional production process analysed, but also new developments that can improve significantly the performance of current commercial production units. In the final sections expected future development for the production processes are presented and discussed.

Finally, the editor wish to thank the authors for their participation, cooperation, help and enthusiasm manifested during the preparation of this book.

J.M.P.Q. Delgado

About the Editor

J.M.P.Q. Delgado

J.M.P.Q. Delgado obtained his PhD in Chemical Engineering in 2002 from Faculty of Engineering of the University of Porto (FEUP), Portugal, where he was Researcher in the Laboratory of Fluid Mechanics /Mass and Heat Transfer/Combustion, at the department of Chemical Engineering. Between 2002 and 2005, he was Pos-Doc researcher, at FEUP, in the Institute of Mechanical Engineering and Industrial Management. Since 2007 he is Auxiliar Investigator in Building Physics Laboratory in the department of Civil Engineering, Faculty of Engineering of the University of Porto, Portugal. Since 2002, he is Assistant Professor of different mathematics subjects, Transport Phenomena, Separation Processes and Air Quality and Noise Pollution. J.M.P.Q. Delgado is been during the last year's organizer of international conferences such as the 5th International Conference on Diffusion in Solids and Liquids (DSL-2009), the 3rd Conference on Building Pathology and Rehabilitation (PATORREB-2009); member of the scientific committee of international symposiums as the International Conference on Diffusion in Solids and Liquids (DSL-2007, DSL-2008) and 20th International Symposium on Transport Phenomena (ISTP-20); organizer and Chairman of special sessions in different international scientific conferences and invited to present various invited lectures. He was co-supervisor of 1 PhD-student and author of 52 publications in reviewed international journals, 3 book chapters and more than 50 other published papers in international conferences.

Contents

Preface   *v*

1. Computational Analysis of the Role of Permeability and Inertia on Fluid Flow through Porous Media   1-19
   A.F. Miguel (Portugal)

2. Modeling of Transport Processes in Solid Oxide Fuel Cells   21-48
   E. Vakouftsia and *F.A. Coutelieris* (Greece)

3. Structure and Properties of Grain Boundaries   49-103
   Vladimir V. Popov (Russia)

4. A Review on Thermal Lattice Monte Carlo Analysis   105-130
   T. Fiedler, I.V. Belova, A. Öchsner and *G.E. Murch* (Australia, Malaysia)

5. Predicting the Effective Thermal Conductivity of Perforated Hollow Sphere Structures (PHSS)   131-151
   S.M.H. Hosseini, A. Öchsner, M. Merkel and *T. Fiedler* (Malaysia, Australia, Germany)

6. Food Dehydration under Forced Convection Conditions   153-177
   A. Mulet, J.A. Carcel, N. Sanjuan and *J.V. Garcia-Perez* (Spain)

7. Dispersion in Porous Media   179-205
   J.M.P.Q. Delgado (Portugal)

8. Nanostructured Materials   207-269
   A. Shokuhfar and *M. Mohebali* (Iran)

9. Modelling and Simulation of Interfacial Heat Transfer   271-291
   M.J.S. De Lemos and *M.B. Saito* (Brazil)

10. Supercritical Fluids and its Applications 293-312
    M. Vázquez Da Silva (Portugal)

11. Biodiesel Production Processes 313-342
    T.M. Mata and *A.A. Martins* (Portugal)

    Index 343-345

1

Computational Analysis of The Role of Permeability and Inertia on Fluid Flow Through Porous Media

A.F. MIGUEL[*]

ABSTRACT

Fluid dynamics is governed by a set of well established differential equations that are based on conservation laws of physics. Advances in computational physics allow simulation of two and three-dimensional flows through complex geometries. These simulations, judiciously employed, can offer much insight into flows through porous media. In particular, Lattice Boltzmann equation method (LBE) is very attractive to investigate flows in complex geometries such as porous media. In this chapter LBE is used to simulate flow in porous media. It is aimed, (i) to investigate the range of validity of Darcy-Forchheimer equation, (ii) to examine the macroscopic transport properties (i.e., the permeability and the inertial parameter), and (iii) to understand the effect of the continuum and the slip-flow regimes on the permeability of porous media.

1. FLUID FLOW IN POROUS MEDIA

A porous medium consists of a solid matrix with void spaces that are in general complicated and distributed throughout the structure. These media are formed in nature as a consequence of many physical, chemical, and physiochemical processes (Nield and Bejan, 1998; Bejan *et al.*, 2004). Flow through porous media has become a very popular research topic due to their wide spectrum of applications (Dullien, 1979; Nield and Bejan, 1998; Bejan *et al.*, 2004). Specially, the prediction of transport properties is a topic of great fundamental and practical significance.

Experiments on water flow through sand led Henry Darcy to formulate a linear relationship between the fluid velocity and the pressure drop (Darcy, 1856). Given a sample of porous medium of length L across which

---

[*] Geophysics Center of Evora, Rua Romao Ramalho 59, 7000-671 Evora, PORTUGAL.
University of Évora, Department of Physics, PO Box 94, 7002-554 Evora, PORTUGAL.
[*] *Corresponding author* : E-mail : afm@uevora.pt

is applied a pressure difference ΔP, the macroscopic flow of a viscous fluid is described by Darcy law (Nield and Bejan, 1998)

$$\frac{\Delta p}{L} = \frac{\mu}{K} u \qquad ...(1)$$

where u is the superficial fluid velocity (the velocity of the fluid in the pores-intrinsic velocity–is related with the superficial fluid velocity through the porosity), μ is the fluid's viscosity and K is the intrinsic permeability. Therefore, this law is used to define the permeability of the medium, provided that the fluid viscosity is known. Whitaker (1986) and Quintard and Whitaker (1994) showed that the Darcy equation can be obtained analytically from averaging the momentum equation.

In spite of its great applicability, the concept of permeability as a global quantity that characterizes the fluid flow, which grounds the validity of Darcy law, holds only (i) in the case of fluid can be regarded as a continuum, and (ii) for low values of the Reynolds number (i.e., for viscous flow).

The value of the Knudsen number determines the degree of rarefaction of the gas and the validity of the continuum flow assumption. For a Knudsen number, defined as the ratio between the mean free path and the characteristic dimension of the flow geometry, less than 0.001 the continuum hypothesis is appropriate. For higher Knudsen numbers the flow is commonly referred to as the slip-flow regime and the intrinsic permeability defined according to Eq. (1) is not anymore an appropriated property (Miguel and Serrenho, 2007).

At sufficiently high Reynolds numbers (and in case of continuum flows), the flow through porous media is well approximated by a nonlinear approach (Dullien, 1979; Nield and Bejan, 1998).

$$\frac{\Delta p}{L} = \frac{\mu}{K} u + \rho \beta u^2 \qquad ...(2)$$

where ρ is the fluid density and β is usually called the inertial parameter. This phenomenological model is known as the as Darcy-Forchheimer equation. Although this equation was proposed from empirical bases, a number of formalizations have been reported (Whitaker, 1996; Miguel et al., 2001). Many attempts have been also made to analytically derive a general equation of motion in porous media. Miguel et al. (2001) present a general equation that is valid in Darcy regime but also at high Reynolds numbers.

Experimental studies have shown that the transition from linear (Darcy flow) to non-linear (Forchheimer or non-Darcy flow) regime occurs gradually as the Reynolds number increases (Ward, 1964). A number of explanations have been given for the onset of non-linearity flow. Early studies have attributed nonlinearity to the occurrence of turbulence (Ward, 1964), and the Reynolds number for identifying turbulent flow in conduits was adapted to describe the onset of nonlinear flow in porous media. Although, deviations from Darcy law have been observed at Reynolds numbers much less than 10, experiments have indicated that the onset of turbulent occurs at higher velocities. Thus, deviations from Darcy's law are not initiated by turbulence. There are experimental evidences that the source of the nonlinear behaviour of flows at high velocities is attributed to inertial effects (Bear and Corapcioglu, 1984; Ma and Ruth, 1993; Miguel, 1998).

The earliest work on the criterion for nonlinear regime in porous media seems to be presented by Chilton and Colburn (1931). Based on the Reynolds number defined in terms of diameter of the beds, d_b,

$$\text{Re}_d = \frac{\rho u d_p}{\mu} \qquad ...(3)$$

and performing experiments performed on packed particles, they observed that the Darcy law is valid when Re_d is less than 40–80. Fancher and Lewis (1933) performed experiments in consolidated sandstones and unconsolidated sands and lead shot. Their experiments show that the nonlinear flow occurs at Re_d in the range of 10–1000 in unconsolidated porous media and at Re_d larger than 0.4–3 in loosely consolidated rocks.

Ergun (1952) proposed the use of the intrinsic velocity, u_i, the porosity, ϕ, and a particle shape factor or "sphericity" factor, c_{ef}, to redefine the Reynolds number

$$\text{Re}_{d\phi} = \frac{c_{ef}\, \rho u_i\, d_p}{\mu\, (1 - \phi)} \qquad ...(4)$$

and suggested that the non-linear flow occurs at $\text{Re}_{d\phi}$ larger than 3–10. Realizing the difficulty of determining the particle diameter, other authors use the permeability to define the Reynolds number. Green and Duwez (1951) redefine the Reynolds number in terms of K and β

$$\text{Re}_{K\beta} = \frac{\rho u K \beta}{\mu} \qquad ...(5)$$

and observed that non-linear flow behaviour starts at $\text{Re}_{K\beta}$ higher than 0.1–0.2.

For convenience, Eq. (2) is often rearranged in dimensionless form

$$f_K = \frac{1}{\text{Re}_K} + \lambda \qquad ...(6)$$

with

$$f_K = \frac{\Delta p}{L}\, \frac{K^{1/2}}{\rho u^2} \qquad ...[7\,(a)]$$

$$\text{Re}_K = \frac{\rho u K^{1/2}}{\mu} \qquad ...[7\,(b)]$$

$$\lambda = \beta K^{1/2} \qquad \ldots[7\ (c)]$$

where f_K is the friction factor, Re_K is the Reynolds number based on permeability and λ is the dimensionless inertia parameter. Experimental data from a variety of porous media has been successfully correlated with this equation (Nield and Bejan, 1998). Besides, experimental measurements presented by Ward (1964) suggest that the Darcy law [i.e., expressed by the first right-hand term of Eq. (2)] is valid when Re_K is less than the 1 to 10 range (see Fig. 1.1).

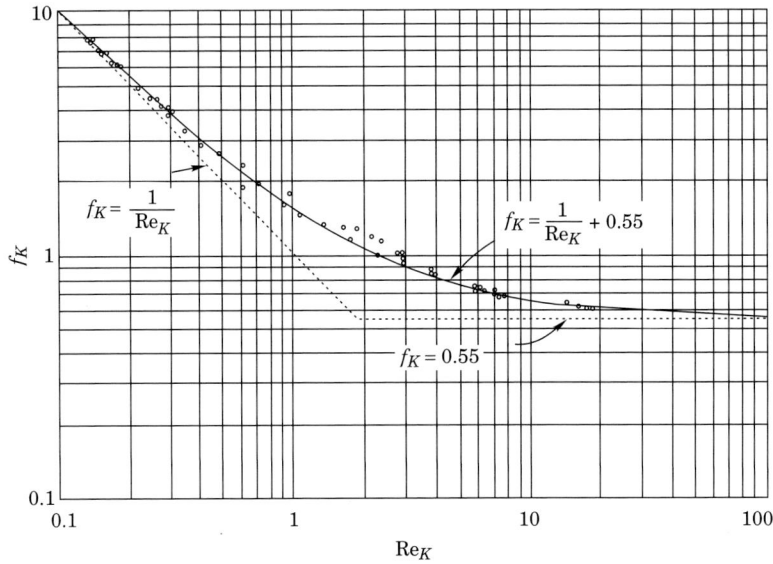

Fig. 1.1. The transition from the Darcy regime to the Forchheimer regime in porous medium (Ward, 1964)

In summary, the criterion for nonlinear (Forchheimer) flow in porous media is based in three different definitions of Reynolds number: Eq. (3) or (4), Eqs. (5) and [7 (b)]. Do these have the same meaning? Which criterion is recommended?

According to Ma and Ruth (1993), the Reynolds number defined by Eqs. (3) and (4) has its root in the similar criterion for flow in ducts. It is a good definition for columns of packed particles, unconsolidated or loosely consolidated sands. According to other authors (Miguel, 1998a; Bejan et al., 2004), Eq. [7 (b)] has also the physical meaning of the Reynolds number because represents the ratio of scaling of the inertia of the flow to the viscous forces in the flow in porous media. In contrast, it has the advantage of easy and wide applicability. In most porous media, pore and bed diameters are very difficult to measure. In fact, several researchers have expressed their preference for using this definition (Bejan et al., 2004).

Ma and Ruth (1993) suggest the definition provided by Eq. (5) has a different meaning of the tradition definition of Re for ducts. They recommended that should be renamed as Forchheimer number. In fact Eq. (5), according to Eqs. [7 (b)] and [7 (c)], is the product of the Reynolds number by the dimensionless inertia parameter. Therefore, the Forchheimer number may be written as

$$Fo_K = \frac{\rho u K \beta}{\mu} = \text{Re}_K \lambda \qquad ...(8)$$

As long as the permeability and inertial parameter can be determined, Eq. (8) constitutes recommended parameter to describe nonlinear flow regime. Therefore, substituting Eq. (8) into Eq. (6) yields

$$f_K = \lambda \left(\frac{1}{Fo_K} + 1 \right) \qquad ...(9)$$

Equation (9) describes the friction factor in terms of Forchheimer number.

2. PERMEABILITY COEFFICIENT AND INERTIAL PARAMETER

Prediction of the permeability and the inertial parameter is a longstanding problem. As mentioned in the last section, the concept of permeability representing a global coefficient for flow, as defined by Eq. (1), is restricted to Darcy regime (creeping flow). The permeability coefficient quantifies the ability of the medium to be crossed by fluid and depends on the properties of the pore space, such as porosity and other structural parameters (Dullien, 1979; Nield and Bejan, 1998). Several empirical and semi-empirical approaches are used to describe the nonlinear dependence of permeability on porosity by power-law or exponential relationships (see Table 1.1).

The best known approach is probably the one first proposed by Kozeny (1927) and later modified by Carman (1937, 1956). The so-called Kozeny-Carman equation is given by

$$K = \frac{1}{c_K S^2} \frac{\phi^3}{(1-\phi)^2} \qquad ...(10)$$

Here ϕ is the porosity, c_K is the Kozeny constant and S is the specific surface (ratio of the pore surface area to the total volume of the sample). An important role of the specific surface is to properly scale data obtained from porous media described by different intrinsic length scales. Usually, S is set equal to $6/d$, where d is a characteristic length of porous media (for example, the mean particle or pore diameter), and values of c_K vary between 2 and 6. A number of similar formulae, derived especially for fibrous porous materials, can be found in the literature (Jackson and James, 1986). While

Kozeny-Carmen relation is very simple and gives satisfactory results for different porous media, it breaks down in the high porosity regime (Koponen et al., 1998; Bejan et al., 2004; Pinela et al., 2005). For high porosity porous media, Koponen et al. (1998) suggest

$$K = \frac{1.39}{\exp[10.1(1-\phi)] - 1} d^2 \qquad \ldots(11)$$

Table 1.1. Some models available in the literature for the intrinsic permeability K

$K\ (m^2)$	Validity	References
$0.0293\phi^{4.57}\, d^2$	$0.4 \leq \phi \leq 0.79$	Adler and Thovert (1998)
$\dfrac{2}{18(1-\phi)} \dfrac{3 - (9/2)(1-\phi)^{1/3} + (9/2)(1-\phi)^{5/3} - 3(1-\phi)^2}{3 + 2(1-\phi)^{5/3}} d^2$	—	Happel and Brenner (1973)
$\dfrac{1}{c_K} \dfrac{\phi^3}{(1-\phi)^2 S^2}$	—	Kozeny (1927) Carman (1937, 1956)
$\dfrac{3.5\phi^3 [1 + 57(1-\phi)^3]}{4(1-\phi)^{1/2}} d^2$	High porosity	Ingmanson et al. (1956)
$\dfrac{3[-\ln(1-\phi) - 0.931]}{80(1-\phi)} d^2$	$0.65 \leq \phi < 1$	Jackson and James (1986)
$\dfrac{1.39}{\exp[10.1(1-\phi)] - 1} d^2$	High porosity	Koponen et al. (1998)
$\dfrac{0.2\phi^3 (\phi - 0.1)}{(1-\phi)^{1.3}} d^2$	$0.65 \leq \phi \leq 0.99$	Miguel (1998a)
$c_{K1} \exp(c_{K2}\phi)\, d^2$	$0.2 \leq \phi \leq 0.99$	Pinela et al. (2005)

A large number of studies (Katz et al., 1959; Geertsma, 1976; Ergun, 1952; Macdonald et al., 1979) have been devoted to the prediction of the inertial parameter as function of the parameters that characterize the pore structure (i.e., porosity, size of pores or permeability). Ward (1964) and Beavers et al. (1973) established that the inertial parameter is inversely proportional to the square root of the permeability. Geertsma (1974); Noman and Archer (1987); Miguel (1998) suggested that β is essentially a function of the porosity and permeability. Recently, Miguel (2010) derived theoretically the inertial parameter of a dendritic flow structure. This study indicates that this parameter is also a function of the pore structure. Some correlations reported in the literature are summarized in the Table 1.2. A detailed review of the published literature on this topic is presented by Nield and Bejan (1999).

Table 1.2. Some models available in the literature for the inertial parameter β

β (1/m)	Validity	References
$\dfrac{1.8\,(1-\phi)}{\phi^3\,d}$	High porosity	Ergun (1952)
$\dfrac{0.550}{K^{0.5}}$	—	Ward (1964)
$\dfrac{0.4}{\phi^{2.13}\,K^{0.5}}$	0.2 < porosity < 0.7	Miguel (1998)
$\dfrac{c_0}{\phi^2\,d}$	—	Blick (1966)

3. COMPUTATIONAL SIMULATION OF FLUID FLOW THROUGH POROUS MEDIA

The increase of computation capacity and the subsequent development of methods of computational fluid dynamics (CFD) have made possible to directly solve many complex fluid-dynamical problems. Since the 1970's, computational modeling of fluid flow through porous media has increased rapidly (Vafai, 2000). Traditionally CFD has concentrated on finding solution to differential continuum equations that govern the fluid flow. Approaches, such as the Navier-Stokes equations, provide a continuum description of macroscopic phenomena provided by partial differential equations. Computational techniques, such as finite-difference and finite-element methods, are then used to transform the continuum description into a discrete one in order to solve the equations numerically on a computer. The results obtained are sensitive, for example, to grid generation (Knupp and Steinberg, 1993). A successfully generated grid is typically an irregular mesh including knotty details that follow the expected streamlines. Recently, lattice-based methods with a regular computational grid have been applied to solve complex fluid flows in porous media. The lattice Boltzmann equation method (LBE) for modelling hydrodynamics has its origins from the lattice gas cellular automata and can be directly derived by discretizing the Boltzmann equation (Succi, 2001; Pan *et al.*, 2006). LBE become very important since it was discovered that very simple models of discrete particles confined to a lattice can be used to solve very complicated flow problems. It is a very simple microscopic, or particle, approach to modelling macroscopic dynamics. The lattice Boltzmann equation method has foundation on kinetic theory.

The LBE has been shown to be equivalent to a finite difference approximation of the incompressible Navier–Stokes equations (Pan *et al.*, 2006). Therefore, it can be viewed as a discrete approximation of the incompressible Navier–Stokes equations, based on kinetic theory rather than continuum theory. The advantages associated to lattice Boltzmann equation method include programming simplicity, intrinsic parallelism, and straightforward resolution of complex solid boundaries and multiple fluid species. LBE method has also some limitations described, for example, by Sankaranarayanan et al. (2003) and Pan et al. (2006). In summary, LBE makes more amenable the computational analysis. Appealing features of the LBE method justify the number of studies devoted to flow simulation in porous media flow simulation (Chen et al., 1991; Koponen et al., 1998; Drazer and Koplik, 2000; Clague et al., 2000; Manwart et al., 2002; Quispe et al., 2005 etc.).

There are several relevant computational studies available in the literature. Vafai and Tien (1981) used a volume-averaged momentum and energy equations to study boundary and inertia effects. Coulaud et al. (1988) carried out simulations for two-dimensional flows across banks of circular cylinders. Larson and Higdon (1989) studied the permeability performing calculations for the Stokes flow through a lattice of spheres. Cancelliere et al. (1990) investigated the permeability of porous media with randomly distributed inclusions. Chen et al. (1991) used the lattice gas automata to study the microscopic behaviour occurring at the pore scale and to obtain volume-averaged parameters from the microscopic point of view. Nakayama et al. (1995) and Lee and Yanh (1997) studied the contributions of both Darcy and Forchheimer regimes to the macroscopic pressure drop in lattices of cubes and cylinders, respectively. Higdon and Ford (1996) used the Stokes flow through ordered fibrous porous media to study the permeability. Koponen et al. (1998) and Clague et al. (2000) carried out studies on the permeability of random fiber webs and on bounded and unbounded fibrous media, respectively. Andrade et al. (1999) studied inertia effects in porous media by means of the control volume finite-difference technique. Drazer and Koplik (2000) studied the transport properties of three-dimensional self-affine rough fractures. Manwart et al. (2002) investigated the flows through a straight rectangular channel and the cubic arrays of spheres in order to estimate the permeability of sandstones. Wang and Liu (2004) studied the scaling relations for the permeability and the inertial parameter in percolation porous media. Quispe et al. (2004, 2005) studied the permeability–porosity relationship of porous media as simple diagenetic or shrinking processes reduces their pore spaces. Pinela et al. (2005) compared simulation resulting from a finite-volume-based technique with some permeability-porosity relationship available in literature. They found that Kozeny-Carman equation is only valid for low porosities. They also suggested that the permeability, for the entire porosity range, is better correlated with the exponential of the porosity. Araujo et al. (2006) studied distributions of channel openings, local fluxes, and velocities in a two-dimensional random medium of non-overlapping disks. They presented theoretical arguments supported by computational data and found scaling laws as function of the porosity. Zobel et al. (2007) studied the influence of orientation distribution of calendared fibrous structures. Recently, Valdes-

Parada et al. (2009) suggest that the local porous medium configuration has an important effect on the permeability value, and that the Kozeny-Carman equation is not a good description of the permeability behaviour in terms of porous media characteristics.

4. LATTICE BOLTZMANN MODEL FOR FLOW THROUGH POROUS MEDIA

In LBM, the fluid is modelled by a single-particle distribution function and the evolution of this function is governed by a lattice Boltzmann equation

$$f_i(x + e_i\delta_t, t + \delta_t) - f_i(x + t) = -\frac{f_i(x + t) - f_i^{eq}(x + t)}{\tau} \qquad ...(12)$$

Here $f_i(x+t)$ is the single-particle distribution function with e_i velocity at x position on the time t, δ_t is the time increment, τ is the non-dimensional relaxation time and f_i^{eq} is the equilibrium distribution function. In LBM, the fluid density, ρ and velocity, v are defined by the single-particle distribution function

$$\rho = \sum_i f_i \qquad ...(13)$$

$$\rho v = \sum_i e_i f_i \qquad ...(14)$$

It can be shown that the Navier-Stokes equations can be derived from Eq. (12) through a Chapman-Enskog expansion procedure in the incompressible limit. A detailed description is provided Quin et al. (1992). Consider a medium with porosity, ϕ which is defined as the fraction of the void space inside the lattice. In the limit of small Mach number

$$\frac{\partial \rho}{\partial t} + \nabla.(\rho v) = 0 \qquad ...(15)$$

$$\frac{\partial (\rho u)}{\partial t} + \nabla \cdot \left(\frac{\rho v v}{\phi}\right) = -\nabla(\varepsilon p) + \nabla.[\mu_e(\nabla v + v\nabla)] \qquad ...(16)$$

where p is the pressure, μ_e is the viscosity ($\mu_e = c_s^2 \rho (\tau - 0.5)\delta_t$) and c_s is the sound speed. If ρ = constant and ϕ = 1, Eq. (16) reduce to the classic Navier-Stokes equation.

In this method the fluid is modelled by particle distributions that move on a lattice divided to solid and fluid points. During one lattice time step, particles propagate to their adjacent lattice points and redistribute their momenta in the subsequent collisions. Each time step involves also the action, e.g., of the external forces on the fluid and the boundary conditions at the solid-fluid interfaces.

Investigations of gas flows in channels showed that when the dimension of a channel is comparable to or smaller than the mean free path of the gas molecules, the gas can no longer be regarded as being in thermodynamic equilibrium and a variety of non-continuum (rarefaction) effects occur (Serrenho and Miguel, 2007). The Knudsen number, Kn, defined as the ratio between the mean free path and the characteristic length of the flow geometry, determines the validity of the continuum flow assumption. For a low Knudsen number (Kn ≤ 0.001), the fluid can be regarded as a continuum. However, for higher Knudsen (*i.e.*, Knudsen numbers between 0.001 and 0.1) the continuum hypothesis is no longer appropriated and the gas flow is commonly referred to as the slip-flow regime. Here, it is assumed that the gas molecules, represented by the particle distribution functions, travel the distance of the lattice mean-free path, l, with a lattice speed, $\delta x / \delta t$, while relax to their equilibrium state in the relaxation time, τ. Therefore, the Knudsen number can be defined as

$$\text{Kn} = \frac{\bar{\delta x} \tau}{d} \qquad \text{...(17)}$$

where d is the characteristic length of the flow geometry. Since the mean free path is inversely proportional to the pressure, then the local Kn is related to the reference values by $\text{Kn}_o \, p_o/p$, where the subscript o means outlet. This constitutes the way to determine the local non-dimensional relation time (Jeong et al., 2006).

Our topological space is based on uniformly distributed cylinders in 2-dimensions for in-line and staggered arrangements (Fig. 1.2). In the simulations the porosity is prescribed (0.15 ≤ φ ≤ 0.95) and Reynolds number changes from 0.004 to 600. Periodic boundary condition is imposed at respective boundaries, while a pressure difference is prescribed between the inlet and outlet. Details of the computational procedure can be obtained in Kim et al. (2001) and Miguel (2010a).

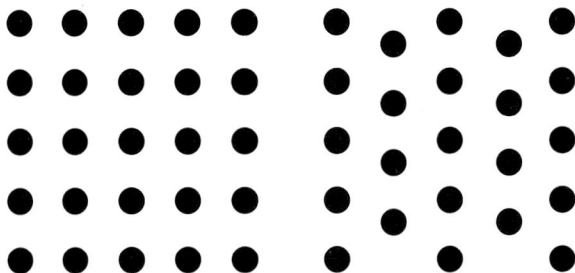

Fig. 1.2. Two dimensional porous space with in-line (left-hand side) and staggered (right-hand side) arrangements

5. CONTINUUM-FLOW REGIME: PERMEABILITY AND INERTIAL PARAMETER

In the following we present and analyze the results of the flow simulations in the geometries depicted in Fig. 1.2.

We apply the dimensionless forms of Darcy-Forchheimer equation [i.e., Eqs. (6) and (9)] as phenomenological models to correlate the variations of the friction factor for different porosities and flow conditions. Based on the velocity and pressure results of the simulations, the permeability and the inertia parameter was estimated and then the friction factor, the Reynolds number and the Forchheimer number calculated. Therefore, these quantities are plotted in terms of the f_K and Re_K or Fo_K. Figures 1.3 and 1.4 show the results of simulations performed with in-line and staggered porous structures for a porosity of 0.3 and 0.6, respectively. In agreement with a large number of experimental and computational studies, we also observe a transition from linear (Darcy) to nonlinear (Forchheimer) regime. Figures 1.3 and 1.4 show that this transition occurs at Re_K in the range of 0.2–0.4 and at Fo_K larger than 0.3–0.7, respectively. In both plots the transition from linear to nonlinear behaviour is gradual and takes place under laminar flow regime. Therefore, Forchheimer regime starts to occur before the existence of turbulence. This means that the inertia effects flatten the f_K curve in a manner reminiscent of the friction factor in turbulent flow over a surface. According to Nield and Bejan (1998) the occurrence of the first eddies in the fluid flow is associated with a local Reynolds number in the range 100 to 300.

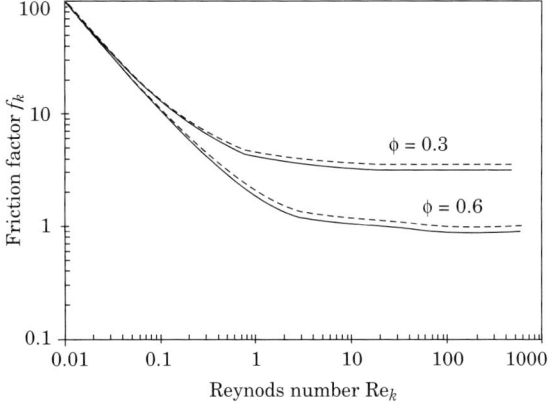

Fig. 1.3. Logarithmic plot showing the dependence of the friction factor f_K on the Reynolds number Re_K:—in-line arrangement and—staggered arrange-ment

Figure 1.3 also reveals that the effect of geometry (i.e., in-line and staggered arrangements) is only significant after the linear (Darcy's regime) flow. This means that, for a uniform distribution of cylinders in a 2D porous structure, the arrangement does not affect the intrinsic permeability (Fig. 1.5). The results of computational simulations depicted in Fig. 1.5 also corroborate numerous studies available in the literature which display a strong dependence of the permeability on the porosity.

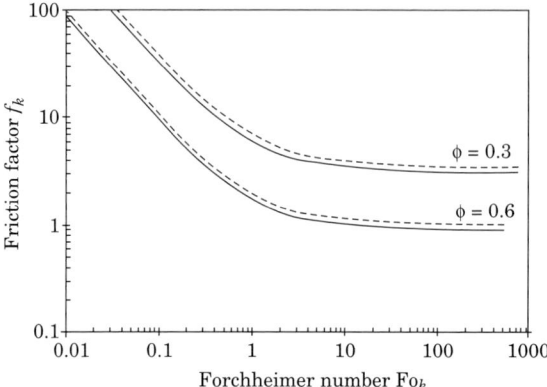

Fig. 1.4. Logarithmic plot showing the dependence of the friction factor f_K on the Forchheimer number Fo_K:—in-line arrangement and—staggered arrangement

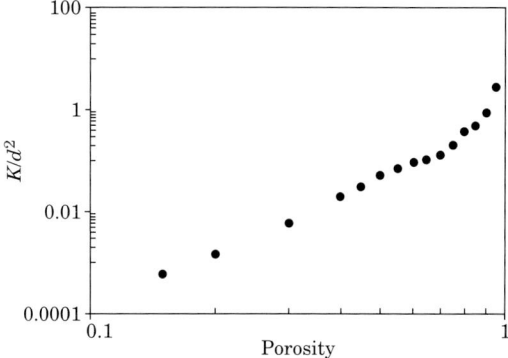

Fig. 1.5. Logarithmic plot showing the dependence of the dimensionless permeability (K/d^2) on the porosity: • in-line arrangement and o staggered arrangement

In Fig. 1.6 we compare the dimensionless permeability generated by the computational simulations with the permeability obtained from Kozeny-Carman (Carman, 1937, 1956); Koponen et al. (1998) and Miguel (1998) correlations (Table 1.1). This plot shows that the Kozeny-Carman correlation is consistent with our data only when the porosity is less than 0.55 (i.e., for larger porosities this correlation overestimates our data). On the other hand, the result obtained from the simulations is significantly different from the values estimated with the correlations presented by Koponen et al. (1998) and Miguel (1998) for low porosities but is in good agreement when the porosity is larger than 0.8 and 0.65, respectively. All correlations present very poor estimation ability for porosities between 0.55 and 0.7. The result depicted in Fig. 1.5 shows that our data approxi-mately coalesces into one exponential curve in the entire porosity domain. Therefore, the points can be fitted with the following equation

$$\frac{K}{d^2} = 0.0003 \exp(9.0956\phi) \quad (r^2 = 0.976) \quad 0.15 \leq \phi \leq 0.95 \quad \ldots(18)$$

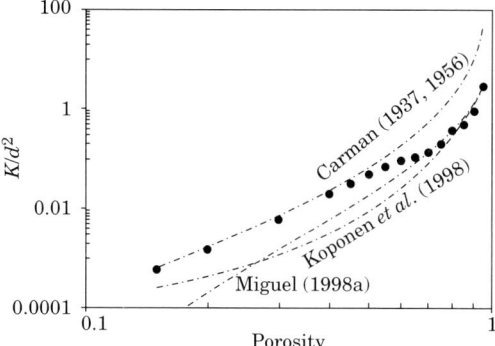

Fig. 1.6. Comparison of K/d^2 obtained from the computational simulations with selected correlations depicted in Table 1.1

This finding is in agreement with the results of early investigations performed by Pinela *et al.* (2005).

Figure 1.7 illustrates the dependence of the dimensionless inertia parameter λ on both the porosity and the solid matrix arrangement. In contrast to the permeability results (Fig. 1.5), λ is higher in the staggered arrangement than in in-line arrangement. Besides, this inertia parameter decreases with the void fraction (porosity). This result corroborates numerous data available in the literature (Ergun, 1952; Geertsma, 1974; Noman and Archer, 1987; Miguel, 1998) which display a strong dependence of porosity.

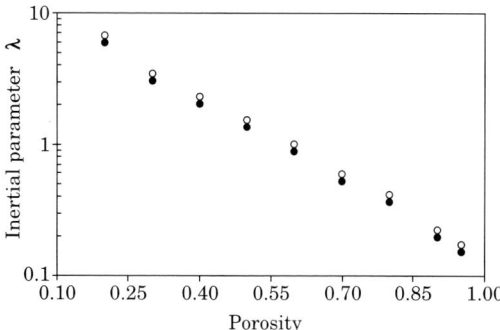

Fig. 1.7. Logarithmic plot showing the dependence of the dimensionless inertial parameter λ on the porosity: • in-line arrangement and o staggered arrangement

The inertia parameter λ obtained in these simulations is compared to the correlations suggested by Ergun (1952) and Miguel (1998) in Fig. 1.8. The plot shows that the Ergun's correlation agrees only for the in-line porous arrangement with porosity less than 0.6. It is also observed that the correlation presented by Miguel (1998) is found to agree only with the staggered porous arrangement for porosities between 0.4 and 0.6. It is noteworthy that both correlations produce very poor estimation of λ for a porosity larger than 0.6, independently of the arrangement. According to Fig. 1.7, the permeability generated by the computational simulations can be fitted with the following equation

$$\lambda = c_\lambda \left(\frac{\phi}{1-\phi}\right)^{-0.867} \quad 0.2 \leq \phi \leq 0.95 \quad \ldots(19)$$

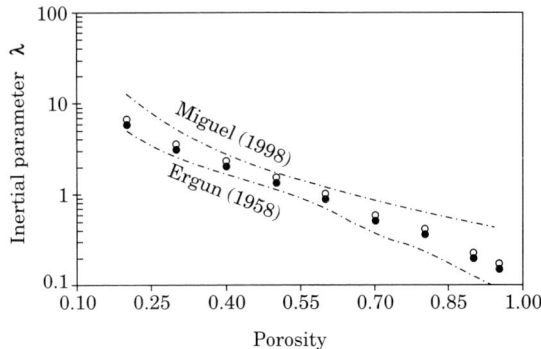

Fig. 1.8. Comparison of l obtained from the computational simulations with selected correlations depicted in Table 1.2: • in-line arrangement and o staggered arrangement

with, $\quad c_\lambda = 1.42\ (r^2 = 0.977)$ staggered arrangement

$c_\lambda = 1.62\ (r^2 = 0.978)$ in-line arrangement

where c_λ is a coefficient that depends on the arrangement of solid matrix. This result shows that the coefficient for a staggered arrangement is 12% higher than for a in-line arrangement.

6. SLIP-FLOW REGIME: PERMEABILITY VERSUS KNUDSEN NUMBER

As mentioned before, the Knudsen number determines the validity of the continuum-flow assumption. For Knudsen numbers ranging from 0.001 to 0.1 the continuum hypothesis is no longer appropriated and the gas flow is commonly referred to as the slip-flow regime.

Figure 1.9 is a summary of the results obtained for the dimensionless permeability in the slip-flow regime. This plot shows that the K/d^2 depends not only on the porosity but also both on the Knudsen number and the arrangement of the solid matrix. Although the porosity dependence is similar for both continuum and slip-flow regimes (i.e., the permeability increases with the porosity), the effect of Knudsen number and solid matrix arrangement are different. For continuum-flow regime, K/d^2 is not dependent of both Kn and matrix arrangement. For slip-flow regime, dimensionless permeability is higher for the in-line arrangement than for the staggered arrangement of the solid matrix. Besides, as the Knudsen number increases the K/d^2 increases too.

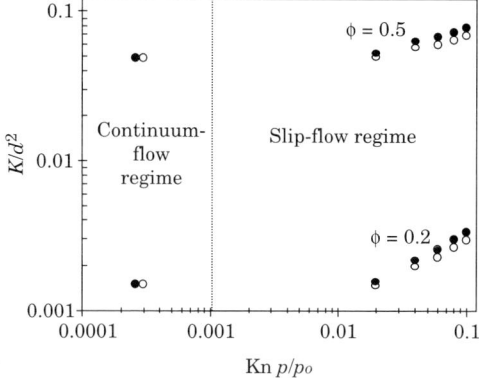

Fig. 1.9. Logarithmic plot showing the dependence of the dimensionless permeability (K/d^2) on the Knudsen number: • in-line arrangement and o staggered arrangement

Notice that both the dimensionless permeability, K/d^2, and the Knudsen number depicted in Fig. 1.9 are dependent on the characteristic dimension of the solid matrix, d. To obtain more insight into these results, the ratio of the intrinsic permeability to the distance of the mean-free path, K/l^2, versus the Knudsen number is depicted in Fig. 1.10. If l = constant, the intrinsic permeability follows an exponential decrease with the Knudson number. This result agrees with the experimental results obtained by Miguel and Serrenho (2007).

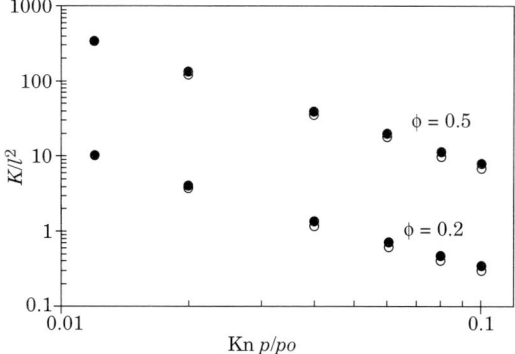

Fig. 1.10. Logarithmic plot showing the dependence of the ratio of intrinsic permeability to distance of the mean-free path (K/l^2) on the Knudsen number: • in-line arrangement and o staggered arrangement

7. FINAL REMARKS

In the development of complex flow systems, the importance of computational simulation is continuously increasing. These simulations produce results that prove highly valuable to science and engineering. This growing importance requires more refined computational techniques. In this chapter, lattice-Boltzmann equation method (LBE) is applied to flow simulation in porous media. The key advantages of lattice Boltzmann equation method to conventional computational techniques include a simple calculation procedure and the possibility to explore the regime where the continuum hypothesis is no longer appropriated-the slip-flow regime.

For continuum-flow regime, the results demonstrate that the permeability is an exponential function of the porosity and do not depends on the solid matrix arrangement. The Kozeny-Carman correlation (Carman, 1937, 1956) agrees only for small porosities and the correlation presented by Koponen et al. (1998) holds well for very high porosities. The inertia parameter is also investigated. It is found that the correlations presented by Ergun (1952) and Miguel (1998) are only suitable for porosities less than 0.6. In addition, it is observed that the transition from linear (Darcy's regime) to nonlinear (Forchheimer regime) flow is gradual and takes place under laminar flow regime.

For the slip-flow regime, the intrinsic permeability depends on the porosity but also both on the solid matrix arrangement and the Knudsen number. The intrinsic permeability follows an exponential decrease with the Knudson number.

ACKNOWLEDGEMENTS

I am especially grateful to Z. Lee for collaborative support on LBE simulations. This work has been partially supported by the Portuguese National Science Foundation (FCT) under contract no. PTDC/EQU-FTT/115509.

REFERENCES

1. Adler, P.M and Thovert, J.F. 1998. Real porous media: local geometry and macroscopic properties, *Appl. Mech. Rev.* **51**: 537–85.
2. Al-Jahmany, Y.Y., Brenner, G. and Brunn, P.O. 2004. Comparative study of lattice Boltzmann and finite volume methods for the simulation of laminar flow through a 4:1 planar contraction, *Int. J. Numer. Meth. Fluids* **46**: 903–20.
3. Andrade Jr., J.S., Costa, U.M.S., Almeida, M.P., Makse, H.A. and Stanley, H.E. 1999. Inertial effects on fluid flow through disordered porous media, *Phys. Rev. Lett.* **82**: 5249–52.
4. Araujo, A.D., Bastos, W.B., Andrade Jr., J.S. and Herrmann, H.J. 2006. The distribution of local fluxes in porous media, *Phys. Rev. E.* **74**: 010401.
5. Bear, J. and Corapcioglu, M.Y. (*Eds.*) 1984. Fundamentals of Transport Phenomena in Porous Media. Martinus Nijhoff, Dordrecht.
6. Beavers, G.S., Sparrow, E.M. and Rodenz, D.E. 1973. Influence of bed size on the flow characteristics and porosity of randomly packed beds of spheres, *ASME J. Appl. Mech.* **40**: 655–60.

7. Bejan, A., Dincer, I., Lorente, S., Miguel, A.F. and Reis, A.H. 2004. Porous and complex flow structures in modern technology. Springer, New York.
8. Blick, E.F. 1966. Capillary orifice model for high speed flow through porous media, *Process Des. Dev.* **1**: 90–4.
9. Cancelliere, A., Chang, C., Foti, E., Rothman, D.H. and Succi, S. 1990. The permeability of a random medium: comparison of simulation with theory, *Phys. Fluids A* **2**: 2085–8.
10. Carman, P.C. 1937. Fluid flow through a granular bed, *Trans. Inst. Chem. Eng. London* **15**: 150–7.
11. Carman, P.C. 1956. Flow of Gases through Porous Media. Butterworths, London.
12. Chen, S., Diemer, K., Doolen, G.D., Eggert, K., Fu, C., Gutman, S. and Travis, B.J. 1991. Lattice gas automata for flow through porous media, *Physica D* **47**: 72–84.
13. Chilton, T.H. and Colburn, A.P. 1931. Pressure drop in packed tubes, *Ind. Engng. Chem.* **23**: 913–9.
14. Chouikh, R., Guizani, A. and Mafilej, M. 1999. Numerical study of laminar natural convection flow around an array of two horizontal isothermal cylinders, *Int. Comm. Heat Mass Transf.* **26**: 329–38.
15. Clague, D.S., Kandhai, B.D., Zhang, R. and Sloot, P.M.A. 2000. Hydraulic permeability of (un)bounded fibrous media using the lattice Boltzmann method, *Phys. Rev. E*. **61**: 616–25.
16. Darcy, H. 1856. Les foutaines publiques de la ville de Dijon. Dalmont, Paris.
17. Drazer, G. and Koplik, J. 2000. Permeability of self-affine rough fractures, *Phys. Rev. E* **62**: 8076–85.
18. Dullien, F.A.L. 1979. Porous Media–Fluid Transport and Pore Structure. Academic Press, New York.
19. Ergun, S. 1952. Fluid flow through packed columns, *Chem. Engng. Prog.* **48**: 89–94.
20. Fancher, G.H. and Lewis, J.A. 1933. Flow of simple fluids through porous materials, *Ind. Engng. Chem.* **25**: 1139–47.
21. Geertsma, J. 1974. Estimating the coefficient of inertial resistance fluid flow through porous media, *SPE* **4706**: 445–50.
22. Geertsma, J. 1976. Discussion of the effects of non-Darcy flow on the behaviour of hydraulically fractured gas wells, *J. Pet. Tech.* **883**: 1178–9.
23. Green Jr., L. and Duwez, P. 1951. Fluid flow through porous metals, *J. Appl. Mech.* **18**: 39–45.
24. Happel, J. and Brenner, H. 1973. Low Reynolds Number Hydrodynamics. Noordhoff International, Leyden.
25. Higdon, J.J.L. and Ford, G.D. 1996. Permeability of three-dimensional models of fibrous porous media, *J. Fluid Mech.* **308**: 341–61.
26. Ingmanson, W.L., Han, S.T., Wilder, H.D. and Myers, W.T. 1956. Resistance of Wire Screens to Flow of Water. Butterworths Scientific Publications, London.
27. Jackson, G.W. and James, D.F. 1986. The permeability of fibrous porous media, *Can. J. Chem. Eng.* **64**: 364–74.
28. Katz, D.L., Cornell, D., Kobayashi, R., Poettmann, P., Vary, J.A., Elenbass, J.R. and Weinaugg, C.F. 1959. Handbook of Natural Gas Engineering. McGraw-Hill, New York.
29. Kim, J., Lee, J. and Lee, K.O. 2001. Nonlinear correction to Darcy's law for a flow through periodic arrays of elliptic cylinders, *Physica A* **293**: 13–20.
30. Knupp, P. and Steinberg, S. 1993. Fundamentals of Grid Generation. CRC Press, Boca Raton.

31. Koponen, A., Kandhai, D., Hellne, E., Alava, M., Hoekstra, A., Kataja, M., Niskanen, K., Sloot, P. and Timonen, J. 1998. Permeability of three dimensional random fiber web, *Phys. Rev. Lett.* **80**: 716–9.
32. Kozeny, J. 1927. Über kapillare leitung des wassers im boden, *Sitzungsber. Akad. Wiss. Wien* **136**: 271–306.
33. Larson, R.E. and Higdon, J.J.L. 1989. A periodic grain consolidation model of porous media, *Phys. Fluids A* **1**: 38–46.
34. Lee, C.K., Sun, C.C. and Mei, C.C. 1996. Computation of permeability and dispersivities of solute or heat in periodic porous media, *Int. J. Heat Mass Transf.* **39**: 661–76.
35. Lee, S.L. and Yang, J.H. 1997. Modeling of Darcy-Forchheimer drag for fluid flow across a bank of circular cylinders, *Int. J. Heat Mass Transf.* **40**: 3149–55.
36. Ma, H. and Ruth, D.W. 1993. The microscopic analysis of high Forchheimer number flow in porous media, *Transport Porous Med.* **13**: 139–60.
37. Macdonald, I.F., El-Sayed, M.S., Mow, K. and Dullien, F.A.L. 1979. Flow through porous media–Ergun equation revisited, *Ind. Eng. Chem. Fundam.* **18**: 199–208.
38. Manwart, C., Aaltosalmi, U., Koponen, A., Hilfer, R. and Timonen, J. 2002. Lattice-Boltzmann and finite-difference simulations for the permeability for three-dimensional porous media, *Phys. Rev. E* **66**: 016702.
39. Miguel, A.F. 1998. Airflow through porous screens: from theory to practical considerations, *Energ. Buildings* **28**: 63–9.
40. Miguel, A.F. 1998a. Transport Phenomena through Porous Screens and Openings. Ph.D. dissertation, Wageningen University, the Netherlands.
41. Miguel, A.F. 2010. Fluid flow in tree-shaped constructal networks: porosity, permeability and inertial parameter. Defect and Diffusion Forum, in press.
42. Miguel, A.F. 2010a. On the estimation of permeability and inertial parameter for porous media using LBE. (Submitted)
43. Miguel, A.F. and Serrenho, A. 2007. On the experimental evaluation of the permeability in porous media using a gas flow method, *J. Phys. D* **40**: 6824–8.
44. Miguel, A.F., van de Braak, N.J., Silva, A.M. and Bot, G.P.A. 2001. Wind-induced airflow through permeable materials: part I, *J. Wind Eng. Ind. Aerodyn.* **89**: 45–57.
45. Nakayama, A., Kuwahara, F., Kawamura, Y. and Koyama, H. 1995. Three-dimensional numerical simulation of flow through a microscopic porous structure, *Proc. ASME/JSME Thermal Engineering Conference* **3**: 313–8.
46. Nield, D. and Bejan, A. 1998. Convection in Porous Media. Springer, New York.
47. Noman, R. and Archer, J.S. 1987. The effect of pore structure on non-Darcy gas flow in some low permeability reservoir rocks, *SPE* **16400**: 103–10.
48. Pan, C., Luo, L.-S. and Miller, C.T. 2006. An evaluation of lattice Boltzmann schemes for porous medium flow simulation. *Computers & Fluids* **35**: 898–909.
49. Pinela, J., Kruz, S., Miguel, A.F., Reis, A.H. and Aydin, M. 2005. Permeability-porosity relationship assessment by 2D numerical simulations. Proc. 16th International Symposium on Transport Phenomena, Prague.
50. Pope, S.B. 2000. Turbulent Flows. Cambridge University Press, Cambridge.
51. Qian, Y., d'Humieres, D. and Lallemand, P. 1992. Recovery of Navier–Stokes equations using a lattice-gas Boltzmann method, *Europhysics Lett.* **17**: 479–84.
52. Quintard, M. and Whitaker, S. 1994. Transport in ordered and disordered porous media II: generalized volume averaging, *Transport Porous Med.* **14**: 179–206.
53. Quispe, J.R., Rozas, R.E. and Toledo, P.G. 2004. Permeability-porosity relationship from a geometrical model of shrinking and flow simulations in two-dimensional pore network. *In*: Reis, A.H. and Miguel, A.F. (*Eds.*), Applications of Porous Media, CGE, Evora, pp. 15–25.
54. Quispe, J.R., Rozas, R.E. and Toledo, P.G. 2005. Permeability–porosity

relationship from a geometrical model of shrinking and lattice Boltzmann and Monte Carlo simulations of flow in two-dimensional pore networks, *Chem. Eng. J.* **111**: 225–36.
55. Sankaranarayanan, K., Kevrekidis, I.G., Sundaresan, S., Lu, J. and Tryggvason, G. 2003. A comparative study of lattice Boltzmann and front-tracking finite–difference methods for bubble simulations, *Int. J. Multiphas. Flow* **29**: 109–16.
56. Succi, S. 2001. The lattice Boltzmann equation for fluid dynamics and beyond. Oxford University Press, Oxford.
57. Vafai, K. 2000. Handbook of Porous Media. Marcel Dekker, New York.
58. Vafai, K. and Tien, C.L. 1981. Boundary and inertia effects on flow and heat transfer in porous media, *Int. J. Heat Mass Transf.* **24**: 195–203.
59. Valdes-Parada, F.J., Ochoa-Tapia, J.A. and Ramires, J.A. 2009. Validity of the permeability Carman-Kozeny equation: A volume averaging approach, *Physica A* **388**: 789–98.
60. Wang, X.H. and Liu, Z.F. 2004. The Forchheimer equation in two-dimensional percolation porous media, *Physica A* **337**: 384–8.
61. Ward, J.C. 1964. Turbulent flow in porous media, *J. Hydraulics Div., Proceedings of AXE* **90**: 1–12.
62. Whitaker, S. 1986. Flow in porous media I: A theoretical derivation of Darcy's law, *Transport Porous Med.* **1**: 3–25.
63. Whitaker, S. 1996. The Forchheimer equation: a theoretical development, *Transport Porous Med.* **25**: 27–36.
64. Zobel, S., Maze, B., Tafreshi, H.V., Wang, Q. and Pourdeyhimi, B. 2007. Simulating permeability of 3-D calendered fibrous structures, *Chem. Eng. Sci.* **62**: 6285–96.

2

Modeling of Transport Processes in Solid Oxide Fuel Cells

E. Vakouftsi[1] and F.A. Coutelieris[2,*]

ABSTRACT

In this chapter two approaches have been adopted to adequately simulate the transport phenomena occurred in a Solid Oxide Fuel Cell (SOFC). The first one is a one dimensional, relatively simple model, providing a rather simple and stable solution by using analytical manipulations and combined numerical techniques. The other is a three-dimensional approach including the coupled solution for flow, heat transfer, mass transport and electrochemistry. It is found that higher fuel consumption corresponds to higher current density values in an equivalent manner. Furthermore, it is clearly depicted that the overpotential increases along the anode, resulted by the consequent increment of the fuel consumption with the distance. Finally, the overpotential increases with current density, which is in accordance with previously presented experimental and theoretical results.

NOTATION

a_a	anodic charge transfer Tafel coefficient [–]
a_c	cathodic charge transfer Tafel coefficient [–]
a_k	concentration exponent used in Eq. (35) [–]
A/V	surface to volume ratio of the catalyst [m^{-1}]
[A/V]$_{eff}$	effective surface to volume ratio of the catalyst [m^{-1}]
b_{PR}	stoichiometric coefficient for products [–]
b_R	stoichiometric coefficient for reactants [–]
$[C_i] = \rho \dfrac{Y_i}{M_i}$	the near wall molar concentration of the i-th reacting species [mol m^{-3}]

---

[1] Department of Mechanical Engineering, University of Western Macedonia, Kozani, GREECE.
[2] Department of Environmental and Natural Resources Management, University of Ioannina, Agrinio, GREECE.
* Corresponding author : E-mail : fkoutel@cc.uoi.gr

$[C_{i,\text{ref}}]$	molar concentration of the species at a reference state at inlet [mol m^{-3}]
C_f	fuel concentration [mol m^{-3}]
$C_{f,\text{ref}}$	uniform reference fuel concentration at the inlet [mol m^{-3}]
C_f^*	dummy unknown for the fuel concentration, given by Eq. (16a) [mol m^{-3}]
D_i	mass diffusion coefficient of species i in the mixture [$m^2\,s^{-1}$]
$D_{i,\text{eff}}$	effective mass diffusion coefficient of species i given by Eq. [31] [$m^2\,s^{-1}$]
$D_{f,\text{eff}}$	effective mass diffusion coefficient of fuel in the mixture [$m^2\,s^{-1}$]
F	Faraday constant [C mol^{-1}]
h_i	enthalpy of species [J kg^{-1}]
h	total enthalpy of the mixture [J kg^{-1}]
I_{cell}	external current density [A cm^{-2}]
i	ionic current density [A cm^{-2}]
i^*	dummy unknown for the ionic current density, given by Eq. (16b) [A cm^{-2}]
$\underline{i}$	ionic current density vector used in Eq. (27) & Eq. (32) [A cm^{-2}]
$\underline{i_F}$	current density vector for the ionic phase [A cm^{-2}]
$\underline{i_S}$	current density vector for the electronic phase [A cm^{-2}]
i_e	index for the electrochemical reactions used in Eq. (35)
J_f	diffusion mass flux of the fuel [kgm$^{-2}\,s^{-1}$]
J_w	diffusion mass flux of the water [kgm$^{-2}\,s^{-1}$]
$\underline{J_i}$	diffusion mass flux vector of the species [kgm$^{-2}\,s^{-1}$]
j_T	transfer current [Am^{-3}]
j_0	exchange current density [Am^{-3}]
k	thermal conductivity [Wm^{-1} K^{-1}]
k_{eff}	effective thermal conductivity of the mixture [Wm^{-1} K^{-1}]
k_F	thermal conductivity of the fluid parts of the porous medium [Wm^{-1} K^{-1}]
k_S	thermal conductivity of the solid parts of the porous medium [Wm^{-1} K^{-1}]
K	coefficient including the geometrical and physical properties of the electrolyte used in Eq. (13), whose units depend on the order of the reaction $\gamma\,[(m^3)^{\gamma-1}\,\text{A mol}^{-\gamma}]$

m	number of hydrogen atoms of the fuel $C_n H_m O_p$ [–]
M_i	molecular weight of the i-th species [kg/mol]
n	number of carbon atoms of the fuel $C_n H_m O_p$ [–]
N	total number of reacting species, used in Eq. (35) [–]
N_g	total number of gas species in the system [–]
p	number of oxygen atoms of the fuel $C_n H_m O_p$ [–]
P	pressure [Pa]
Q	thermal sources due to exothermic/endothermic reactions [Wm^{-2}]
$R_{F,\,\text{eff}}$	effective ionic resistance in the catalyst layer [Ω m^{-1}]
$R_{S,\,\text{eff}}$	effective resistance in the solid phase [Ω m^{-1}]
R	universal gas constant [J mol^{-1} K^{-1}]
Sc	Schmidt number [–]
T	temperature [K]
$\underline{U}$	velocity vector [ms^{-1}]
x	spatial coordinate
Y_i	mass fraction of the ith chemical species [–]
$Y_{p,i}$	mass fraction in the pore fluid of the ith chemical species [–]

GREEK LETTERS

γ	order of reaction [–]
δ	diffusion length scale [m]
ε	porosity [–]
η	electrode overpotential [V]
η_a	anodic overpotential [V]
κ	permeability [m^2]
μ	viscosity [kgm^{-1} s^{-1}]
ρ	mass density of mixture [kgm^{-3}]
σ_F	ionic phase electrical conductivity [Ω^{-1} m^{-1}]
σ_S	solid phase electrical conductivity [Ω^{-1} m^{-1}]
τ	tortuosity [–]
$\overline{\overline{\tau}}$	shear stress tensor [Nm^{-2}]
φ_F	ionic potential of the fluid phase in the porous medium [V]
φ_S	electric potential of the fluid phase in the porous medium [V]
ω_i	production/destruction rate of the i-th chemical species in gas phase due to homogeneous and/or heterogeneous reactions [kgm^{-3} s^{-1}]

1. INTRODUCTION

1.1. General

During last decades globally, the gradually increasing energy demands, the rising environmental and health concerns as well as the depletion of natural recourses, intensified the problem of discovering new energy production technologies that would utilize current fuels more effectively by raising the efficiency of the production energy units and be more environmentally friendly reducing emissions to the atmosphere. Fuel cells seem to be one potential answer to this goal and offer to the energy policy makers the alternative choice of having energy independence under fully environmental-friendly conditions. At the same time, used autonomously or even be connected in parallel with the electrical grid, they offer stability and energy production with zero carbon dioxide emissions compared to the conventional energy systems. It should be clarified that, fuel cells can be considered as renewable energy sources applications on the understanding that they will operate with fuel produced by alternative energy sources and would be stored efficiently. In the long term, it will be possible for fuels cells to run on environmentally neutral fuels produced by using sustainable and/or renewable energy sources such as solar and wind power[1].

Fuel cells are electrochemical devices that directly convert the chemical energy of the gaseous fuels into electricity, overcoming Carnot limitations. Theoretically, they can reach high efficiencies and unlike batteries, are able to provide continuous supply of electric power when replenished with fuel. Furthermore, fuel cells are very compact units without moving parts, assuring therefore silent operation, fewer materials design constraints and quite large portability. A typical fuel cell stack consists of an anode and a cathode compartment between which a catalyst (electrolyte) layer exists. The atmospheric air flows in one gas channel, while the other is continuously supplied by a feeding mixture (usually a hydrogen-rich mixture containing some higher hydrocarbons as well as carbon monoxide and dioxide). The principle operation of a typical fuel cell, in most cases, incorporates the formation of water from hydrogen and oxygen and the production of electricity and heat. The basic types of fuel cells categorized according to the electrolyte used are as follows: (i) the Alkaline Fuel Cell (AFC); (ii) the Proton Exchange Membrane Fuel Cell (PEMFC); (iii) the Phosphoric Acid Fuel Cell (PAFC); (iv) the Molten Carbonate Fuel Cell (MCFC) and (v) the Solid Oxide Fuel Cell (SOFC).

The first three types operate at relatively low temperatures (AFC at about 333–500 K, PEMFC at about 353–393 K and PAFC at about 150–220 K). It should be outlined that, fuel processor and cleaning systems should be incorporated in such units despite their high cost since direct supply with pure hydrogen is imperative (for example AFCs are sensitive to carbon dioxide and PEMFCs to carbon monoxide). Unlike low temperature fuel cells, MCFC and SOFC operate at high temperatures, varying between 873–973 K and 873–1273 K for MCFC and SOFC, respectively, and as a result the use of precious metal electrocatalysts is redundant[2, 3].

Among the other types of fuel cells, SOFCs present considerable advantages due to their high operating temperature, which favors electrochemical reaction kinetics permitting the use of low cost metal catalysts while it promises higher efficiencies when recovering the high energy waste heat (electrical efficiencies of 45–50% can be achieved). Furthermore, SOFC operation allows external or even internal reforming reaction to occur, thus fuels such as carbon monoxide and hydrocarbon fuels that are considered as poison to low temperature fuel cells, can be used with minimal fuel processing[4–10]. However, apart from the benefits mentioned above, there are some drawbacks regarding the materials used caused mainly by the high temperature, such as thermal expansion of the different fuel cell components and stability issues due to mechanical deficiencies and catalyst deactivation that need to be confronted. A lot of effort has been made in fabricating new, cost effective materials with improved physical and electrochemical characteristics which would enhance both SOFC performance and stability[11].

Owing to the variety of SOFCs advantages, they are considered as ideal candidates for various applications. As mentioned previously, they operate at high temperatures and the waste thermal energy (waste heat) can be recovered in combined heat and power systems (CHP) for production of electricity and power in large scale distributed power generation systems as well as in small-scale domestic heat and power production units, taking full advantage of the fuel used. At the same time, SOFCs are considered to be useful as auxiliary power units (APUs) for various electrical systems in cooking and transportation, such as in vehicles air conditioning supplies and for portable electronics i.e., cell phones and personal computers. In addition, SOFC can be used for chemical cogeneration of electricity and chemical compounds with the use of appropriate materials. Furthermore, they can be the alternative choice for remote distributed power generation either in areas that there is no grid supply, such as in isolated islands, or when local power production is necessary, such as in small power units[12–17].

A typical configuration of a SOFC as well as the fundamental processes taking place within it, are schematically presented in Fig. 2.1. The basic components of a typical Solid Oxide Fuel Cell are the anode and the cathode electrode separated by a dense electrolyte, which are usually referred to as PEN (Positive electrode/Electrolyte/Negative electrode). Fuel is supplied at the anode electrode, while air is introduced at the cathode electrode respectively. The electrolyte presents high ionic and minimum electronic conductivity and at the same time it prevents fuel crossover and mixing of the relevant gas supplies. The fuel is electrochemically oxidized by the oxygen ions, O^{2-}, formed at the cathode electrode by oxygen reduction reaction and migrate through the electrolyte to the anode. It is evident that the same number of electrons must be transferred through the electrolyte in order to preserve equilibrium between the electrodes. The electrons released at the anode compartment flow via an external electrical circuit.

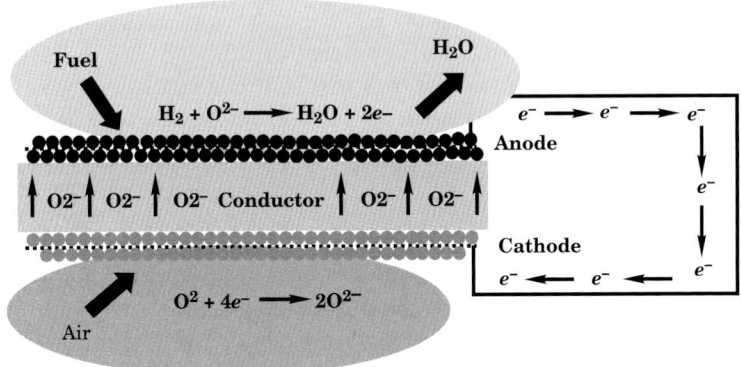

Fig. 2.1. Fuel cell configuration

SOFCs can be classified according to their temperature operation in low-, intermediate- and high-temperature SOFCs with operating temperature between 773 K–823 K, 823 K–1073 K and 1073 K–1273 K, respectively. Regarding geometrical aspects and designs, one of the most commonly used geometries is the tubular one, initially introduced by Siemens-Westinghouse[18]. The principle of operation is that the tubular fuel cell is comprised of a unique tube and the air is supplied internally, while the fuel is thought to flow at the surroundings. Another design is the planar geometry in which the PEN assembly is surrounded by the collectors[19, 20]. According to the flow direction of air and fuel it could be subdivided into (*i*) co flow planar SOFC, where the air and the fuel have the same direction, (*ii*) counter-flow, where they have opposite directions and (*iii*) cross-flow, where they have perpendicular flows to each other[21]. The tubular and planar geometries can be found as electrolyte or electrode (anode or cathode) supported[22, 23].

1.2. About Modeling

In general, it is widely known that mathematical modeling is considered as a useful, time and cost saving tool compared to experimental procedures, which is quite demanding in terms of human potential, equipment needed and money spent. In particular, studying the effect of operational variables such as velocity, temperature and pressure on fuel cell performance, would contribute in better understanding and deep knowledge of the processes occurring, on prevailing mechanisms as well as on their interactions.

Additionally, the influence of both geometric characteristics, such as length of the fuel cell and the width of the diverse PEN components, and material microstructure properties, such as porosity, surface to volume area, tortuosity, porous diameter, thermal and electronic conductivity, etc. would not only provide information about the fuel cell behavior under different load conditions, but would reveal crucial detail for future optimization as well. Finally, innovative geometries could be created and examined, being probably indicative for future construction in real scale units[24–26].

There are several studies in the relative literature that examine SOFCs through modelling, focused either on the microstructure behaviour of the system in terms of material properties (microscopic modelling) or on general operational quantities and conditions such as velocity and temperature (macroscopic modelling).

On the first category, one dimensional model approaches, usually attempt to describe reaction and mass transport phenomena inside the PEN assembly. Costamagna et al.[27] created a theoretical one-dimensional model, taking into account electronic and ionic transport together with the electrochemical reactions. They examined the influence of morphology on electric resistance and the effect of electrode thickness on fuel cell performance. On their further work at the same year[28], they discussed some optimization considerations as well. Later, Chan et al.[29] adapted the model of Costamagna et al. to study the effects of all forms of polarization on anode electrodes by using the molecular and Knudsen diffusion. They found out that polarization reaches a minimum for certain particle size and that there is a strong link between particle size and anode thickness, while it was outlined that hydrogen concentration influences polarization significantly.

Based on the assumptions made for an anode supported SOFC in the work of Yakabe et al.[30], Suwanwarangkul et al.[31] creating an one dimensional SOFC model evaluated concentration overpotential using various mass transport models, i.e., the Ficks's model, the dusty-gas model and the Stefan-Maxwell model. Achmann et al.[32] focused on mathematical modeling of mass and heat transport phenomena in a 2-D planar SOFC incorporating an anode and a cathode substrate. They used the mean transport pore model (MTPM) to describe the mass transport phenomena in the porous electrodes, which was thought to be advantageous compared to the approach proposed by Yakabe et al.[30]. Chan et al.[33] incorporated mass transport phenomena and cathode electrode microstructure for examining cathode overpotential under different operational conditions. Shi et al.[34] developed a 2-D isothermal model which was validated in regard to the separate losses of each PEN component. Presvytes and Vayenas in 2007[35] presented a model describing surface diffusion reaction mechanisms taking into account that reaction zone is extended on the Ni catalytic surface over several hundreds of Å and it is not limited on the three phase boundary of the anode electrode. Their model was solved for various sphere-based Ni particle geometries and showed good agreement with experimental results.

Apart from the microscopic models analyzed previously, a great number of theoretical investigations are dedicated to macroscopic simulations. One of the very first attempts in fuel cell modeling was that of Debenetti and Vayenas[36], where a single-cell was simulated as a solid state electrocatalytic reactor. Later on, Vayenas et al.[37] proceeded with a two dimensional model in order to adequately represent the variation of crucial parameters inside the cell and overcome the 1-D imposed limitations. They created a 2-D cross-flow monolithic fuel cell taking into account the transport of oxygen ions through the catalyst layer.

Achenbach et al.[38], performed a time-dependent mathematical simulation for a planar SOFC and studied the effect of different flow regimes on temperature and current density distributions, taking into account internal methane steam reforming and water gas shift reaction, both in equilibrium. One year later, Bessette et al.[39] proposed a three-dimensional model for a tubular SOFC, where quite good agreement with experimental results for both temperature and current was achieved. Ferguson et al.[40], produced a mathematical code flexible in geometries, by using C programming language and finite volume approach. This code was validated against planar, cylindrical and tubular geometry as well as under hydrogen or methane feedstock supply.

Costamagna[24] presented in 1997 a new concept for a rectangular SOFC integrated with an air pre-heater where the two dimensional time dependent model predicted power and efficiency as well as current density and temperature. One year later, Costamagna and Honegger[41] extended this work by introducing circular shaped cell stacks operating under high fuel utilization.

In addition, Lehnert et al.[42] developed a 1-D model analysing the transport of gases in the porous anode, based on three microstructural material properties i.e., porosity, tortuosity and mean value of pore diameter. Yakabe et al.[30,43] calculated the thermal stresses occurring in a 3-D planar SOFC by taking into account transport phenomena and reactions implemented by the commercial STAR-CD program. They concluded that co-flow is beneficial in overcoming the adverse effects that steep thermal stresses, developed by the internal reforming reaction, have on fuel cell performance. Finally, Aguir et al., model[44] investigates the mismatch between the thermal load associated with the rate of steam reforming reaction and the local amount of heat available from the fuel cell reactions. Both counter and co flow configurations have been examined under different operational conditions. At the same year Racknagle et al.[45] developed a simulation tool for a three dimensional SOFC using the commercial software

STAR CD, where they outlined that the co-low case performed better in term of thermal stability.

Recently, the work of Yakabe and Sakurai[46] focused on the current path inside the cell components considering the geometric characteristics of the planar SOFC. They pointed out the existence of diagonal electric current in the electrolyte and in the interconnects. The species transport inside a porous anode taking into account Stefan-Maxwell model and Knudsen diffusion and assuming that the reactions occurred at a thin zone layer inside the porous media was investigated by Hussain et al.[47] for the calculation of molar fractions and concentration overpotential values.

A 2-D model assuming isothermal conditions for a syngas fed SOFC implying the water gas shift reaction and other electrochemical reactions was performed and validated against experiments produced by Suwanwarankul et al.[48]. The work of Hwang et al.[49] aims in investigating the species electrochemical characteristics and mass transport in a free breathing cathode of a planar SOFC, incorporating the combinatory effect of the coupled transport phenomena (mass, momentum, species and charged species) using the finite volume method. The works of Paloukis and Neophytides[26] and Zhu et al.[50] both account for cogenetarion of syngas and electricity in SOFC. Applying their new concept, Paloukis and Neophytides managed to increase the overall efficiency of a methane-fed co-generated unit of synthesis gas production and electricity by periodically inversing the flow through the reactor and taking into account methane oxidation.

Coutelieris in 2008[51] developed a simplified 1-D model and a more complicated 2-D, one in order to accurately simulate the transport phenomena occurred in the anode of a fuel cell and predict current densities and overpotential values. The comparison between these models underlined the necessity of multi-dimensional modelling. A finite element based algorithm was applied by Arpino and Massaroti[52] for the modeling of transport phenomena in a SOFC analyzing the effect of inlet fuel velocity and current density on the fuel cell performance. Suzuki et al.[53] focused on methane fed anode supported SOFCs and claimed that increasing the anode pore size can improve output voltage. Apart from the three dimensional model, they created a simplified simulation code which could be applied at complex geometries and promises to predict output voltage, temperature and current density eliminating the computational cost. Hussain et al.[54] also aimed in modeling an anode supported SOFC for various fuel feedstocks. Their concept of treating the electrodes as two distinct areas (i.e., the backing layer and the reaction zone layer where electrochemical reactions take place) allowed the incorporation of microscopic characteristics into their model which discovered that anode overpotential is determinative for the potential losses in the fuel cell.

1.3. Content Guide

The most significant aspects of transport processes occurring in a typical SOFC are examined here in terms of mathematical modeling. More precisely, the present work is organized as follows:

Firstly, a rather simple one-dimensional approach is used in order to develop a predictive mathematical model for the transport phenomena occurring along the axis of a generic typical fuel cell anode. A numerical solution for the mass and charge transport equations was obtained by using specific numerical methods. To obtain this solution, extensive assumptions and simplifications (like isothermal operation, pre-defined current at outlet, etc.) should be involved, therefore it significantly diverges from reality in terms of quantitative results. Macroscopic electrochemical quantities, such as current densities and overpotential values have been estimated throughout that model. Obviously, significant effort to overcome some of the above mentioned deficiencies has been incorporated here.

Further to this simple modeling approach, more detailed three-dimensional simulations have been also implemented for the same processes in order to overcome superficial assumptions and mathematical manipulations. For this purpose, averaged equations are applied in both flow regions and porous media taking into account convective and diffusion effects. As a result, the three dimensional steady state model produced is capable of predicting macroscopic properties such as velocity, temperature and species distribution profiles as well as the above mentioned electrochemical characteristics.

2. THEORETICAL BACKGROUND: ELECTROCHEMICAL AND SURFACE REACTIONS

The most significant advantage of SOFC compared to other fuel cell types, as mentioned previously, is that they allow wide fuel flexibility due to high operational temperature and the materials used. As a result, apart from hydrogen (H_2), SOFC can be fed with carbon monoxide (CO) and synthesis gas (H_2 and of various ratios), as well as with natural gas and hydrocarbon fuels (coal fuels). The electrochemical reactions are taking place at the interface between catalyst and ionic conductor and the gas phase, the so called three-phase boundary layer (TPB). Precisely, at the anode electrode the electrochemical oxidation reactions of H_2 and CO can be described as

$$H_2 + O^{2-} \rightarrow H_2O + 2e^- \qquad \text{...[1 (a)]}$$
$$CO + O^{2-} \rightarrow CO_2 + 2e^- \qquad \text{...[1 (b)]}$$

Experiments showed that the electrochemical oxidation rate of is significantly higher (1–2.5 times) than that of CO[55–57], thus a large number of models in existing bibliography neglect the contribution of in the production of current density.

With the use of appropriate anode cermets, such as Cu-YSZ and instead of Ni-YSZ, the direct oxidation of hydrocarbons and/or alcohols is possible in SOFCs[58–60]. In general, assuming that fuel can be expressed as $C_n H_m O_p$ where $p = 0$ or $p = 1$, denoting hydrocarbon or alcohol fuels, respectively, the electrochemical oxidation reactions can be expressed as

$$C_n H_m + (2n + 0.5m) O^{2-} \to nCO_2 + 0.5m H_2O + (4n + m)e^- \text{ where } p = 0$$
...[2 (a)]

$$C_n H_m O + (2n + 0.5m - 1) O^{2-} \to nCO_2 + 0.5m H_2O + (4n + m - 2) e^-$$
where $p = 1$. ...[2 (b)]

It is common that atmospheric air (humidified or not) and scarcely pure O_2 is introduced in the cathode gas channel. At the cathode electrode, the electrochemical oxygen reduction is taking place, written as,

$$0.5 O_2 + 2e^- \to O^{2-} \qquad ...(3)$$

and in case of fuels in the form $C_n H_m O_p$ is accordingly

$$(n + 0.25m) O_2 + (4n + m) e^- \to (2n + 0.5m) O^{2-} \text{ where } p = 0 \quad ...[4 (a)]$$

$$(n + 0.25m - 0.5) O_2 + (4n + m - 2) e^- \to (2n + 0.5m - 1) O^{2-}, \text{ where } p = 1.$$
...[4 (b)]

In practice, apart from the electrochemical reactions, numerous chemical reactions homogeneous or heterogeneous ones may appear in a SOFC. The most commonly considered in simulation models are presented below. In general, the water gas shift reaction (WGS),

$$CO + H_2O \leftrightarrow CO_2 + H_2 \qquad ...(5)$$

is applied when both hydrogen and carbon monoxide are at least present and is usually assumed to be in equilibrium.

For effective use of hydrocarbon fuels they should be partially reformed to H_2 and through internal or external steam reforming reaction as follows[5]

$$C_n H_m + n H_2O \to nCO + (0.5m + n) H_2 \text{, where } p = 0 \quad ...[6 (a)]$$

$$C_n H_m O + (n - 1) H_2O \to nCO + (0.5m + n - 1) H_2 \text{, where } p = 1.$$
...[6 (b)]

Additionally, the presence of hydrocarbons may lead to carbon formation and in order to study the adverse effects on catalyst deactivation the Boudouard reaction[61–65]

$$2CO \leftrightarrow C + CO_2 \qquad ...(7)$$

as well as the cracking reaction (for those hydrocarbons it is applicable)

$$C_n H_m \leftrightarrow nC + 0.5\, mH_2 \qquad ...(8)$$

may be introduced.

3. MODELING

As mentioned before, modeling is a powerful tool for the study of fuel cells. Hereafter, we are going to present a simplified one-dimensional model, where all the transport processes is assumed to take place only in the axial direction, and a more complicated 3-D one, where flow, heat transfer, mass transport and electrochemistry are taken into account.

3.1. One Dimensional Model

For the one dimensional model, a typical 2-D cut of a fuel cell containing the anode area is schematically presented in Fig. 2.2. In particular, the anode is divided into three regions: an electrolyte region, an active catalyst region (catalyst layer) that provides a catalytic site for the oxidation of fuel and a diffusion region (diffusion layer) composed of highly porous and electronically conductive material. The electrolyte is assumed to be dense, while the diffusion layer is located adjacent to the flow channel and favoring the distribution of reactants to the catalyst layer. From modeling point of view, the area of interest is the catalyst layer. The fuel fed to the anode compartment of the SOFC is assumed to be a hydrogen rich mixture ($H_2 > 90\%$). It is also assumed that hydrogen reacts only partially in the catalyst/electrolyte interface; therefore a remaining portion follows the stream to the outlet and/or crossovers to the cathode compartment.

Fig. 2.2. Schematic diagram of the simulated (anode) area

By assuming that the pressure in both anode and cathode compartments is constant and that the prevailing transport processes occur mainly at the direction parallel to the cell axis, the mass flux in the catalyst layer for steady-state isothermal conditions can be expressed as follows

$$J_f = -D_{f,\text{eff}} \frac{dC_f}{dx} + C_f J_w \qquad ...(9)$$

where C_f is the fuel concentration, $D_{f,\text{eff}}$ is the effective diffusivity of fuel in the catalyst layer and J_f, J_w are the mass fluxes of the fuel and the water, respectively. In the above equation, the convection term has been neglected since the velocity is very low within the catalyst layer[66, 67]. By taking into account the material balance for the water, its mass flux is given as

$$J_w = \frac{I_{\text{cell}} - i}{4F} \qquad ...(10)$$

where $\dfrac{I_{\text{cell}} - i}{4F}$ is the flux of the produced water mass. I_{cell} and i denote the cell and ionic current density, respectively, while F represents the Faraday's constant.

It is worth noticing that the oxidation of fuel along the axis of the fuel cell is linear with the ionic current density, thus the variation of mass flux is analogous to the spatial evolution, x, of current density, i.e.,

$$\frac{dJ_f}{dx} = -\frac{1}{4F}\frac{di}{dx} \qquad ...(11)$$

By differentiating Eq. (9) and using Eq. (11), the governing differential equation for mass transfer within the catalyst layer of a SOFC becomes as follows

$$D_{f,\text{eff}}\frac{d^2 C_f}{dx^2} = \frac{I_{\text{cell}} - i}{4F}\frac{dC_f}{dx} + \frac{I_{\text{cell}}}{4F}(1 - C_f)\frac{di}{dx} \qquad ...(12)$$

Since the last equation combines the mass transport with the local current density, it is necessary to develop the relative expression for i. Starting from the well known Butler-Volmer equation and after same simplifications, a Tafel-type equation can be obtained

$$\frac{di}{dx} = K(C_f)^\gamma e^{\frac{4a_a F}{RT}\eta_a} \qquad ...(13)$$

where γ is the order of the reaction, a_a is the anodic charge transfer coefficient, η_a is the anodic overpotential, R is the universal gas constant, T is the temperature and K is a coefficient including the geometrical and physical properties of the electrolyte. By involving the local potentials at catalyst layer and at ionomeric phase as well as the Ohm's low, the variation of the overpotential down within the catalyst layer is given as[68, 69]

$$\frac{d\eta}{dx} = (R_{F,\text{eff}} + R_{S,\text{eff}})i - R_{S,\text{eff}} I_{\text{cell}} \qquad ...(14)$$

where $R_{F,\text{eff}}$ is the effective ionic resistance in the catalyst layer and $R_{S,\text{eff}}$ is the effective resistance in the solid phase.

Again, by differentiating Eq. (13) and using Eq. (14), it is easy to obtain the governing differential equation for current density within the catalyst layer as follows

$$\frac{d^2 i}{dx^2} = \left\{\frac{\gamma}{C_f}\frac{dC_f}{dx} + \frac{4a_a F}{RT}\left[(R_{F,\text{eff}} + R_{S,\text{eff}})i - R_{S,\text{eff}} I_{\text{cell}}\right]\right\}\frac{di}{dx} \qquad ...(15)$$

In order to numerically integrate the system of Eqs. (12) and (15), it is necessary to decrease the order by introducing new unknowns C_f^* and i^* (downgrade technique), thus a system of four equations is produced as follows:

$$\frac{dC_f}{dx} = C_f^* \qquad \text{...[16 (a)]}$$

$$\frac{di}{dx} = i^* \qquad \text{...[16 (b)]}$$

$$\frac{dC_f^*}{dx} = \frac{1}{4FD_{f,\text{eff}}} \left[(I_{\text{cell}} - i) C_f^* + I_{\text{cell}} (I_{\text{cell}} - C_f) i^* \right] \qquad \text{...[16 (c)]}$$

$$\frac{di^*}{dx} = \left\{ \frac{\gamma}{C_f} C_f^* + \frac{4a_a F}{RT} \left[(R_{F,\text{eff}} + R_{S,\text{eff}}) i - R_{S,\text{eff}} I_{\text{cell}} \right] \right\} i^* \qquad \text{...[16 (d)]}$$

The boundary conditions accompanying the above system are as follows

$$C_f^*(x=0) = 0 \qquad \text{...[17 (a)]}$$

$$C_f(x=0) = C_{f,\text{ref}} \qquad \text{...[17 (b)]}$$

$$i(x=0) = 0 \qquad \text{...[17 (c)]}$$

$$i^*(x=0) = 0 \qquad \text{...[17 (d)]}$$

where is the uniform reference fuel concentration at the inlet. Although it is rather obvious that

$$i(x = \text{cell length}) = I_{\text{cell}} \qquad \text{...(18)}$$

the analysis here does not use the relation above for the description of the boundary condition as in[66], because it is better for the accurate convergence of a first order system of differential equations to be accompanied by initial conditions at the same point of the domain[70]. Therefore, the above Eq. (18) is considered as one of the criteria against which the solution of the system must be evaluated.

3.2. Three Dimensional Model

3.2.1. Transport Phenomena in Gas Channel

For the three dimensional model, the fundamental transport phenomena occurring are the flow and heat transfer, the mass transport as well as the charge transfer, which are described by the equations of continuity, momentum and species (neutral or charged) conservation equation. Therefore, the velocity field, the temperature profile, the gas composition and the electric potential distribution of the fuel cell can be calculated. For better insight on the processes appearing, the fuel cell geometry can be divided in two general regions: one is the non porous region, which refers to the gas channels and the other includes all the porous parts of the fuel cell i.e., anode and cathode electrode, the dense electrolyte and even the current collectors. A two dimensional cut of the fuel cell is depicted in Fig. 2.3, where the dimensions have been chosen in accordance with Ramakrishna et al.[25].

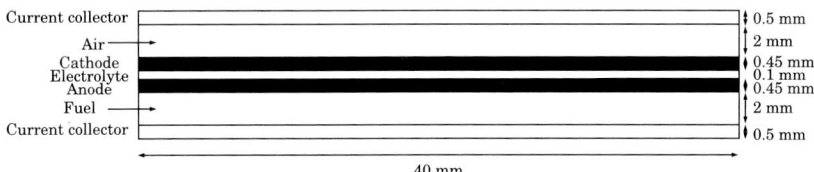

Fig. 2.3. Simulated SOFC geometry

First of all, as far as the non porous regions are concerned, the mass conservation equation (continuity equation) can be written as

$$\frac{\partial \rho}{\partial t} + \nabla \cdot (\rho \underline{U}) = 0 \qquad \ldots(19)$$

where $\underline{U}$ is the velocity vector, ρ is the fluid density and t is the time.

The velocity in fuel cell gas channels is usually quite low, so the assumption of laminar flow can be considered accurately. As a result, for compressible fluids and by neglecting gravitational effects, the momentum equation is

$$\frac{\partial}{\partial t}(\rho \underline{U}) + \nabla \cdot (\rho \underline{U}\underline{U}) = -\nabla P + \nabla \cdot \overline{\tau} \qquad \ldots(20)$$

where P is the pressure and $\overline{\tau}$ is the shear stress tensor.

By neglecting radiation, the energy conservation equation becomes

$$k\frac{\partial}{\partial t}(\rho h) + \nabla \cdot (\rho \underline{U} h) = \nabla \cdot (k\nabla T) + \frac{\partial P}{\partial t} + \dot{Q} \qquad \ldots(21)$$

where h is the total enthalpy of the fluid, k is the thermal conductivity of the mixture and $\dot{Q}$ describes the additional thermal sources due to exothermic/endothermic reactions.

The mass fraction of each gas species can be computed by the species conservation equation as follows

$$\frac{\partial}{\partial t}(\rho Y_i) + \nabla \cdot (\rho \underline{U} Y_i) = \nabla \cdot \underline{J_i} + \dot{\omega}_i \qquad \ldots(22)$$

where Y_i are the mass fractions of the i-th chemical species, $\dot{\omega}_i$ is the production/destruction rate of the ith chemical species in gas phase due to homogeneous reactions, and $\underline{J_i}$ is the species mass diffusion flux.

In literature, there are several approaches used to calculate the mass diffusion flux, but a detailed description and discussion of them is out of the scope of this analysis. The simplest model used to calculate the species diffusion flux is the Fick's law

$$\underline{J_i} = \rho D_i \nabla Y_i \qquad \ldots(23)$$

where D_i is the mass diffusion coefficient of species i in the mixture.

Although several models for the estimation of diffusivity have been presented in the relative literature, in the present study the mass diffusion coefficient for gas channels is obtained trough Schmidt number, Sc, according to the relation

$$D_i = \frac{\mu}{Sc} \qquad \ldots(24)$$

where μ is the viscosity of gas species.

Obviously, all the above differential equations are strongly coupled and should be integrated along with the appropriate boundary conditions, which are application depended.

3.2.2. *Transport Phenomena in Porous Media*

Porous materials are widely used in SOFC applications and particularly at the anode and cathode electrodes, since they present large catalytic surface areas. All the transport phenomena described above by the Eqs. (20)–(22) have to be slightly modified to incorporate the porous nature of the electrodes and simulate the transport processes there since the porous material structure implies mass transport limitations.

Analytically, the mass conservation equation in a porous medium can be written as

$$\frac{\partial}{\partial t}(\varepsilon \rho) + \nabla \cdot (\varepsilon \rho \underline{U}) = 0 \qquad \ldots(25)$$

where ε is the porosity of the medium representing the volume occupied by the pores to the total volume of the porous media.

In porous regions, the momentum conservation equation becomes

$$\frac{\partial}{\partial t}(\varepsilon \rho \underline{U}) + \nabla \cdot (\varepsilon \rho \underline{UU}) = -\varepsilon \nabla P + \nabla \cdot (\varepsilon \overline{\tau}) + \frac{\varepsilon^2 \mu}{\kappa} \underline{U} \qquad \ldots(26)$$

where μ is the viscosity of the fluid and κ is the permeability representing the square of the volume to surface area ratio of the porous material. Note that the last term of the above equation represents the Darcy's law and describes the superficial velocity in the porous medium[71].

The energy conservation equation, inside a porous medium, written in terms of enthalpy takes into account thermal convection and conduction, species diffusion contributions as well as chemical reaction (neutral or electroctrochemical) effects on temperature. It can be written as

$$\frac{\partial}{\partial t}(\varepsilon \rho h) + \nabla \cdot (\varepsilon \rho \underline{U} h) = \nabla \cdot \left(k_{\text{eff}} \nabla T + \sum_{i=1}^{N_g} \underline{J}_i h_i \right) + \varepsilon \overline{\tau} \nabla \underline{U}$$

$$+ \varepsilon \frac{dP}{dt} - j_T \eta + \frac{|\underline{i}|^2}{\sigma} \qquad \ldots(27)$$

where k_{eff} is the effective thermal conductivity of the mixture, N_g is the total number of gas species in the system and h_i is the enthalpy of the i-th species. The temperature gradient is significantly affected inside the porous material by the electrochemical reactions and consequently both Joule heating and electrical work affect the energy transfer. This effect is described by the last two terms in the energy transfer equation.

By taking into account the combination of porous and solid parts of the porous medium, the effective thermal conductivity of this medium can be defined as[72]

$$k_{\text{eff}} = -2k_S + \frac{1}{\dfrac{\varepsilon}{2k_S + k_F} + \dfrac{1-\varepsilon}{3k_S}} \qquad \ldots(28)$$

where k_F is the thermal conductivity of fluid parts of the porous medium and k_S is the thermal conductivity of solid parts of the porous medium.

The conservation equation of the i-th gas species is given by the relation

$$\frac{\partial}{\partial t}(\varepsilon \rho Y) + \nabla \cdot (\varepsilon \rho \underline{U} Y_i) = \nabla \cdot \underline{J}_i + \dot{\omega}_i \qquad \ldots(29)$$

where the Fick's model can be also applied, given as

$$\underline{J}_i = \rho D_{i,\text{eff}} \nabla Y_i \qquad \ldots(30)$$

where $D_{i,\text{eff}}$ is the effective mass diffusion coefficient of species i.

In order to take into account the porosity, ε, and the tortuosity, τ, of the porous medium, a number of correlations can be found in literature such as the Daggan model, but the most applicable one is the Bruggeman correlation, defined as[73]

$$D_{i,\text{eff}} = \varepsilon^\tau D_i \qquad \ldots(31)$$

For the calculation of species diffusion fluxes, apart from the Fick's model implemented in the present study, several models such as Stefan-Maxwell model for multi component systems and Dusty Gas model (DGM), which incorporates both the Stefan-Maxwell formulation and the Knudsen diffusion can be applied[74, 75].

Regarding the charge conservation, in conducting materials, the sum of all current flows should be zero based on elecro-neutrality. Thus, the current conservation equation is

$$\nabla \cdot \underline{i} = 0 \qquad \ldots(32)$$

where $\underline{i}$ is the current density vector.

However, based on the analysis of Newman and Tabias[76], the charge transport consists of electronic and ionic phase transports. During electrochemical reactions, electrons are either transferred from the pores (ionic phase) to the solid region (electronic phase) or vice versa, i.e., the electron transfer is expressed as the transfer current, j_T, where

$$-\nabla \cdot \underline{i}_F = \nabla \cdot \underline{i}_S = j_T \qquad \ldots(33)$$

where i_F is the current density vector flowing through the pores (ionic phase) and i_S is the current density vector flowing through the solid parts of the porous medium (electronic phase). By applying the Ohm law, the transfer current yields

$$\nabla \cdot (\sigma_F \nabla \varphi_F) = -\nabla \cdot (\sigma_S \nabla \varphi_S) = j_T \qquad \ldots [34\,(a)]$$

where σ_F and σ_S are the ionic phase conductivity and the solid phase conductivity, respectively, while φ_F and φ_S are the ionic potential of the fluid and of the electric potential of the solid, respectively.

For the non-conducting electrolyte the above equation Eq. [34 (a)] becomes

$$\nabla \cdot (\sigma_F \nabla \varphi_F) + \nabla \cdot (\sigma_S \nabla \varphi_S) = j_T \qquad \ldots [34\,(b)]$$

and the current transfer j_T can be defined through the Butler-Volmer equation as

$$j_T = \frac{j_0 (A/V)}{\prod_{i=1}^{N} [C_{i,\text{ref}}]^{a_{i_e}}} \left[\exp\left(\frac{a_a F}{RT}\eta\right) - \exp\left(-\frac{a_c F}{RT}\eta\right) \right] \prod_{i=1}^{N} [C_i]^{a_{i_e}} \qquad \ldots(35)$$

where i_e indexes the electrochemical reactions, j_0 is the exchange current density, a_a and a_c are the anodic and cathodic charge transfer coefficients, respectively, determined by the Tafel plots, is the total number of reacting species, $[C_i]$ is the near wall molar concentration of the ith reacting species or expressed in mass fractions is $[C_i] = \rho \dfrac{Y_i}{M_i}$, where M_i is the molar weight of the i-th species, $[C_{i,\text{ref}}]$ is the molar concentration at a reference state at inlet and a_k is the concentration exponent. The overpotential, η, can be expressed as the potential difference between the solid phase and the porous phase potential,

$$\eta = \varphi_S - \varphi_F \qquad \ldots(36)$$

At surface reactions occurring at porous/catalyst interfaces, the volumetric reaction rate (production/destruction of a gas species i) can be computed assuming a balance between the reaction flux and the diffusion flux on the surface[77]. This rate is given as

$$\dot{\omega} = \rho D_i \frac{Y_i - Y_{P,i}}{\delta} [A/V]_{\text{eff}} \qquad \ldots(37)$$

where $Y_{P,i}$ is the mass fraction of the i-th species in the pore fluid, $[A/V]_{\text{eff}}$ denotes the effective surface to volume ratio of the catalyst and represents the catalyst load and δ is the diffusion length scale. For the electrochemical reactions, the production/destruction rate is expressed through the current transfer by the relation

$$\dot{\omega} = (b_{PR} - b_R) \frac{j_T}{F} \qquad \ldots(38)$$

where b_{PR} and b_R are the stoichiometric coefficients of the products and reactants, respectively.

4. RESULTS AND DISCUSSION

4.1. Results for One Dimensional Model

The numerical solution of the systems of differential Eqs. [16 (*a*)], [16 (*b*)], [16 (*c*)] and [16 (*d*)] was achieved with the use of a non-linear shooting scheme in conjunction with the multidimensional Newton algorithm[78]. The space was discretized through a constant-step mesh, while the resulting nonlinear system of ordinary differential equations was solved by using the 4th order Runge-Kutta method. A FORTRAN code[51] was developed to implement the above mentioned numerical scheme, while the values used for the constants/parameters involved in the system are given in the following Table 2.1. The computational time in a simple home PC was of order of seconds, depended on the initial guesses for the Newton method.

Table 2.1. Parameter values for the one-dimensional model

a_a	0.7[25]
T (K)	1073
γ	0.25[66]
$C_{f,\text{ref}}$ (kg)	3.069 × 10^{-7}
$R_{f,\text{eff}}$ ($m^2 s$)	1.9 × 10^{-10}[79]
R (Jmol^{-1} K^{-1})	8.314
F (Cmol^{-1})	96484
$R_{F,\text{eff}}$ (Ωm^{-1})	31250[80]
$R_{S,\text{eff}}$ (Ωm^{-1})	26455030[81]

The hydrogen concentration obtained by the one-dimensional model as function of dimensionless normalized length is presented in Fig. 2.4 for various cell current densities, I_{cell}. It is clear that hydrogen concentration

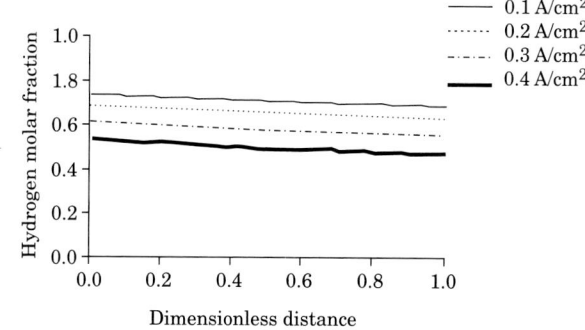

Fig. 2.4. Hydrogen concentration within the catalyst layer for various I_{cell} values

decreases with the current density, because increasing consumption corresponds to higher current density values. It is also depicted that the concentration profiles are nearly linear for low current densities, because diffusion dominates at these conditions, therefore only a small portion of hydrogen is consumed in the catalyst layer. The ionic current density in the catalyst layer as calculated by the one-dimensional model is presented in Fig. 2.5 for various I_{cell} values. A slight increase is observed near the outlet area, following the behavior of hydrogen consumption with the distance. Furthermore, the ionic current density increases slowly with I_{cell}.

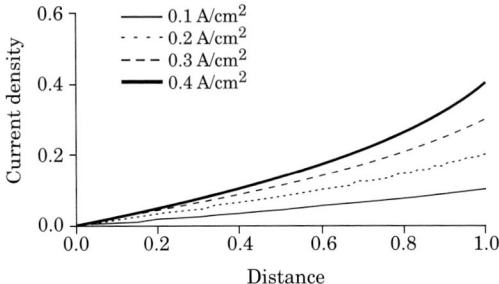

Fig. 2.5. Ionic current density within the catalyst layer

Figure 2.6 presents the overpotential in the catalyst layer. As above, several values of external current density have also been considered. A rather smooth increment from the inlet to the outlet is observed (note the scale of y-axis), resulted by the consequent increment of the hydrogen consumption with the distance, as it has been previously shown. Furthermore, the overpotential increases with current density, which is in accordance with previously presented experimental and theoretical results[66, 67].

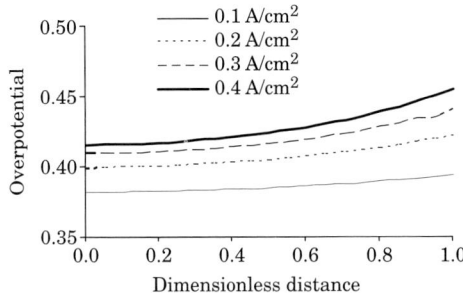

Fig. 2.6. Overpotential within the catalyst layer for various values

4.2. Results for Three Dimensional Model

For the three dimensional approach, a planar SOFC geometry was used, which consists of seven separated volumes: the fuel and air channel where the corresponding mixtures were introduced, the porous anode and cathode electrodes where the reactions occurred, the dense electrolyte through which oxygen ions migrate to reach anode electrode and finally the anode and cathode contacts. In the present study, both H_2 and CO were assumed to react with oxygen ions, in accordance with Eqs. [1 (a)] and [1 (b)] and the kinetics of the Butler-Volmer equation [Eq. (35)] are listed in Table 2.2, where the exchange current density, j_0, and the charge transfer coefficients (a_a and a_c) for both electrodes can be seen. It was considered that the electrochemical reactions took place only in the porous electrodes and precisely on the three-phase boundary layer, while no other reaction occurred either on porous interface or in bulk phase.

Table 2.2. Electrochemical kinetics for the three-dimensional model

	Anode	Cathode
j_0 [Am^{-3}]	10^{11}	10^{10}
a_a [–]	0.7	0.7
a_c [–]	0.7	0.7

The feeding stream was assumed to be a reformate mixture consisting of 9.6% H_2, 26.0% CO, 2.16% CO_2 and 42.8% H_2O on w.t.% basis, while typical composition for the atmospheric air at the cathode channel has been applied (23.3% O_2 and 76.7% N_2 on w.t. basis).

Regarding the boundary conditions, the mass flow rates for the anodic and the cathodic mixtures were assumed to be constant and equal to 3.0×10^{-7} kg/s and 4.0×10^{-6} kg/s for anode and cathode respectively, while pressure of 1 atm was set at the inlets and outlets. Considering no accumulation, zero mass flux was set at the walls and at the outlets as well. Additionally, preheated mixtures at 1173 K entered the fuel cell, while zero heat flux was set at all the other boundaries. Finally, constant value of zero overpotential was set to the anode contact, while the overpotential of cathode contact was set to –0.7 V, as well.

Table 2.3. Porous media properties

	Anode	Electrolyte	Cathode
ε [–]	0.40	0.01	0.50
κ [m^2]	1e^{-12}	1e^{-18}	1e^{-12}
k_S [Wm^{-1} K^{-1}]	6.23	2.7	9.6
σ_S [Ωm^{-1}]	100000	1e^{-20}	7700
σ_F [Ωm^{-1}]	10	10	10

Both electrodes were modeled as isotropic porous media and their physical characteristics are listed in Table 2.3[25, 82]. Furthermore, it was assumed that the gases followed the ideal gas law of gases for the calculation of density, ρ, and the kinetic theory of gases was used to estimate viscosity, μ. The specific heat, c_p, derived by fittings to the experimental JANNAF curves and the mass diffusivity was calculated by imposing 0.7 to the Schmidt number.

Assuming steady state conditions, the numerical solution for the equations described at the relative section §3.2 was obtained by the commercial package CFD-ACE+ by ESI-Group©, which is based on the finite volume method, together with all the appropriate boundary conditions in order to achieve residual values for all the quantities less than 10^{-4}. The three dimensional fuel cell was discretized in space by structured grid consisting of 33516 cells[83].

All the typical distributions for the main physical quantities (velocity, temperature mass fractions and overpotential) obtained for constant gas inlet mixture and temperature for both fuel and air mixtures are presented below in contour plot in a two dimensional cut in the middle plane of the fuel cell.

Figure 2.7 shows the developed velocity profile, which is obviously parabolic for both flow channels, satisfying the non-slip conditions applied on the walls. Due to the mass flow rates imposed, high velocity values can be seen in the air channel and lower in the fuel channel, while in the porous media zero velocities are observed. The higher value observed is approximately 5 m/s, which can be considered as quite high, being necessary to satisfy the energetic demands of the system in terms of cooling as well as of sufficient amount to keep the chemical reactions active.

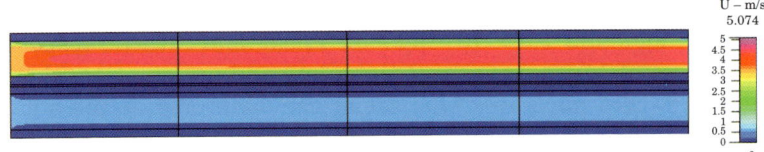

Fig. 2.7. Velocity profile

The molar fractions of the major reactants and products along the fuel cell are presented in Figs 2.8 and 2.9. As far as hydrogen [Fig. 2.8 (a)] and carbon monoxide [Fig. 2.8 (b)] are concerned, it is rather clear that they

Fig. 2.8. Mass fraction profile for (a) hydrogen and (b) carbon monoxide

are consumed at the anode channel and the anode electrode. Furthermore, their fractions decrease with the distance, due to the electrochemical reaction controlling the system. The consumption of both species is accompanied by relative production of both CO_2 [Fig. 2.9 (a)] and H_2O [Fig. 2.9 (b)], in accordance with the relative reaction rates. At the same time, a small depletion of oxygen can be seen in the cathode and in the cathode channel which is attributed to its participation in the electrochemical reactions (see Fig. 2.10). The O_2 gradient is quite small since the air mass flow rate is high enough to avoid oxygen depletion that would reduce fuel cell's performance, while the highly convective regime could be considered as an extra barrier to the extent of the electrochemical reactions.

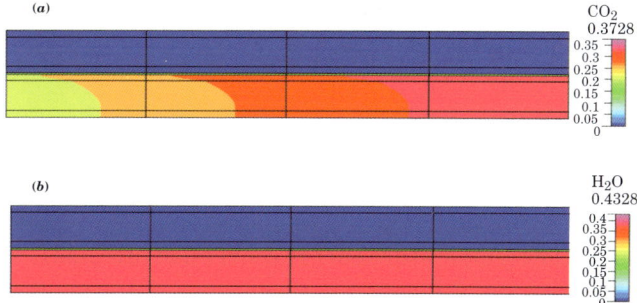

Fig. 2.9. Mass fraction profile for (a) carbon dioxide and (b) steam

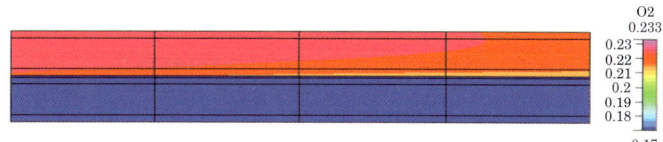

Fig. 2.10. Mass fraction profile for oxygen

As can be seen in Fig. 2.11, the temperature varies along the fuel cell length from the value of 1173 K imposed at the inlet boundary to 1256 K which is reached at the outlet. This temperature variation is strongly affected by the electrochemical reactions occurring at the anode electrode. Hydrogen and carbon monoxide electroxidation reaction is exothermic, thus a result certain amount of heat is released in the cell and transferred locally

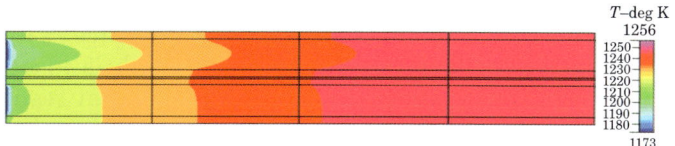

Fig. 2.11. Spatial distribution for temperature

due to convection and conduction. Consequently, temperature increase can be observed along the cell's length and it is evident that convectional effects prevail at air channel due to the developing velocities. Finally, the overpotential profile is depicted in Fig. 2.12, where it is clear that overpotential at anode electrode is higher compared to the one observed at the cathode electrode due to fuel depletion at anode channel. It should be also mentioned that overpotential increases along the anode electrode and this raise is followed by the reduction of active species mass fractions *i.e.*, H_2 and CO.

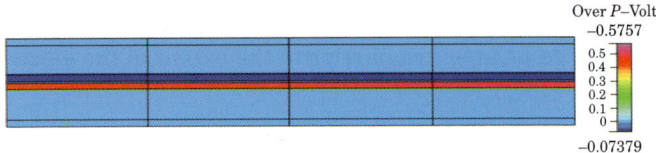

Fig. 2.12. Overpotential profile

5. CONCLUSIONS

In general, the use of fuel cells presents advantages because of direct conversion of chemical to electrical energy where the efficiency is not limited by the Carnot cycle. Unlike batteries, fuel cells are continuous flow systems, operating under low temperature energy conversion processes. Among other types of fuel cells, Solid Oxide Fuel Cells present special interest for application because of their high operation temperature, which allows flexibility in fuel choice. Finally, it is worth noticing that modeling could be considered as a very powerful tool for the deeper understanding of the fundamental phenomena occurred in the cells as well as for the successful design and operational management of such systems.

In the present work, two approaches have been adopted to adequately simulate the transport phenomena occurred in a SOFC. The first one is a one dimensional, relatively simple model, providing a rather simple and stable solution by using analytical manipulations and combined numerical techniques. To overcome deficiencies arising from one-dimensional simulation, another three-dimensional approach has also been developed, including the coupled solution for flow, heat transfer, mass transport and electrochemistry.

It is found that higher fuel consumption corresponds to higher current density values in an equivalent manner. Furthermore, it is clearly depicted that the overpotential increases along the anode, resulted by the consequent increment of the fuel consumption with the distance. Finally, the overpotential increases with current density, which is in accordance with previously presented experimental and theoretical results. From qualitative point of view, both models describe with sufficient accuracy the operation of a fuel cell in terms of feedstock concentration and overpotential produced.

REFERENCES

1. Larminie, J. and Dicks, A. 2002. Fuel cell systems explained. John Wiley & Sons Ltd., West Sussex, UK.
2. Vielstich, W., Lamm, A. and Gasteiger, H.A. (*Eds.*) 2003. Handbook of fuel cells fundamentals technology and applications, John Wiley & Sons Ltd, Chichester, UK.
3. EG&G Services Parsons Inc. 2000. Fuel Cell Handbook. 6th edn. 2002. U.S. Department of Energy, National Energy TechnologyLaboratory. Produced under contract DE-AM26–99FT40575. Norgantown, West.
4. Coutelieris, F.A., Douvartzides, S. and Tsiakaras, P. 2003. The importance of the fuel choice on the efficiency of a solid oxide fuel cell system, *J. Power Sources* **123**: 200–5.
5. Achenbach, E. and Reinsche, E. 1994. Methane/Steam reforming kinetics for solid oxide fuel cells, *J. Power Sources* **52**: 238–99.
6. Grgicak, C.M., Green, R.G. and Giorgi, J.B. 2008. SOFC anodes for direct oxidation of hydrogen and methane fuels containing H_2S, *J. Power Sources* **179**: 317–28.
7. Kim, T., Ahn, K., Vohs, J.M. and Gorte, R.J. 2007. Deactivation of ceria-based SOFC anodes in methanol, *J. Power Sources* **164**: 42–48.
8. Jamsak, W., Assabumrungrat, S., Douglas, P.L., Laosiripojana, N., Suwanwarangkul, R., Charojrochkul, S. and Croiset, E. 2007. Performance of ethanol-fuelled solid oxide fuel cells: Proton and oxygen ion conductors, *Chem. Eng. J.* **133**: 187–94.
9. Ni, M., Leung, D.Y.C. and Leung, M.K.H. 2008. Mathematical modeling of ammonia-fed solid oxide fuel cells with different electrolytes, *Int. J. Hydrogen Energ.* **33**: 5765–72.
10. Fuerte, A., Valenzuela, R.X., Escudero, M.J. and Daza, L. 2008. Ammonia as efficient fuel for SOFC, *J. Power Sources* (in press).
11. Tsipis, E.V. and Kharton, V.V. 2008. Electrode materials and reaction mechanisms in solid oxide fuel cells: a brief review II. Electrochemical behavior vs. materials science aspects, *J. Solid State Electr.* **12**: 1367–91.
12. Rienshe, E., Achenbach, E., Froning, D., Haines, M.R., Heidug, W.K., Lokurlu, A. and von Andrian, S. 2000. Clean combined-cycle SOFC power plant-cell modelling and process analysis, *J. Power Sources* **86**: 404–10.
13. Palsson, J., Selimovic, A. and Sjunnesson, L. 2000. Combined solid oxide fuel cell and gas turbine systems for efficient power and heat generation, *J. Power Sources* **86**: 442–8.
14. Costamagna, P., Magistri, L. and Massardo, A.F. 2001. Design and part-load performance of a hybrid system based on a solid oxide fuel cell reactor and a micro gas turbine, *J. Power Sources* **96**: 352–68.
15. Pimentel, D., Rodrigues, G., Wang, T., Abrams, R., Goldberg, K., Staecker, H., Ma, E., Brueckner, L., Trovato, L., Chow, T.C., Govindarajulu, U. and Boerke, S. 1994. Renewable energy: Economic and environmental issues, *Bioscience* **44**: 536–47.
16. Lu, N., Li, Q., Sun, X. and Khaleel, M.A. 2006. The modeling of a standalone solid-oxide fuel cell auxiliary power unit, *J. Power Sources* **1661**: 938–48.
17. La O', G.J., In, H.J., Crumlin, E., Barbastathis, G. and Shao-Horn, Y. 2007. Recent advances in microdevices for electrochemical energy conversion and storage, *Int. J. Energ. Res.* **31**: 548–75.
18. Singhal, S.C. 2000. Advances in solid oxide fuel cell technology, *Solid State Ionics* **135**: 305–13.
19. Hirschenhofer, J.H., Stauffer, D.B., Engleman, R.R. and Klett, M.G. 1997. Fuel Cell Handbook. 4th edn., Business/Technology Books, Orinda, USA.

20. Nagel, F.P., Schildhauer, T., Biollaz, S.M.A. and Wokaun, A. 2008. Performance comparison of planar, tubular and Delta8 solid oxide fuel cells using a generalized finite volume model, *J. Power Sources* **184**: 143–64.
21. Ferguson, J.R., Fiard, J.M. and Herbin, R. 1996. Three-dimensional numerical simulation for various geometries of solid oxide fuel cells, *J. Power Sources* **58**: 109–22.
22. Aguiar, P., Adjimana, C.S. and Brandona, N.P. 2004. Anode-supported intermediate temperature direct internal reforming solid oxide fuel cell. I: model-based steady-state performance, *J. Power Sources* **138**: 120–36.
23. Suwanwarangkul, R., Croiset, E., Pritzker, M.D., Fowler, M.W., Douglas, P.L. and Entchev, E. 2007. Modelling of a cathode-supported tubular solid oxide fuel cell operating with biomass-derived synthesis gas, *J. Power Sources* **166**: 386–99.
24. Costamagna, P. 1997. The benefit of solid oxide fuel cells with integrated air preheater, *J. Power Sources* **69**: 1–9.
25. Ramakrishna, P.A., Yang, S. and Sohn, C.H. 2006. Innovative design to improve the power density of a solid oxide fuel cell, *J. Power Sources* **158**: 378–84.
26. Paloukis, F. and Neophytides, S.G. 2007. Numerical simulation of methane fuelled cogenerative SOFCs for the production of synthesis gas and electrical energy, *Chem. Eng. Sci.* **62**: 3868–81.
27. Costamagna, P., Costa, P. and Antonucci, V. 1998. Micro-modelling of solid oxide fuel cell electrodes, *Electrochim. Acta* **43**: 375–94.
28. Costamagna, P., Costa, P. and Arato, E. 1998. Some more considerations on the optimization of cermets solid oxide fuel cell electrodes, *Electrochim. Acta* **43**: 967–72.
29. Chan, S.H. and Xia, Z.T. 2001. Anode micro model of solid oxide fuel cell, *J. Electrochem. Soc.* **148**: A388–A394.
30. Yakabe, H., Ogiwara, T., Hishinuma, M. and Yasuda, I. 2001. 3-D model calculation for planar SOFC, *J. Power Sources* **102**: 144–54.
31. Suwanwarangkul, R., Croiset, E., Fowler, M.W., Douglas, P.L., Entchev, E. and Douglas, M.A. 2003. Performance comparison of Fick's, dusty-gas and Stefan–Maxwell models to predict the concentration overpotential of a SOFC anode, *J. Power Sources* **122**: 9–18.
32. Ackmann, T., de Haart, L.G.J., Lehnert, W. and Stolten, D. 2003. Modeling of mass and heat transport in planar substrate type SOFCs, *J. Electrochem. Soc.* **150**: A783–A789.
33. Chan, S.H., Chen, X.J. and Khor, K.A. 2004. Cathode micromodel of solid oxide fuel cell, *J. Electrochem. Soc.* **151**: A164–A172.
34. Shi, Y., Cai, N., Li, C., Bao, C., Croiset, E., Qian, J., Hu, Q. and Wang, S. 2007. Modeling of an anode-supported Ni–YSZ|Ni–ScSZ|ScSZ|LSM–ScSZ multiple layers SOFC cell Part I. Experiments, model development and validation, *J. Power Sources* **172**: 235–45.
35. Presvytes, D. and Vayenas, C.G. 2007. Mathematical modelling of the operation of SOFC Nickel-cermet anodes, *Ionics* **13**: 9–18.
36. Debenedetti, P.G. and Vayenas, C.G. 1983. Steady-state analysis of high temperature fuel cells, *Chem. Eng. Sci.* **38**: 1817–29.
37. Vayenas, C.G., Debenedetti, P.G., Yentekakis, I. and Hegedus, L.L. 1985. Cross-flow solid-state electrochemical reactors: a steady-state analysis, *Ind. Eng. Chem. Res. Fundam.* **24**: 316–24.
38. Achenbach, E. 1994. Three-dimensional and time-dependent simulation of a planar solid oxide fuel cell stack, *J. Power Sources* **49**: 333–48.
39. Bessette, N.F., Wepfer, W.J. and Winnick, J. 1995. A mathematical model of a solid oxide fuel cell, *J. Electrochem. Soc.* **142**: 3792–3800.

40. Ferguson, J.R., Fiard, J.M. and Herbin, R. 1996. Three-dimensional numerical simulation for various geometries of solid oxide fuel cells, *J. Power Sources* **58**: 109–22.
41. Costamagna, P. and Honegger, K. 1998. Modeling of solid oxide heat exchanger integrated stacks and simulation at high fuel utilization, *J. Electrochem. Soc.* **145**: 3995–4007.
42. Lehnert, W., Meusinger, J. and Thom, F. 2000. Modeling of gas transport phenomena in SOFC anodes, *J. Power Sources* **87**: 57–63.
43. Yakabe, H., Hishinuma, M., Uratani, M., Matsuzaki, Y. and Yasuda, I. 2000. Evaluation and modelling of performance of anode-supported solid oxide fuel cell, *J. Power Sources* **86**: 423–31.
44. Aguiar, P., Chadwick, D. and Kershenbaum, L. 2002. Modelling of an indirect internal reforming solid oxide fuel cell, *Chem. Eng. Sci.* **57**: 1665–77.
45. Recknagle, K.P., Williford, R.E., Chick, L.A., Rector, D.R. and Khaleel, M.A. 2003. Three-dimensional thermo-fluid electrochemical modeling of planar SOFC stacks, *J. Power Sources* **113**: 109–14.
46. Yakabe, H. and Sakurai, T. 2004. 3D simulation on the current path in planar SOFCs, *Solid State Ionics* **174**: 295–302.
47. Hussain, M.M., Li, X. and Dincer, I. 2006. Mathematical modeling of transport phenomena in porous SOFC anodes, *Int. J. Therm. Sci.* **161**: 1012–22.
48. Suwanwarangkul, R., Croiset, E., Entchev, E., Charojrochkul, S., Pritzker, M.D., Fowler, M.W., Douglas, P.L., Chewathanakup, S. and Mahaudom, H. 2006. Experimental and modeling study of solid oxide fuel cell operating with syngas fuel, *J. Power Sources* **161**: 308–22.
49. Hwang, J.J. 2006. Species-electrochemical modeling of an air-breathing cathode of a planar fuel cell, *J. Electrochem. Soc.* **153**: A1584–A1590.
50. Zhu, H., Kee, R.J., Pillai, M.R. and Barnett, S.A. 2008. Modeling electrochemical partial oxidation of methane for cogeneration of electricity and syngas in solid-oxide fuel cells, *J. Power Sources* **183**: 143–50.
51. Coutelieris, F.A. 2008. Modeling of transport phenomena in a fuel cell anode, *Defect Diffus. Forum*, pp. 273–6; 820–8.
52. Arpino, F. and Massarotti, N. 2008. Numerical simulation of mass and energy transport phenomena in solid oxide fuel cells, *Energy* (in press).
53. Suzuki, M., Shikazono, N., Fukagata, K. and Kasagi, N. 2008. Numerical analysis of coupled transport and reaction phenomena in an anode-supported flat-tube solid oxide fuel cell, *J. Power Sources* **180**: 29–40.
54. Hussain, M.M., Li, X. and Dincer, I. 2009. A general electrolyte-electrode-assembly model for the performance characteristics of planar anode-supported solid oxide fuel cells, *J. Power Sources* (in press).
55. Sukeshini, A.M., Habibzadeh, B., Becker, B.P., Stoltz, C.A., Eichhorn, B.W. and Jackson, G.S. 2006. Electrochemical oxidation of H_2, CO and CO/H_2 mixtures on patterned Ni anodes on YSZ electrolytes, *J. Electrochem. Soc.* **153**: A705–A715.
56. Matsuzaki, Y., Baba, Y. and Sakurai, T. 2004. High electric conversion efficiency and electrochemical properties of anode-supported SOFCs, *Solid State Ionics* **174**: 81–86.
57. Costa-Nunes, O., Gorte, R.J. and Vohs, J.M. 2005. Comparison of the performance of Cu–CeO_2–YSZ and Ni-YSZ composite SOFC anodes with H_2, CO and syngas, *J. Power Sources* **141**: 241–9.
58. Park, S., Cracium, R., Vohs, J.M. and Gorte, R.J. 1999. Direct oxidation of hydrocarbons in a solid oxide fuel cell I. Methane oxidation, *J. Electrochem. Soc.* **146**: 3603–5.
59. Zhu, H., Kee, R.J., Pillai, M.R. and Barnett, S.A. 2008. Modeling electrochemical partial oxidation of methane for cogeneration of electricity and syngas in solid–oxide fuel cells, *J. Power Sources* **183**: 143–50.

60. Irvine, J.T.S. and Sauvet, A. 2001. Improved oxidation of hydrocarbons with new electrodes in high temperature fuel cells, *Fuel Cells* **1**: 205–10.
61. Xu, J. and Froment, G.F. 1989. Methane steam reforming and water-gas shift: I Intrinsic Kinetics, *AICHE J.* **35**: 88–96.
62. Ahmed, K. and Foger, K. 2000. Kinetics of internal steam reforming of ethane on Ni/YSZ based anodes for solid oxide fuel cells, *Catal. Today* **63**: 479–87.
63. He, H. and Hill, J.M. 2007. Carbon deposition on Ni/YSZ composites exposed to humidified methane, *Appl. Catal. A-Gen* **317**: 284–92.
64. Lin, Y.B., Zhan, Z.L., Liu, J. and Barnett, S.A. 2005. Direct operation of solid oxide fuel cells with methane fuel, *Solid State Ionics* **176**: 1827–35.
65. Walters, K.M., Dean, A.M., Zhu, H.Y. and Kee, R.J. 2003. Homogeneous kinetics and equilibrium predictions of coking propensity in the anode channels of direct oxidation solid-oxide fuel cells using dry natural gas, *J. Power Sources* **123**: 182–9.
66. Jeng, K.T. and Chen, C.W. 2002. Modeling and simulation of a direct methanol fuel cell anode, *J. Power Sources* **112**: 367–75.
67. Jeng, K.T., Kuo, C.P. and Lee, S.F. 2004. Modeling the catalyst layer of a PEM fuel cell cathode using a dimensionless approach, *J. Power Sources* **128**: 145–51.
68. Hammer, B. and Norskov, J.K. 1995. Why gold is the noblest of all metals, *Nature* **376**: 238–40.
69. Kim, H., Lu, C., Worrell, W.L., Vohs, J.M. and Gorte, R.J. 2002. Cu-Ni cermet anodes for direct oxidation of methane in solid-oxide fuel cells, *J. Electrochem. Soc.* **149**: A247–A250.
70. Ames, W.F. 1977. Numerical Methods for Partial Differential Equations, Academic Press. New York, USA.
71. Dullien, F.A. 1979. Porous media: Fluid transport and pore structure. Academic Press, California, USA.
72. Gurau, V., Liu, H. and Kakac, S. 1998. Two-dimensional model for proton exchange membrane fuel cells, *AIChE J.* **44**: 2410–22.
73. Pulkrabek, W.W. 2003. Engineering Fundamentals of the Internal Combustion Engine. 2nd edn., Pearson Prentice Hall, New Jeresey, U.S.A.
74. Tseronis, K., Kookos, I.K. and Theodoropoulos, C. 2008. Modelling mass transport in solid oxide fuel cells anodes: a case for multicomponnet dusty gas model, *Chem. Eng. Sci.* **63**: 5626–38.
75. Bird, R.B., Stewart, W. and Lightfoot, E.N. 1960. Transport Phenomena. Wiley, New York, USA.
76. Newman, J.S. and Tabias, C.W. 1962. Theoretical analysis of current distribution in porous electrodes, *J. Electrochem. Soc.* **109**: 1183–91.
77. Jaouen, F., Lindbergh, G. and Sundholm, G. 2002. Investigation of mass-transport limitations in the solid polymer fuel cell cathode, *J. Electrochem. Soc.* **149**: A437–A447.
78. Keller, H.B. 1968. Numerical methods for two-point boundary-value problems. Blaisdel, Waltham, USA.
79. Perry, R.H. and Green, D.W. 1977. Chemical Engineers Handbook. 7th edn., McGraw-Hill, New York, USA.
80. Scott, K., Taama, W. and Cruickshank, J. 1997. Performance and modelling of a direct methanol solid polymer electrolyte fuel cell, *J. Power Sources* **65**: 159–71.
81. Baxter, S.F., Battaglia, V.S. and White, R.E. 1999. Methanol fuel cell model: Anode, *J. Electrochem. Soc.* **146**: 437–47.
82. Klein, J.-M., Bultel, Y., Pons, M. and Ozil, P. 2007. Modeling of a Solid Oxide fuel cell by methane: Analysis of carbon deposition, *J. Fuel Cell Tech.* **7**: 425–34.
83. ESI-Group. 2003. CFD-RC, CFD-ACE+ User Manual, Huntsville, USA.

3

Structure and Properties of Grain Boundaries

VLADIMIR V. POPOV[*]

ABSTRACT

This chapter is a brief review on the state of affairs in studies of structure and properties of grain boundaries (GBs) in metals. It is demonstrated that GBs structure and properties can be understood only based on simulation combined with a number of various experimental techniques, such as high-resolution electron microscopy (HREM), diffusion investigations, emission Mossbauer spectroscopy, etc. Models of structure and energy of equilibrium and non-equilibrium grain boundaries are considered, specific features of various point defects formed in GBs are analyzed, diffusion properties of GBs in coarse-grained, submicrocrystalline and nano-structured materials are described. A new method of investigation of grain boundaries based on the use of the emission Mossbauer spectroscopy is described in detail, and its capabilities are analyzed. It is demonstrated that the results of the Mossbauer studies do not completely agree with Fisher's model of grain boundary diffusion, and a new recently suggested model which may be considered as the Fisher's model modification is described. This model is in agreement both with the results of traditional diffusion experiments and with the NGR data. The results of the Mossbauer studies of GBs corresponding to determination of mechanisms of GB diffusion, evaluation of segregation factors, estimation of an extent of non-equilibrium state of GBs, etc. are discussed. Further goals and nearest projects in studies of GBs are considered.

1. INTRODUCTION

The majority of materials, especially metals, are used in form of polycrystals, an integral part of which are grain boundaries (GBs) appreciably affecting many practically important properties of polycrystalline materials. Modern conceptions of GBs internal arrangement and properties have been established relatively recently and are generalized in a number of monographs and reviews including[1–6].

---

[1] Institute of Metal Physics, Russian Academy of Sciences-RUSSIA.
[*] Corresponding author : E-mail : vpopov@imp.uran.ru

A grain boundary is defined as an area between two perfect single-phase crystals (or grains) of different crystallographic orientation contacting with each other. According to modern conceptions, a GB proper is an area in which the arrangement of atoms differs from that in a perfect lattice. In metals the width of this area is not more than 2–3 inter-atomic spaces due to strong screening[7]. GB properties must differ from that of the bulk because a specific structure different from that of a main crystal is formed at an interface of two grains. That is why a GB may be considered as an independent phase distinguishing from a bulk in structure and properties.

A GB state essentially determines properties of metals, particularly such processes as grain-boundary squeezing, super-plasticity, inter-granular fracture, etc. The role of grain boundaries is especially high in nanostructured materials in which a GB fraction can be as high as several tens percents[8].

Before early eighties mainly equilibrium GBs were under study, but at present non-equilibrium GBs especially characteristic of nano- and sub-microcrystalline materials obtained by severe plastic deformation (SPD) attract the growing attention of the materials science researchers[9]. A great number of publications deal with GB studies, and it is next to impossible to review them in one relatively short chapter. For example, the volume of a monograph by Sutton and Balluffi "Interfaces in Crystalline Materials"[6] is 819 pages. That is why in this chapter we restrict our consideration to grain boundaries in metals paying main attention to GB structure, defects, energy and GB diffusion.

2. GEOMETRICAL CHARACTERISTICS OF A PLANE BOUNDARY

Misorientation of grains, *i.e.*, orientation of two neighboring crystallites relative to each other, is commonly described as a turn of a crystallite relative to another one around their common crystallographic axis $\vec{u}$ by an θ angle, θ being a misorientation angle and $\vec{u}$ a misorientation axis. Vector is chosen of a unit length and is given by two components. A plane boundary is given by $\vec{n}$, a unit normal to it. Only two of three components of vector $\vec{n}$ are independent, as the third one is obtained from a normalization condition, *i.e.*, two degrees of freedom are connected with an interface orientation. Thus, in macroscopic description a plane boundary is given by five independent parameters three of which determine crystallites mis-orientation (turn angle and axis) and the other two—boundary orientation. These characteristics to a great extent determine GB physical properties and serve as a basis for their classification. If $\vec{u} \perp \vec{n}$, *i.e.*, turn axis is in a boundary plane, it is a tilt boundary, and if $\vec{u} \parallel \vec{n}$ it is a twisting boundary. At other relative orientations of $\vec{u}$ and $\vec{n}$ a boundary is referred to as a mixed one.

When a grain boundary is formed, two parts of a crystal may not only turn relative to each other, but shift as well. From a microscopic point of view, *i.e.*, when a discrete crystal arrangement is taken into account, a shift of one crystal relative to another results in a new boundary structure with the same macroscopic parameters. That's why in a microscopic description of a GB one should add three components of a vector characterizing rigid shift of neighboring crystals to the five macroscopic parameters determining crystallite misorientation and boundary plane[10].

3. GRAIN BOUNDARY STRUCTURE MODELS

All the modern models of GB structure are based on a conception that a boundary is not an amorphous layer of a substance between two crystals but a "crystalline" formation. The first attempt to associate an arrangement of atoms at an interface with lattices of crystallites forming a boundary was done by Kronberg and Willson[11] who used a conception of a coincidence site lattice (CSL), independently suggested earlier by several researchers for a GB model. The CSL model is a basis of all modern conceptions of boundary structure.

If a bicrystal containing a boundary is formed by two crystal lattices, L_1 and L_2, turned by a θ angle relative to each other, and a boundary plane is determined by normal $\vec{n}$, then at definite values of θ a part of L_1 and L_2 sites will coincide. Such values of θ and corresponding GBs are referred to as special ones. A CSL is characterized by an inverse density of coinciding sites Σ, equal to the number of sites of one of the lattices in a unit cell of CSL. As an example Fig. 3.1 demonstrates a projection of coincidence of two cubic lattices misoriented by an angle of $\theta_0 = 36.87°$ [001] corresponding to $\Sigma = 5$ on a (001) plane.

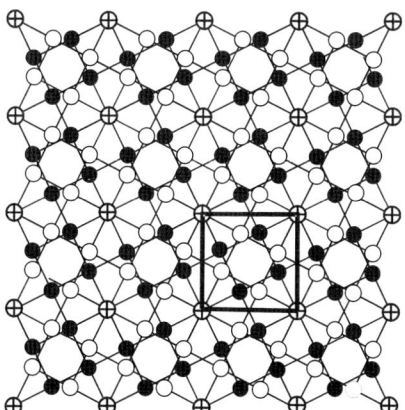

Fig. 3.1. Superposition of pure cubic lattices L_1 (white circles) and L_2 (black circles), turned relative to each other by $\theta = 36.87°$ around the [001] axis ($\Sigma = 5$). Circles with crosses are coinciding sites, and bold lines denote an elementary CSL unit

To build a GB in CSL an area is separated out, in one side of which L_1 sites are filled with atoms, and in another side L_2 sites (Fig. 3.2).

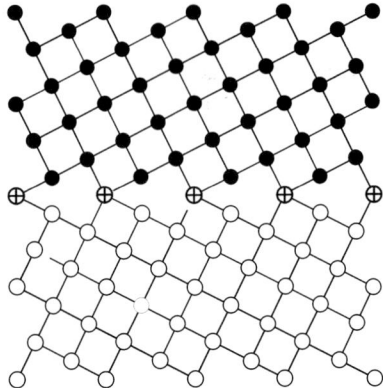

Fig. 3.2. Schematic view of a special tilt boundary with $\theta = 36.87^0$, [001] ($\Sigma = 5$)

At special misorientations θ_0 [hkl] coinciding sites are quite frequently observed. The special misorientations are of specific interest, as special boundaries, *i.e.*, boundaries between grains misoriented by special angles, possess peculiar physical properties. Boundaries with small Σ are commonly characterized with low mobility, low grain-boundary diffusion coefficient and other specific features.

From a geometrical point of view boundaries with as high as one likes (but definite) value of Σ may be considered as special ones. However from a physical point of view special boundaries are only those the properties of which really differ from that of boundaries of an ordinary type. At very high values of Σ a boundary periodicity will only slightly affect its properties.

A list of special misorientation angles $\theta_{0,i}$ for cubic lattices corresponding to $\Sigma < 50$ is given in Table 3.1 according to the data of[12]. A method of analytical construction of CSL for cubic lattices is described in[13]. A technique of special angles determination for hexagonal crystals with rational axes ratio of c/a is given in[14], and for other lattice types–in[15–18].

Table 3.1. Special misorientation angles ($\Sigma < 50$) for cubic lattices[12]

Σ	*hkl*	θ_0	Σ	*hkl*	θ_0	Σ	*hkl*	θ_0	Σ	*hkl*	θ_0
3	111	60.00	21	221	44.41	33	110	58.99	43	210	27.91
5	100	36.86	23	311	40.45	35	211	34.05	43	322	60.77
7	111	38.21	25	100	16.26	35	331	43.23	45	311	28.62
9	110	38.94	25	331	51.68	37	100	18.92	45	211	36.87
11	110	50.47	27	110	31.59	37	310	43.14	45	221	53.13
13	100	22.61	27	210	35.43	37	111	50.57	47	331	37.07
13	111	27.79	29	100	43.60	39	111	32.20	47	320	43.66
15	220	48.18	29	221	46.40	39	321	30.13	49	111	43.57
17	100	28.07	31	111	17.90	41	100	12.68	49	511	43.57
19	110	26.52	31	320	54.50	41	210	40.89	49	322	49.23
19	111	46.82	33	110	20.05	41	110	55.88			
21	111	27.78	33	311	33.56	41	111	15.18			

Note: The second column indicates Miller's indexes of a turn axis.

A CSL conception allows classifying different type boundaries. A boundary containing only coinciding sites of $L_1 L_2$ superposition is referred to as a symmetrical one if it coincides with one of $L_1 L_2$ superposition symmetry planes, otherwise it is called a quasi-symmetrical. A boundary containing non-coinciding sites along with the coinciding sites of L_1 or L_2 lattices is referred to as an asymmetrical.

Various auxiliary lattices are used in description of a GB structure, the main of which are the above-described coinciding sites lattice and a displacement shift complete (DSC) lattice. The DSC lattice was suggested by Bollman[19-20]. DSC lattice sites are formed by all vectors of $\Delta \vec{r} = \vec{r}^{(1)} - \vec{r}^{(2)}$, where $\vec{r}^{(1)}$ and $\vec{r}^{(2)}$ are radius-vectors of L_1 and L_2 lattice sites. An ensemble of sites occupied by atoms of both lattices at various translations of $\Delta \vec{r}$ retaining a superposition picture forms a DSC lattice. A net in Fig. 3.3 demonstrates a DSC lattice in case of $\Sigma = 5$. An important property of DSC lattice is that at L_1 translations relative to L_2 by difference vectors of $\Delta \vec{r} = \vec{r}^{(1)} - \vec{r}^{(2)}$ a CSL is retained, a part of L_1 sites transferring to L_2 sites and former coincident sites separating. As shifts by vectors of DSC lattice don't change crystal-geometry of superposition of disoriented lattices, complete grain-boundary dislocations must have Burgers vectors equal to DSC translations.

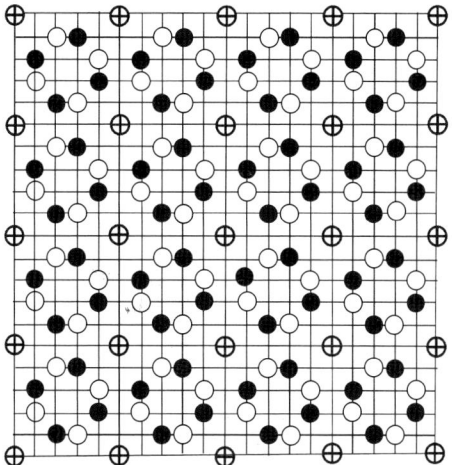

Fig. 3.3. A displacement shift complete lattice for pure cubic lattices L_1 (white circles) and L_2 (black circles) turned relative to each other by $\theta = 36.87^0$ around [001] axis ($\Sigma = 5$). Circles with crosses are coinciding sites

Formally any GB may be described as some plane dislocation pileup. Dislocation models adequately describe orientation dependence of GB energy up to misorientation angles of about 5^0[21].

At high misorientation angles a nearest special boundary described by CSL with small value of Σ may be chosen as a reference structure[22]. Then a GB is represented in form of a net of dislocations superposed on a special boundary. A dislocation net serves for an additional turn by an angle of $\Delta\theta = \theta - \theta_0$ from a special θ_0 to a given θ angle. The smallest Burgers vectors of full grain-boundary dislocations are equal to basis vectors of a DSC lattice constructed for a misorientation angle of θ_0. Complete grain-boundary dislocations in boundaries close to special ones are referred to as secondary grain boundary dislocations (GBDs) contrary to primary GBDs which are lattice dislocations captured by a boundary[20].

Numerous observations of secondary GBD nets confirm the prediction of the DSC lattice theory that the Burgers vector of these dislocations equals to DSC vector of special boundaries[23]. At high deviations of θ from θ_0 this model is not valid because of grain-boundary dislocation cores overlapping.

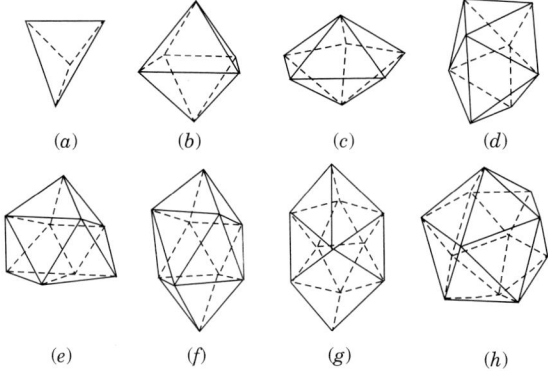

Fig. 3.4. Structural units of grain boundaries[24]: (a) tetrahedron, (b) octahedron, (c) pentagonal bi-pyramid, (d) tetragonal dodecahedron, (e) capped trigonal prism; (f) archimedean anti-prism, (g) octadecahedron, (h) regular icosahedron

As shown in[24–25], a GB atomic structure may be represented as a dense stacking of polyhedrons which are called Voronoy-Bernal polyhedrons[26].

Using definite type polyhedrons in tops of which equal radius spheres are located one can describe a boundary structure considering polyhedrons as structural units. All faces of these polyhedrons are equilateral triangles the edges of which equal to doubled spheres radii so that the spheres touch each other. There exist eight polyhedrons like that (Fig. 3.4), the simplest ones being a tetrahedron and an octahedron [Fig. 3.4 (a), (b)].

In early eighties Sutton and Vitek carried out simulation of a large number of GBs with [001], [110] and [111] misorientation axes and big periods up to CSL with $\Sigma = 491$[27–30]. The main result of these publications is a discovery of two main classes of grain boundaries. It was found that for each tilt axis under consideration there is a definite set of short period GBs containing sequences of only one type structural units. These boundaries were referred to as preferential ones. Any boundary with a misorientation angle intermediate between misorientation angles of two preferential boundaries consists of their structural units taken in definite proportion and sequence. For example, Fig. 3.5 demonstrates schematic images of symmetrical tilt boundaries [110] $\Sigma = 1$ (0°), $\Sigma = 27$ (31.59°) and $\Sigma = 83$ (17.86°) in FCC crystals taken from[31]. A boundary in Fig. 3.5 (a) represents purely a perfect crystal which is used to demonstrate an origin of A element. Figure 3.5 (b) demonstrates a preferential boundary with a tilt angle of 31.59° formed by B elements. Figure 3.5 (c) shows a boundary with a misorientation angle of 17.86° falling between misorientation angles of preferential A and B boundaries. It is seen that the structure of AAB boundary consists of A and B units. Thus, it is possible to predict theoretically the structure of any symmetrical tilt boundary and to confirm it by computer simulation.

Note that the sequence of structural units arrangement is as follows: the units present in a boundary in minority are separated as far as possible, i.e., they are located in a boundary as uniformly as possible[27].

High-resolution electron microscopic (HREM) studies of tilt boundaries structure demonstrate validity of the structural units model[32–33], though there are some deviations, particularly the presence of the third type of structural units[34]. It should be noted that in theoretical calculations of GB structure it was also found that in some boundaries with high Σ values there are three types of structural units[31].

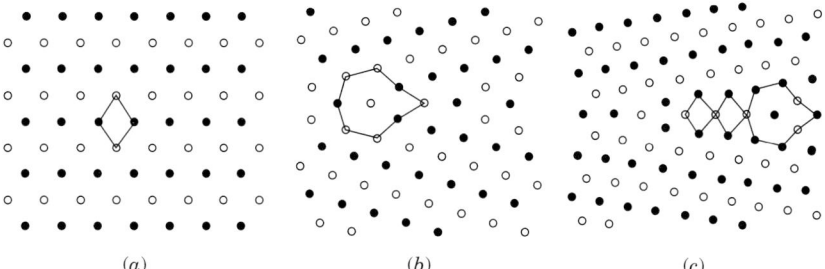

Fig. 3.5. Structure of tilt boundaries [110] in FCC crystals: (a) $\Sigma = 1$, $\theta = 0°$ (A); (b) $\Sigma = 27$, $\theta = 31.59°$ (B); (c) $\Sigma = 83$, $\theta = 17.86°$ (AAB)

The structural units model is closely connected with a GB dislocation model. Analyzing structures of arbitrary GBs Sutton and Vitek showed in[27], [35]–[36] that structural units present in a boundary in minority serve as grain-boundary dislocation cores for which a reference structure is a preferential boundary composed of structural units present in majority. For example, in a boundary with $\Sigma = 89$ the structure of which was calculated in[35] [see Fig. 3.6 (a)], structural units B are the cores of grain-boundary dislocations with Burgers vector of $b = \frac{2}{27}$ [11$\bar{5}$]. Extra semi-planes connected with these dislocations are shown in Fig. 3.6 (b).

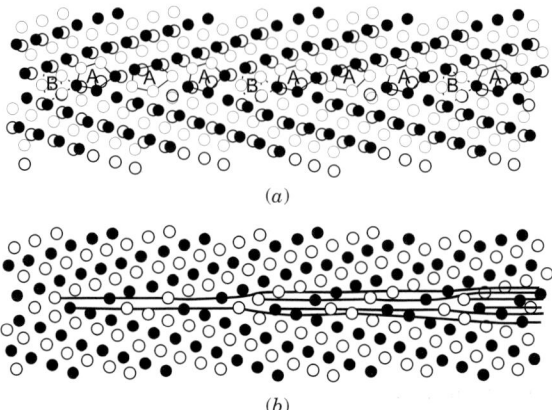

Fig. 3.6. Atomic structure of a tilt boundary $[1\bar{1}0]$ $\Sigma = 89$ (145.1°) (a) and structural dislocations in it (b)

As one can directly obtain a GB dislocation structure from the structural units model, this model was further referred to as a model of structural units/grain-boundary dislocations (SU/GBD)[37]. One of possible experimental proofs of ordered GB structure and validity of the SU/GBD model is an observation of dislocations in arbitrary boundaries.

Balluffi with coauthors carried out detailed electron-microscopic studies of GB dislocation structure in aluminum and showed that even at temperatures up to $0.96T_m$ dislocations in special and arbitrary boundaries are still localized[38].

The structural units model however has a limited range of applicability. As shown in[39–40], this model in its present form is effective for low-index misorientation axes such as [001], [110] and [111] and for pure tilt or twisting boundaries. In case of more complicated axes and mixed GBs the prediction capability of the model decreases as the number of preferential boundaries is increasing and intervals between them are decreasing[39].

To avoid difficulties connected with the GB dislocation model Li suggested a disclination model[41–42], according to which a tilt boundary with a misorientation angle of θ, intermediate between two special misorientations of θ_A and θ_B, may be represented in form of alternating sections of special boundaries A and B. In this case partial wedge disclinations with the power of $\pm \Delta\theta = \pm (\theta_2 - \theta_1)$ are formed at interfaces of the sections, i.e., a boundary may be considered as a wall of disclination dipoles. Both sections of boundary A, the length of L_A, and of boundary B, the length of L_B, may be equally taken as disclination dipoles, a period of a disclination dipole wall being $H = L_A + L_B$. However the disclination model suggested in[41–42] has not found practical application, the reasons of which according to[43] are as follows. Firstly, in[41–42] the disclination cores contribution to GB energy is not taken into account, which makes impossible a self-consistent determination of a disclination wall period. Secondly, the principle of special boundaries selection used as basic for the description of boundary properties in all misorientation range is not determined. Thirdly, in a boundary microstructure consideration in Li's model one more contribution to energy connected with structural elements distortions must be added. The fact is that in the disclination model generally not a whole number of these elements is put in one period of a boundary, and they must be either deformed or torn in the sites where disclinations are located. That's why the Li's disclination model is valid only for misorientation angles close to the special ones which makes it essentially only one of versions of the grain-boundary dislocations model. A more general disclination-structural model is suggested in[43–44] enabling to describe the structure and properties of high-angle boundaries in all the range of misorientation angles.

Let's consider this model in more detail following[44]. As mentioned above, for every crystal and every GB type there is a definite set of preferential boundaries consisting of one type structural units, while the rest (not preferential) boundaries consist of structural elements of two preferential boundaries taken in a respective proportion and sequence (structural units model). Consider a tilt boundary $\Sigma = 169$ [001](1250) as an example. Elementary cell of this boundary contains three structural elements of a (210) boundary (A elements) and two elements of (310) boundary (B elements) [Fig. 3.7 (a)]. Arbitrary boundaries with other misorientations may be represented analogously in form of a sequence of structural elements of preferential boundaries. As in junction lines of different type structural elements misorientation angle changes from one preferential angle to another, these lines may be considered as partial wedge disclinations the power of $\pm \theta = \pm (\theta_2 - \theta_1)$. As a result a boundary is represented in form of a wall of disclination dipoles in which the dipole role is played by one-type structural elements [Fig. 3.7 (b)]. A period of disclination dipole wall equals to a boundary elementary cell in the direction normal to tilt axis, and dipole arrangement in this period may generally be non-uniform. Structural elements comprising an arbitrary boundary differ from that in preferential boundaries by some compression strain, but the energy of this strain is negligibly small compared to other contributions in boundary energy, as in the model of structural units a boundary period and the number of structural elements in it are consistent.

As shown in[43], [44], in the framework of this model one can calculate energy curves of equilibrium boundaries versus misorientation angle for any tilt boundary using a small number of experimentally measured parameters (preferential boundaries energies).

According to[45] the most accurate results in GB energy calculations may be obtained by combination of disclination-structural model with the results of computer simulation of atomic structure of a boundary from the first principles. Capabilities of modern computers enable to calculate accurately structures and energies of pure boundaries with small periods (*i.e.*, preferential boundaries). Using these energies in the disclination model one can calculate boundaries energies in all the misorientation angles range.

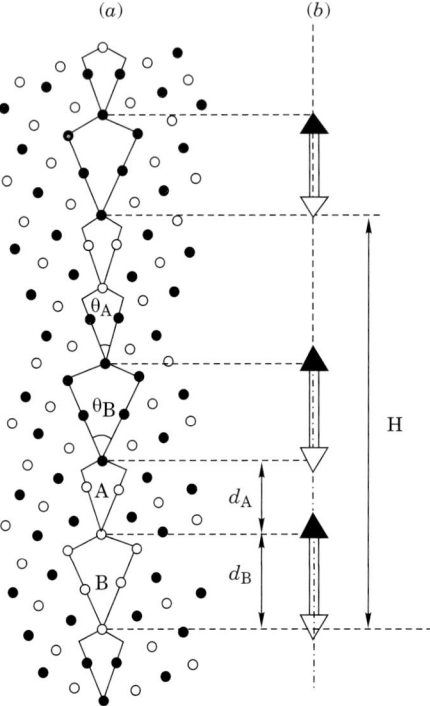

Fig. 3.7. Structure of $\Sigma = 169[001]$ (1250) tilt boundary in form of a sequence of structural elements of (210) and (310) preferential boundaries (*a*) and in disclination-structural model (*b*)[43]

Many atomic configurations revealed in the framework of geometrical models may appear to be energy-unfavorable. A well-founded analysis of grain boundaries is possible only if the GBs atomic structure is known, and this opportunity appeared recently due to numerous data of computer simulation. In computer simulation a boundary is considered as an isolated superlattice (periodical in two dimensions), and an effect of grains is taken into account only by boundary conditions. The result of computer simulation is an equilibrium structure in which all atoms of boundary area occupy positions corresponding to a system free energy minimum.

4. POINT DEFECTS IN GRAIN BOUNDARIES

In a GB as well as in a regular lattice, point defects of two types, vacancies and interstitials, may be present. As shown in[46–48], formation energy of point defects in GBs may be appreciably lower than in a regular lattice, the former being highly non-uniform. For instance, formation energy of vacancies in GBs may be only 20% of its bulk value, but in some positions it may be higher than in the volume, all deviations of vacancy formation energy being highly localized in a relatively narrow area of a GB ($\pm$ 0.5 nm). Beyond this region it practically coincides with the bulk value. One more peculiarity of point defects in grain boundaries is that formation energies, and consequently concentrations of vacancies and interstitials in a GB are close to each other, whereas in a regular lattice formation energy of vacancies is much lower than of interstitials, and thus interstitials concentration is negligible, and they don't play any essential role. In[49–51] Ma et al. and Nomura and Adams estimated migration energies of interstitials and jumps in a GB by a molecular statics method and concluded that diffusion mechanisms connected with interstitial sites in some cases may be as important as a vacancy mechanism.

More detailed studies of point defects formation in GBs were carried out in[46–48] by atomistic simulation methods by the example of tilt boundaries in Cu. The structure of GBs containing point defects was studied thoroughly in these papers from a viewpoint of possible delocalization and instability of point defects in a GB, the following observations being made. As a rule, vacancies induce relatively small relaxations of neighboring atoms and remain well localized in those sites where they have arisen. In some cases, however, the relaxation is so strong that vacancies delocalize in a GB structure. An example of vacancy delocalization is shown in Fig. 3.8. If vacancies arise in position 2 in $\Sigma = 9$ ($1\bar{2}2$) [011] GB, then an atom initially located in 1' position relaxes to a middle point between 1' and 2 positions which results in vacancy delocalization between these two positions. Absolutely the same situation takes place when a vacancy is formed at 1' site. Thus, considering the final configuration in Fig. 3.8 one cannot say in what site a vacancy has formed (1' or 2). According to Suzuki and Mishin[47], vacancy delocalization is a general phenomenon which may take place in many GBs and cores of other extended defects.

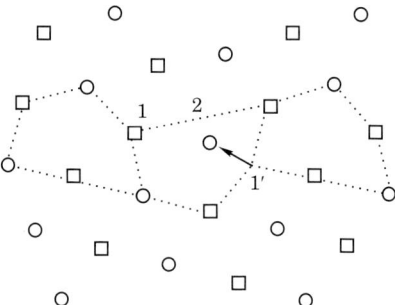

Fig. 3.8. Example of vacancy delocalization in Cu grain boundaries[47]. Vacancy at site 2 in the $\Sigma = 9$ ($1\bar{2}2$) GB. Atom 1' strongly relaxes towards the vacancy

Another feature of vacancies formed in GBs is their instability in some sites of a GB. It is illustrated in Fig. 3.9 demonstrating examples of unstable vacancies formation. It is shown in Fig. 3.9 (a) that a vacancy formed at site 6 in Σ = 5(310)[001] GB is filled by atom 1 in the process of relaxation. In other words, the vacancy in site 6 is absolutely unstable and it spontaneously transfers into a vacancy at site 1. Analogously, a vacancy formed at site 6 in Σ = 9 (1$\bar{2}$2) [011] GB [Fig. 3.9 (b)] relaxes into a vacancy delocalized between 1' and 2 sites. As in the former case, site 6 does not keep a stable vacancy and is filled with atom 2 by an athermic way.

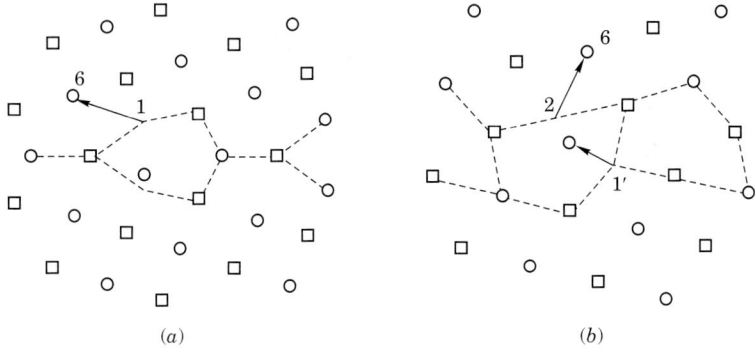

Fig. 3.9. Examples of vacancy instability in Cu grain boundaries[47]. (a) Unstable vacancy at site 6 in the Σ = 5 (310) [001] GB. The atom initially at site 1 fills the vacancy. (b) Unstable vacancy at site 6 in the Σ = 9 (1$\bar{2}$2) [011] GB. The atom initially at site 2 fills the vacancy

Interstitials in GBs were found in three structural forms: (1) localized in a relatively open region ("pore"), (2) delocalized over a relatively large area, and (3) interstitial dumbbells[46]–[48]. Examples of these structural forms are shown in Fig. 3.10. Figure 3.10 (a) demonstrates a preferential interstitial site in Σ = 5 (210)[001] GB. It is a center of a triangle formed by 2, 2' and 3 atoms. Localized interstitials can form at this site which gives only small relaxations to neighboring atoms. Figure 3.10 (b) illustrates a delocalized interstitial Σ = 5 (310)[001] GB. Atomic relaxations induced by interstitials extend to two structural elements which results in a configuration with two I and $\tilde{I}$ interstitials and a vacancy distributed between $\underline{1}$, 2 and $\underline{6}$ sites. Alternatively this may be considered as an interstitial at $\tilde{I}$ site and a Frenkel's couple formed by interstitial I and a vacancy delocalized between 2 and $\underline{6}$ sites. Interstitials can also form a configuration of a split dumbbell oriented parallel or normal to a tilt axis. A dumbbell oriented normally to a tilt axis in Σ = 7 (2$\bar{3}$1) [111] GB is shown in Fig. 3.10 (c).

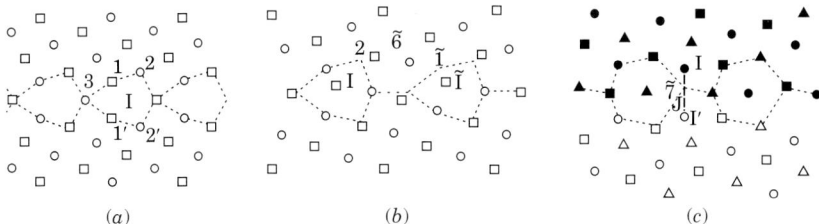

Fig. 3.10. Examples of structural interstitials in GBs[47]: (a) preferential interstitial site 1 in the $\Sigma = 5\,(210)[001]$ GB; (b) delocalized interstitial in the $\Sigma = 5\,(310)[001]$ GB; (c) interstitial dumbbells in the $\Sigma = 7\,(2\bar{3}1)\,[111]$ GB. An interstitial atom introduced near site I forms a dumbbell with an atom initially at site $\tilde{7}$

5. NON-EQUILIBRIUM STATE OF GRAIN BOUNDARIES

Non-equilibrium state of GBs is most typical of the majority of polycrystalline materials. This state is characteristic of boundaries in deformed materials especially in those subjected to SPD, the non-equilibrium structure remaining after short annealing, quenching or irradiation damage. Specific properties of boundaries in a non-equilibrium state are responsible for changing the character of deformation and recrystallization processes, and they also affect interaction of boundaries with impurities. The term "non-equilibrium boundaries" was suggested by Grabski and Korski as early as in 1970[52], but it came into use in scientific literature much later[53–57], a great variety of GB states being designated by it. For example, this term is used to denote boundaries with a non-equilibrium concentration of point defects[53, 57], boundaries with a curved surface[53], boundaries containing captured lattice and inserted grain-boundary dislocations[53–57], etc. One should keep in mind that any boundary proper is a non-equilibrium defect in a crystal, and that is why a conception of thermodynamic equilibrium of grain boundaries is to a certain extent conditional. More detailed conception of non-equilibrium boundaries was given by Valiev and Nazarov with coauthors[4, 58–60].

According to[4], an equilibrium state of a GB structure may be characterized by minimal free energy at given crystal-geometrical parameters and external conditions (temperature, pressure, concentration of impurities). If one considers grain-boundary defects of only dislocation nature, then the above-given definition of grain structure equilibrium is equivalent to a requirement that a boundary must not have long-range stress fields and its structure must be stable[61–62]. A GB equilibrium structure is as a rule characterized by a periodical distribution of structural units, disclination dipoles or grain-boundary dislocations. Consequently, interaction forces between elements of its structure are mutually balanced, and no strain arises in crystallites separated by a equilibrium boundary. An equilibrium state of a boundary can be attained as a result of a prolonged annealing at high enough temperatures.

A non-equilibrium state of a GB structure is characterized by an enhanced energy, and it results from the presence of defects in the structure. GBs are effective barriers for lattice dislocations movement, and hence at external impacts, *e.g.* at plastic deformation or recrystallization annealing, lattice dislocations captured by boundaries are accumulated[4]. Numerous electron-microscopic studies of structure of deformed polycrystals prove this general property of high-angle GBs[63–66]. The density of dislocations captured by a boundary increases with the growth of deformation degree, and at as low strain as 2% dislocation density reaches the value of about $10^7 - 10^8$ m^{-1} [67–68]. Boundary captured dislocations have a lattice Burgers vector of one of the grains which may be geometrically represented as a sum of vectors of a DSC lattice of this boundary. That's why a lattice dislocation may split into grain-boundary dislocations with Burgers vectors equal to that of DSC lattice which are inserted or primary GB dislocations[2, 4, 58]. Non-dissociated captured grain-boundary dislocations represent a special case of primary GB dislocations. Generally, when Burgers vector of a captured grain-boundary dislocation has components both in the boundary plane and normal to it its dissociation results in formation of glide (with Burgers vector parallel to a boundary plane) and sessile (with Burgers vector perpendicular to a boundary plane) primary GB dislocations. Hence, for splitting it's necessary to have a high enough temperature and an opportunity for sessile primary GB dislocations to climb.

The main property of primary GB dislocations is the presence of non-screened stress fields. At complete or partial absence of stress field screening of primary GB dislocations a GB possesses long-range stress fields. Hence, GBs containing primary GB dislocations have an excess energy of the elastic origin compared to equilibrium boundaries.

Thus, lattice dislocations capture by a boundary results in the formation of a system of primary GB dislocations in this boundary, and the latter is out of equilibrium. In this state long-range stresses and excess energy are practically completely determined by inserted dislocations, their type, density and distribution.

Components of non-equilibrium structure of grain boundaries were in detail analyzed by Nazarov in[60]. This analysis shows that a periodical wall formed by edge dislocations with Burgers vector normal to a GB plane has an equilibrium structure. On the contrary, a system of glide dislocations in a GB with a sum Burgers vector not equal to zero is always a source of long-range stresses, *i.e.*, it is non-equilibrium. Another type of a non-equilibrium state results from non-uniform dislocation density along a boundary. A non-uniform distribution of GB dislocations at any orientation of Burgers vector is a source of long-range stresses. Dislocation glide is generally non-uniform from a microscopic viewpoint and occurs along more or less randomly distributed glide planes. Consequently, location of dislocations captured by boundaries at plastic deformation will be random. As a result, the presence of such non-equilibrium ensembles of primary GB dislocations must be characteristic of GB structure in strained polycrystals.

As grain boundaries are elements of structure of a polycrystal with a net of GBs, defects untypical of separate grains may form in these boundaries. A detailed analysis of defective structures of that type was made by Rybin with co-authors[69–71], and it was shown that at plastic deformation a jump of deformation tensor appears at an interface of two neighboring crystallites because of their different orientation. In its turn accumulation of plastic deformation jump at a GB results in the formation of the rotation type defects in triple junctions which are represented in form of a combination of a wedge disclination with an axis coinciding with a junction line and a more complex defect. In case of deformation of a model two-dimensional polycrystal a junction defect will be represented only in form of a junction disclination. As shown by Rybin[72], the junction disclinations formation is the most typical phenomenon at plastic deformation of both polycrystals and fragmented materials, and it accounts for important peculiarities of large plastic deformations.

Thus, analyzing the available publications one can distinguish according to[59–60] the following three types of non-equilibrium dislocation structures which may form at plastic deformation of polycrystals (Fig. 3.11):

1. Disordered ensembles (nets) of dislocations resulting from microscopic non-uniformity of glide in grains.

2. Junction disclinations accumulating due to incompatibility of plastic deformation tensors of grains.

3. Ensembles of glide dislocations arising because of an arbitrary orientation of Burgers vectors of primary GB dislocations.

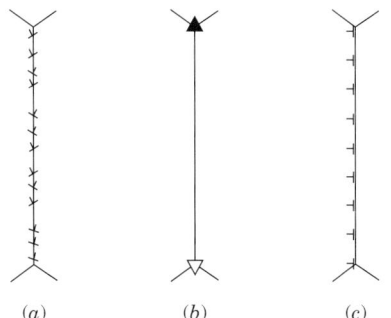

Fig. 3.11. Components of a GB non-equilibrium dislocation structure forming at plastic deformation[60]: (a)–a disordered dislocation wall; (b)–a dipole of junction disclinations; (c)–an ensemble of tangential inserted GB dislocations

The physical meaning of dividing of the non-equilibrium dislocation structures in GBs into three types consists in the fact that their contributions in polycrystal properties are in most cases well distinguished and additive.

Various methods are used to determine a non-equilibrium state of GBs. It's most simply to judge about non-equilibrium state of boundaries by the presence of long-range elastic stresses in near-boundary areas. This effect may be noticed by specific contrast in electron-microscopic images. In the very first electron-microscope studies of materials obtained by severe plastic deformation a specific appearance of their grain boundaries compared to that of ordinary annealed materials was observed[73]. Particularly, a considerable broadening of thickness extinction contours was noticed. Later in[74–75] the physical nature of this broadening was analyzed, and it was shown to be due to a high level of the internal stresses and distortions near GBs of samples subjected to SPD. Based on this analysis a technique was suggested of determination of elastic strains as a function of a distance from a GB. Using this technique elastic strains near GBs of Cu obtained by equal-channel angle pressing (ECAP) were studied, and it was shown that the elastic strains distribution is not uniform, namely, it has a maximum in a near-boundary area and an exponential drop at a distance of about 5 nm from a boundary[75]. Such distribution of elastic strains near a GB can be described by a configuration of edge glide dislocations in a GB [Fig. 3.11 (c)].

HREM gives certain opportunities for observation of non-equilibrium state of grain boundaries. This method allows observation of big amount of various defects, such as steps, facets and dislocations, the presence of which gives rise to elastic distortions near boundaries, and the latter can also be revealed by HREM. Besides, it's possible to discover a non-equilibrium state of grain boundaries using Mössbauer spectroscopy.

6. GRAIN BOUNDARY ENERGY

Specific energy of a GB (E_b) is determined as the energy of a boundary surface unit. In many cases it's convenient to consider a boundary as a surface with a certain surface tension (γ_b) which is determined as a constant of proportionality in an expression of work (δW) required for an increase of surface area by δA:

$$\delta W = \gamma_b \delta A \qquad \ldots(1)$$

If temperature T, system volume V and the number of particles of all components n_i are constants, then

$$(\delta W)_{T,V,n_i} = \delta F \qquad \ldots(2)$$

where F is the Helmholtz energy.

By definition the Helmholtz energy change is equal to

$$\delta F = \delta E - T\delta S \qquad \ldots(3)$$

where S is an entropy.

From Eq. (1)–(3) one obtains

$$\gamma_b = \frac{\delta E}{\delta A} - T\frac{\delta S}{\delta A} \qquad ...(4)$$

or
$$\gamma_b = E_b - TS_b \qquad ...(5)$$

where S_b is a specific entropy of a GB.

One can see from expression (5) that generally a GB specific energy is not equal to the surface tension. At low temperatures (≤ 10 K) the second term in the right part of (5) is small, and then $\gamma_b \approx E_b$. Theoretical calculations of boundary energy are usually carried out for $T = 0$ K, and they give $\gamma_b = E_b$, while experimentally the surface tension is measured which may be much lower than E_b.

The surface tension of GBs is most often determined by measuring an angle in a groove of thermal etching forming in the place of a boundary exit to the surface at heating in vacuum up to high temperatures close to the melting temperature (Mullinz method[76], Fig. 3.12).

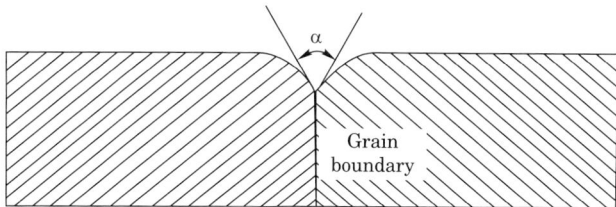

Fig. 3.12. Schematic image of a bicrystal section for the comparison of surface tension of a GB and a free surface

Mullins equation describes equilibrium at a thermal etching groove mouth in a GB:

$$\cos(\alpha/2) = \gamma_b/2\gamma \qquad ...(6)$$

where γ is surface tension of a free surface.

As noted above, special boundaries with small values of Σ possess specific physical properties. Particularly, their energy is as a rule considerably lower than that of ordinary boundaries because special boundaries have a high density of coinciding sites, and atoms located in the coinciding sites cause the smallest distortions of a crystal lattice. Near special boundaries inter-atomic spaces, the number and direction of inter-atomic bonds are retained to a greatest extent. One can see cusps on the experimental dependences of GB energy on misorientation angles, which is characteristic of both twisting and tilt boundaries (Figs. 3.13 and 3.14).

Figure 3.13 demonstrates the measured variation of a GB specific energy with twist angle, θ, for <100> twist boundaries in Al and Cu. One can clearly see a number of cusps on $E_b(\theta_{twist})$ curves for twist angles corresponding to small values of Σ. An analogous picture is observed for tilt boundaries. As seen from Fig. 3.14, there are cusps on $E_b(\theta_{tilt})$ curves at angles corresponding to small Σ.

A formula directly connecting a GB energy with Σ was suggested in[77]:

$$E_b(\Sigma) = \bar{E}_b (1 - \Sigma^{-1}) \qquad ...(7)$$

where $\bar{E}_b$ is an average GB energy.

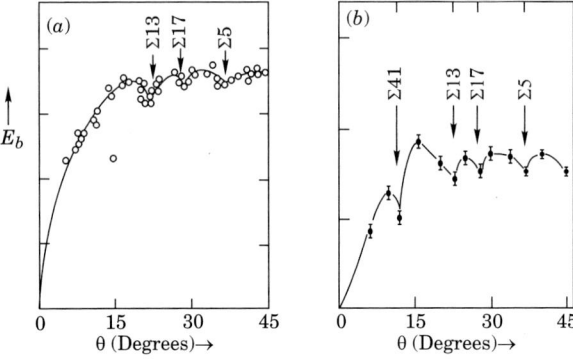

Fig. 3.13. Measured variation E_b with twist angle, θ, for $\langle 100 \rangle$ twist boundaries in Al (a) and Cu (b)[6]

It should be mentioned however that the experimental results analysis demonstrates an absence of linear dependence of E_b on Σ^{-1}. For example, in the experiments made in[78] boundaries corresponding to $\Sigma = 83$ appeared to be energetically more favorable than boundaries with $\Sigma = 9$ and $\Sigma = 33$. An analogous result was obtained in[79–80] as well as in calculations of boundaries energy[2]. Thus, one should keep in mind that the density of coinciding sites in a boundary, Σ_b^{-1}, has a physical meaning while the volume fraction of coinciding sites, Σ^{-1}, is purely a conventional parameter. Consequently, grain-boundary energy must correlate with Σ_b^{-1} but not with Σ^{-1}. For example, it was shown in[2] that for the surface tension experimental points with high accuracy fall to a line

$$\gamma_b = \bar{\gamma}_b (1 - \Sigma_b^{-1}) \qquad ...(8)$$

quite the same as (7).

Introduction of excess defects in a boundary changes its atomic structure and gives additional energy ΔE_b. As a result GB energy in a non-equilibrium state may be expressed as

$$E_b = E_b^0 + \Delta E_b \qquad ...(9)$$

where E_b^0 is a GB energy in an equilibrium state.

Measurements of a GB excess energy in deformed polycrystals were for the first time made by Grabski and Korski[52]. Their experiments have shown that at plastic deformation up to 15% a GB energy in Ni increases from its equilibrium value of $0.3E_s$ (E_s is a surface energy) to $0.6E_s$, i.e., it increases by a factor of two.

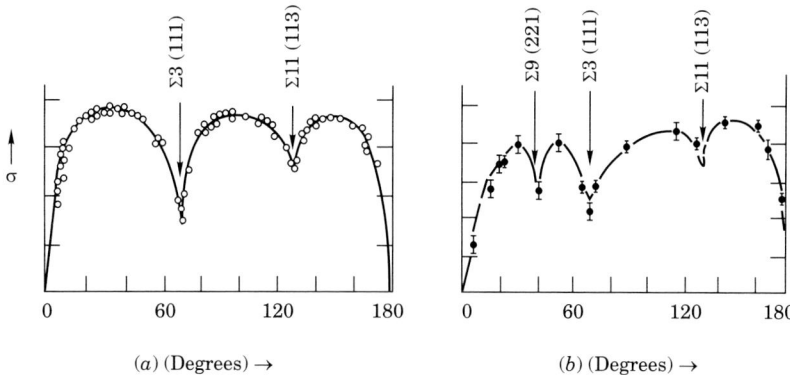

Fig. 3.14. Measured variation of E_b with tilt angle, θ, for $\langle 110 \rangle$ symmetric tilt grain boundaries in Al (a) and NiO (b)[6]

There exist two approaches to estimation of non-equilibrium boundaries energy. The first one is a thermodynamic approach in which the main parameter characterizing GBs structure is a free volume, and it was developed by Chuveldiev in[81–82]. Following a known idea of Mott[83] about an islet-like arrangement of grain boundaries, it is suggested that an arbitrary GB consists of a mixture of islets of two different phases, one of them being equated to a liquid with a relative free volume of 5%, and another is identified as a crystalline phase with a free volume close to zero. An average free volume of a GB is assumed to be 1–2%. A GB free energy is calculated as a sum of energies of phases and interfaces. According to[81–82], lattice dislocations absorbed by a boundary increase its free volume which results in growth of energy. As follows from calculations[81], a GB free energy growth corresponding to a free volume increase by 0.5% equals to 25% of an equilibrium free energy. In an approach suggested in[81–82] the determining role of dislocation cores is assumed.

Another approach to an assessment of non-equilibrium GBs was developed by Nazarov et al.[60, 84–86]. The difference of this approach from the previous one is that the role of dislocation cores is neglected, and elastic fields of extrinsic dislocations is considered in an assumption that their role is dominating in an excess energy of boundaries. It is supposed that GBs in a non-equilibrium state may contain three types of non-equilibrium structures forming at plastic deformation, namely, disordered dislocation nets, junction disclinations and ensembles of glide dislocations. As mentioned above, such division of non-equilibrium dislocation structures in GBs is physically based because of their well distinguished and additive contributions into properties of polycrystalline materials.

It was shown in[60] that individual dislocations with a non-screened stress field cannot exist in large crystals as their energy is proportional to a square of a characteristic size of a body. That is why dislocation bonding into multi-pole configurations is energy favorable. In grains dislocation glide is limited, which results in the formation of dislocation dipoles with a lever approximately equal to grain size [Fig. 3.15 (a)]. Consequently, one should consider not individual boundaries with random dislocation walls but a pair of boundaries with opposite dislocation signs. Analogously, dislocations formed by disclinations entering GBs from one grain appear to be bonded into disclination quadrupoles [Fig. 3.15 (b)], whereas glide GB dislocations form closed systems with a sum Burgers vector equal to a zero [Fig. 3.15 (c)].

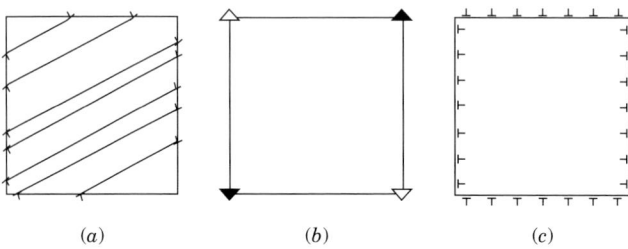

Fig. 3.15. Formation of dislocations in GBs at glide in a grain (a), disclination quadrupole (b) and a closed system of tangential dislocations (c)[60]

As shown in[87], a total energy of random nets of GB dislocations is just a sum of energies of constituent dipoles as these nets consist of independent dislocation dipoles. Therefore a system energy per one dipole is equal to an intrinsic energy of a dislocation dipole:

$$\Delta E = \frac{Gb^2}{4\pi (1-\nu)} \ln \frac{L}{b} \qquad ...(10)$$

where G is a shear modulus; b is a dislocation Burgers vector; ν is a Poisson's ratio; and L is an average grain size.

With an account of (10), an excess energy of a GB per unit area is

$$\Delta E_b = \frac{Gb^2 \rho}{4\pi (1-\nu)} \ln \frac{L}{b} \qquad ...(11)$$

where ρ is dislocation density.

An excess energy of a GB unit area caused by disclinations is described by formula[86]:

$$\Delta E_b = \frac{W}{2L} = \frac{G \langle \Omega^2 \rangle L \ln 2}{16\pi (1-\nu)} \qquad ...(12)$$

where W is an average energy of one quadrupole, and $\langle \Omega^2 \rangle$ is a dispersion of power of partial junction disclinations in grain junctions.

For the third component of GB defects, *i.e.*, for an ensemble of extrinsic GB dislocations a two-dimensional model was considered in[85] based on which the following expression was obtained:

$$\Delta E_b = \frac{GL \langle \beta^2 \rangle (\pi - 2\ln 2)}{2\pi (1-\nu)} \qquad ...(13)$$

where $\langle \beta^2 \rangle$ is a dispersion of a total density of Burgers vectors of extrinsic dislocations in grain boundaries.

The results obtained in[60] show that all three components of non-equilibrium dislocation structure make commensurable contributions into a GB excess energy, and an estimation of total excess energy agrees with the available experimental data.

Conclusions on the presence of long-range elastic stress fields caused by non-equilibrium state of grain boundaries may be done based on X-ray and dilatometric studies, investigation of thermal properties, measurements of density, etc. However all these are indirect methods, and the effects revealed with their application in some cases may be caused not by the non-equilibrium state of grain boundaries, but by other factors. The only methods enabling to characterize directly the state of GBs and near-boundary areas are GB diffusion studies considered in the next section and emission Mössbauer spectroscopy described in the last section of this chapter.

7. DIFFUSION ALONG GRAIN BOUNDARIES

Diffusion along grain boundaries plays a key role in many processes occurring in commercial materials at enhanced temperatures, as grain boundaries are the paths of accelerated diffusion. As the GB diffusion coefficient D_b is higher than the volume diffusivity D_V, and the activation energy is lower, $Q_b < Q_V$, GB diffusion is of importance not only at high temperatures, but at low and even room temperatures as well. For example, exploitation period of many products of microelectronics is limited by room temperature diffusion. Especially high is the role of GB diffusion in nanocrystalline materials.

The majority of mathematical descriptions of grain-boundary diffusion are based on the model of an isolated boundary firstly suggested by Fisher[88]. In this model it is assumed that a grain boundary is a semi-infinite isotropic layer of constant thickness, δ, with high diffusivity, D_b, enclosed in a semi-infinite perfect crystal with low diffusivity, D_V, the grain boundary being perpendicular to a surface on which a diffusant is deposited (Fig. 3.16). Concentrations of diffusing atoms in volume and grain boundary are denoted as $C_V(x, y, t)$ and $C_b(y, t)$, respectively.

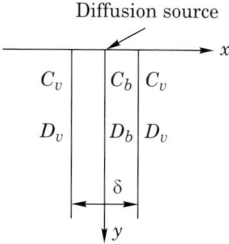

Fig. 3.16. Fisher's model of grain-boundary diffusion

In a diffusion experiment a layer of impurity atoms or isotope atoms of the same element is deposited on a specimen's surface, and then the specimen is annealed at constant temperature T for a period of t. At annealing atoms diffuse from the specimen's surface into the bulk by two paths, directly into grains and much faster along GBs. Atoms diffusing along GBs in their turn leave them and diffuse into lattice areas neighboring to a boundary forming a zone of volume diffusion around the boundary. Mathematically this diffusion process is described by a system of two equations

$$\frac{\partial C_V}{\partial t} = D\left(\frac{\partial^2 C_V}{\partial x^2} + \frac{\partial^2 C_V}{\partial y^2}\right), \quad \text{where } |x| > \delta/2 \qquad ...(14)$$

$$\frac{\partial C_b}{\partial t} = D_b \frac{\partial^2 C_b}{\partial y^2} + \frac{2 D_V}{\delta}\left(\frac{\partial C_V}{\partial x}\right)_{x=\delta/2} \qquad ...(15)$$

These equations describe volume and grain-boundary diffusion, respectively. Here $C_V(x, y, t)$ is concentration of the diffusing atoms in the volume and $C_b(y, t)$ is their concentration in a boundary. The second term in the right part of Eq. (15) accounts for the atom flux from a boundary into the volume.

Boundary conditions at GB/surrounding matrix interfaces for isotropic self-diffusion into a pure metal are expressed purely as concentration continuity at an interface

$$C_b(y, t) = C_V(x = \pm\delta/2, y, t) \qquad ...(16)$$

For an impurity diffusion in a pure metal the corresponding condition is expressed through an equilibrium segregation factor(s):

$$C_b(y, t) = s C_V(x = \pm\delta/2, y, t) \qquad ...(17)$$

The latter equation is based on the assumption that impurity atoms in a grain boundary are in a local thermodynamic equilibrium with the impurity atoms in lattice areas neighboring to a boundary, and, besides, it is assumed that grain-boundary segregation obeys Henry's law according to which

$$s = s_0 \cdot \exp\left(-\frac{E_s}{RT}\right) \qquad ...(18)$$

where E_s is the segregation energy.

The main equations of Fisher's model, Eqs. (14) and (15), can be solved analytically. Integral presentation of an accurate solution for a constant and instant source was derived by Whipple[89] and Suzuoka[90–91], respectively. Unfortunately, practical value of all such solutions is quite limited due to their too intricate mathematical form.

In grain-boundary diffusion studies an experiment is usually carried out as follows. A diffusant is deposited on a specimen's surface, and after a constant temperature annealing the diffusant concentration distribution is studied in the specimen's depth. When self-diffusion is studied, the same element isotopes are used, and for impurity diffusion another element atoms are taken. The classical radiotracer method is most often used to determine diffusing impurity distribution. Besides, the secondary-ion mass spectrometry, Auger electron spectrometry, autoradiography and other methods are applied. The methods of grain-boundary diffusion investigation are reviewed in a monograph by Kaur, Mishin and Gust[92]. Most of GB diffusion measurements were made using the radioisotope and sequent sections techniques. The parameter measured in such experiments is an average diffusant concentration in a layer, $\bar{C}$, versus penetration depth y. This function is referred to as a concentration, penetration or diffusion profile, and it gives information on GB diffusion parameters. Many studies deal with an obtaining of approximate solutions for Fisher's model and simple ways of the profiles treatment. Some of them are considered below according to[92, 93–94].

Fisher considered a simplified form of Eqs. (14) and (15) and obtained the following approximate solution for a constant source

$$\bar{C} \propto C_0 \exp(-\pi^{-1/4} w) \qquad \text{...(19)}$$

where

$$w = \frac{y}{(s\delta D_b)^{1/2}} \left(\frac{4D}{t}\right)^{1/4} \qquad \text{...(20)}$$

Equation (19) assumes that a penetration profile built as $\bar{C}$ versus y tends to a straight line. From this line tilt $\partial \ln \bar{C}/\partial y$ one can determine the product $s\delta D_b$

$$s\delta D_b = 1.128 \, (D_V/t)^{1/2} \, (-\partial \ln \bar{C}/\partial y)^{-2} \qquad \text{...(21)}$$

The volume diffusivity is taken known from independent measurements.

Exponential Fisher's solution (19) is not accurate. If a profile calculated using an accurate analytical Whipple's solution[89] is built as $\bar{C}$ versus y^n with various n, the best straight line is obtained with $n = 6/5$. It was also found that

$$-\frac{\partial \ln \bar{C}}{\partial w^{6/5}} \approx 0.78 \qquad \text{...(22)}$$

Thus, knowing the tilt of a linear part of the $\ln \bar{C}$ versus $y^{6/5}$ curve, one can calculate the $s\delta D_b$ product as

$$s\delta D_b = 1{,}322 \, (D/t)^{1/2} \, (-\partial \ln \bar{C}/\partial y^{6/5})^{-5/3} \qquad \text{...(23)}$$

for a constant source. Equation (23) may be used only in the following two conditions. Parameter β, determined as

$$\beta = \frac{s\delta D_b}{2D \, (Dt)^{1/2}} \qquad \text{...(24)}$$

must be large enough (for practical purposes $\beta > 10$), and parameter

$$\alpha = \frac{s\delta}{2(Dt)^{1/2}} \qquad \ldots(25)$$

must be small enough (for practical purposes $\alpha < 0.1$).

Similarly, for the diffusion from an instant source it was found that

$$-\frac{\partial \ln \overline{C}}{\partial w^{6/5}} \approx 0.775 \qquad \ldots(26)$$

When α is small enough and $\beta > 10^4$, the product $s\delta D_b$ may be determined from an expression

$$s\delta D_b = 1{,}308 \, (D/t)^{1/2} \, (-\partial \ln \overline{C}/\partial y^{6/5})^{-5/3} \qquad \ldots(27)$$

If β is smaller than 10^4, then the right part of Eq. (26) is no longer constant and is a function of β (and therefore of time as well). For $\beta < 10^4$ Eq. (27) takes the form:

$$s\delta D_b = 1{,}206 \, (D^{0{,}585}/t^{0{,}605})^{1/1{,}19} \, (-\partial \ln \overline{C}/\partial y^{6/5})^{-5/2{,}975} \qquad \ldots(28)$$

when $10^2 < \beta < 10^4$, and

$$s\delta D_b = 1{,}084 \, (D^{0{,}91}/t^{1{,}03})^{1/1{,}94} \, (-\partial \ln \overline{C}/\partial y^{6/5})^{-5/2{,}91} \qquad \ldots(29)$$

when $10 < \beta < 10^2$.

Equations (23) and (27)–(29) are the basic ones used in the everyday practice for GB diffusion measurements.

Note that these equations give only the triple product of $s\delta D_b$, while separate values of s, δ and D_b remain unknown. Even for self-diffusion in pure metals ($s = 1$) one can determine only the δD_b product.

Besides GB diffusion proper, diffusion along grain boundaries includes the direct volume diffusion from a surface, diffusion outflow from a boundary into the volume and diffusion in the volume adjacent to the boundary. Depending on annealing temperature and duration, grain size and other conditions, this or that process may dominate and determine the total rate of the whole process. For proper determination of the diffusion parameters one should know the controlling process, because the concentration profile form depends to a great extent on the dominating kinetic regime. Moreover, diffusion characteristics which can be obtained from the penetration profile are generally different for different regimes and must be identified *a priori*.

The first and the most widely used classification of diffusion regimes in polycrystals is that for a system of uniformly distributed GBs suggested by Harrison[95]. It suggests diffusion behavior in three regimes denoted as type *A*, *B* and *C* (see Fig. 3.17).

Regime *A* is observed at high temperatures (usually higher than $(0{,}7 \div 0{,}8)\,T_m$) and/or at prolonged annealing and/or fine grains. At such conditions the volume diffusion path $(D_V t)^{1/2}$ is much greater than the grain size, L, and diffusion fluxes from neighboring boundaries overlap. The condition of regime A realization is:

$$(D_V t)^{1/2} \gg L \qquad \ldots(30)$$

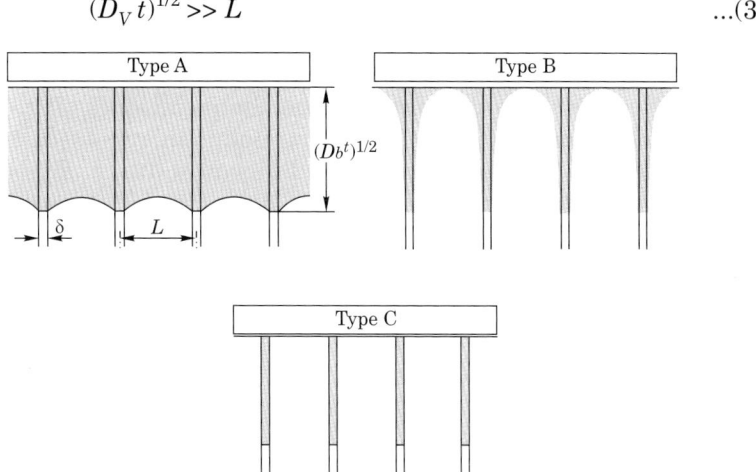

Fig. 3.17. Diffusion kinetic regimes of A, B and C types in a polycrystal with uniformly distributed grain boundaries

The condition of regime B realization is expressed as an inequality:

$$\delta \ll (D_V t)^{1/2} \ll L \qquad \ldots(31)$$

In this case diffusion fluxes from neighboring boundaries don't meet, and grain boundaries are isolated from each other. Regime B is realized at lower temperatures than regime A and/or shorter diffusion annealing and/or coarser grains. For reasonable (about dozens of hours) times and grain sizes (1–10 µm) regime B covers the widest and most frequently occurring temperature interval of $0.4 \div 0.7 T_m$. The only parameter which can be determined in regime B is the product of $s\delta D_b$. In this regime one may use Eqs. (23) and (27)–(29) for a profile treatment. Conditions of regime B are most commonly encountered in the diffusion measurements.

Condition of regime C is determined by the relationship

$$(D_V t)^{1/2} \ll \delta \qquad \ldots(32)$$

This inequality shows that regime C is realized if $\alpha \gg 1$. For practical purposes a regime is considered as C at $\alpha > 10$. In this case volume diffusion is "frozen", and diffusion flux from a GB into the volume may be neglected. Diffusion proceeds only along grain boundaries. Regime C is usually realized at temperatures lower than $0.4 T_m$. Concentration profile in regime C for an instant source is a Gauss function

$$\overline{C} \propto \exp(-y^2/4 D_b t) \qquad \ldots(33)$$

or error function for a constant concentration source

$$\overline{C} \propto \mathrm{erfc}\,(y/2(D_b t)^{1/2}) \qquad \ldots(34)$$

As noted above, GB diffusion experiments are usually made in regime B, and a penetration profile is treated using $y^{6/5}$ [Eqs. (23) and (27)–(29)]. For impurity diffusion along grain boundaries this method gives a triple product $s\delta D_b$ and for self-diffusion δD_b.

From measurements in regime C at relatively low temperatures one can directly determine D_b. Combining the obtained values of D_b with δD_b extrapolated from self-diffusion measurement in regime B at higher temperatures one can determine a GB diffusion width, δ. This approach was used in[96] and the diffusion width of a GB was found to be about 0.5 nm. This value agrees with a traditional assumption that $\delta = 0.5$ nm suggested by Fisher[88]. This value of δ is also in agreement with a GB width assessment by HREM, field ion microscopy and other techniques[6,92].

Using an assumption that $\delta = 0.5$ nm, one can determine the grain-boundary segregation factor based on the results of measurements of GB impurity diffusion in B and C regimes. This approach is realized in Munster's school studies (see e.g.[97–100]). The most complete data on grain-boundary diffusion parameters are given in[101].

Thus, using Fisher's model one can determine GB diffusion parameters based on the analysis of concentration profiles. However the Fisher's model is an idealized one and does not have a direct association with a GB structure, and a number of attempts to modernize it have been undertaken.

One of the main assumptions of Fisher's model is that a GB is uniform and isotropic. Several attempts to reject this assumption were made[102–107], various types of spatial non-uniformity of a GB being considered. For example, there are attempts to solve the problem of GB diffusion in case of GB non-uniformity[102–104]. These models don't contradict the results of GB diffusion studies by the method of layers removal, as for real polycrystals and real annealing times they predict the same concentration dependence on penetration depth as the Fisher's model. Their main result is confirmation of validity of standard Fisher's solution with the only difference that in case of a non-uniform GB the solution parameters become effective quantities with intricate dependence on microscopic diffusion characteristics of GBs.

In[102, 105] the effect of accelerated outflow paths of a diffusant atoms from a GB into the volume such as dislocations, sub-grains and others were taken into account. In[105] a special diffusion regime D was suggested when dislocation outflow prevails over a common volume pumping-out. In low-temperature diffusion experiments one maybe should take into account the accelerated outflow of atoms from GBs, though there are no unambiguous experimental evidences of a marked effect of this process on GB diffusion for equilibrium boundaries of recrystallization origin. However, it's quite possible that its contribution is high in strained materials, especially in that subjected to SPD.

In[106–108] Fisher's model modification was suggested to explain the presence of two lines in grain-boundary emission Mössbauer spectra (see Section 8). It was assumed that between a grain boundary and crystallites volume there exists a so-called pumping zone, characterized by a high concentration of point defects and consequently having a relatively high diffusivity, which makes possible the diffusing atoms transition from GBs to adjacent areas of a matrix at relatively low temperatures when volume diffusion is suppressed. This model was widely used to interpret the results of the Mössbauer investigations (see e.g.[106–110]). However it does not agree with direct diffusion experiments as it predicts an existence of such diffusion kinetic regimes which are not observed experimentally.

Thus, up to now Fisher's model has been the main instrument for diffusion experiments treatment. Nevertheless, there is some internal contradiction in the modern interpretation of this model. It's usually assumed that is case when volume diffusion is frozen, *i.e.*, regime C is realized, the diffusing impurity atoms locate only in GBs and don't transfer into crystallites volume. On the other hand, Fisher's model assumes that on a surface dividing a GB and the surrounding matrix there is a local equilibrium given by Eq. (17), but for this equilibrium to establish some amount of the diffusant must transfer into near-boundary areas. An attempt to solve this contradiction was undertaken in[111, 112], in which a modification of Fisher's model is suggested. This model is considered in detail in the last section of this chapter.

Grain-boundary diffusion mechanisms are discussed in many monographs and reviews[6, 92, 113–115]. Up to recently it was assumed by analogy with volume diffusion that the main way of GB diffusion is the vacancy mechanism. The majority of the available experimental data, including measurements of diffusion activation volume which appears to be approximately equal to an atomic one[115], and of isotopic effect[116] testify in favor of the vacancy mechanism.

However, the problem of determining the mechanism (or mechanisms) of fast diffusion of substitutional atoms in grain boundaries has proven to be much more difficult and is still not well resolved. The reasons for this state of affairs are clear: grain boundary structures are generally complex and irregular, and there exist an infinite number of different types of them. Various mechanisms may therefore be dominant in various boundaries. In addition, the thin nature of GBs precludes many types of critical experiments which have been used successfully to study diffusion mechanisms in bulk crystal lattices. One of a few methods enabling to make conclusions about a dominating GB diffusion mechanism is the emission Mössbauer spectroscopy[117] (see Section 8).

In recent years the main efforts in GB diffusion mechanisms studies have been connected with atomistic simulation of diffusion and defects by methods of molecular dynamics, molecular statics, Monte-Carlo kinetics, etc. In the very first works on GB diffusion simulation[49–51] a great role of interstitials was revealed, and it was shown that in definite conditions grain-boundary diffusion may proceed through interstitial sites. In[6] possible mechanisms of GB diffusion are considered based mainly on geometrical representation, but it is not taken into account that point defects in GBs markedly differ from that in a regular lattice.

A comprehensive analysis of possible mechanisms of GB diffusion was done in[46–48] by an example of tilt boundaries in Cu. The results of these publications concerning GB diffusion mechanisms are briefly considered below. It is shown that vacancies are typically found to move in GBs by exchanges with individual atoms, just as they do in the bulk (Fig. 3.18). However, in contrast to bulk diffusion, vacancies in GBs can also induce collective jumps of two or more atoms. Such "long" vacancy jumps always involve sites that cannot support a stable vacancy. For example, the unstable vacancy at site 6 in the $\Sigma = 5(310)$ [001] GB is responsible for the long vacancy jump $1' \to 6' \to \tilde{4}$ involving two atoms as well as for the $1 \to 6 \to \tilde{2} \to \tilde{1}$ jump involving three atoms (Fig. 3.18).

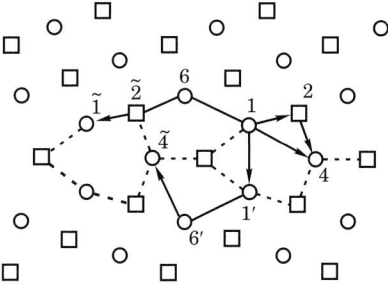

Fig. 3.18. Vacancy jumps in the $\Sigma = 5$ (310) [001] GB in Cu[48]

Localized interstitials can move by two mechanisms, direct and indirect. Under the direct mechanism an interstitial atom travels along a GB by jumping between neighboring interstitial positions. While energetically unfavorable in $\Sigma = 5$ GBs[46], this mechanism was found to operate in some other GBs.

Under the indirect mechanism, an interstitial atom displaces a neighboring regular atom to another interstitial position and takes its place. This process occurs by a simultaneous jump of both atoms. Furthermore, an interstitial atom can initiate a chain of atomic displacements and push out the last atom of the chain into another interstitial position which can be well separated from the initial one. All atoms involved in this process move in a concerted manner and not one after another. Figure 3.19 illustrates this mechanism for the diffusion perpendicular to the tilt axis. Notice that the same interstitial jump (e.g., $I \to \tilde{I}$ in the $\Sigma = 5$ (310) [001] GB) can be implemented in several different ways, some of them involving more atoms than others. The four-atom jump $I \to 2 \to \tilde{6} \to \tilde{2} \to \tilde{I}$ in Fig. 3.19 actually happens in two steps, with the formation of an intermediate delocalized interstitial configuration.

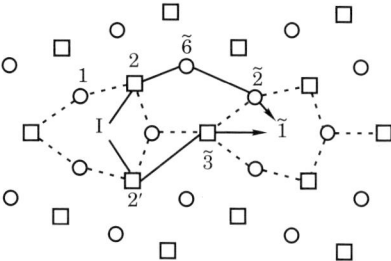

Fig. 3.19. Interstitial jumps in the $\Sigma = 5$ (310) GB in Cu[48]

An interstitial dumbbell always moves by a collective jump of three or more atoms. Finally, the highly delocalized interstitial I can migrate along the tilt axis with an extremely small activation barrier. Each step of this motion is accompanied by a very small displacement of each individual atom, yet the entire relaxation zone translates by one period.

This consideration shows that even for equilibrium special GBs various diffusion mechanisms can be realized including intricate multi-step ones. One can expect that in boundaries of common type, and especially in the non-equilibrium ones the situation is even more complicated, but diffusion in such boundaries is quite poorly studied by now.

In recent years an avalanche growth of publications dealing with nano-materials is observed. Particularly, great attention is paid to diffusion studies in nanocrystalline (NC) and nanostructured (NS) materials[118–120]. Table 3.2 demonstrates comparison of diffusion coefficients in NC and NS materials (D_n) with GB diffusion coefficients in polycrystalline metals (D_b). It can be seen that diffusivities in nano-materials are by several orders of magnitude higher than GB diffusion coefficients at the same temperatures. Based on these results Gleiter assumed an existence in nano-materials of an "amorphous" GB phase which can range up to a half of their total mass and is characterized by an absence of not only long-range, but even a short-range order in atoms arrangement[118]. This assumption contradicts firstly with experimental data which testify that the diffusion rate of substitutional impurities in amorphous metal alloys is essentially lower that the GB diffusion rate in polycrystals, and, secondly, with the results of GBs studies by HREM which demonstrate that GBs in nanocrystalline as well as in polycrystalline materials are layers the thickness of 2–3 inter-atomic spaces with periodical arrangement of atoms[121, 122].

Klotsman[130] cast doubt on Gleiter's conclusion[118] about considerably higher GB diffusion coefficients in NC materials, as the latter used a diffusion model not adequate to the experiment because the experimental conditions did not correspond to regime C. It should be noted however that in[119] it is shown that from all possible solutions corresponding to an experimentally observed dependence of layer concentration on depth the lowest value of a true GB diffusion coefficient will be determined in regime C, and consequently Gleiter's conclusion[118] on a considerable increase of GB diffusion rate in NC materials still remains valid.

Table 3.2. Diffusion coefficients in nano- and polycrystalline metals

Solute	L, nm	Solvent	T, K	Method	Diffusivity, m²/sec		Ref.
					D_n	D_b	
1	2	3	4	5	6	7	8
NC Cu	10	^{67}Cu	293	RT	$2.6 \cdot 10^{-20}$	$4.8 \cdot 10^{-24}$	[123]
	10	Ag	293	EPMA	$4.8 \cdot 10^{-17}$	$8.1 \cdot 10^{-22}$	[124]
	10	Au	293	AES	$4.8 \cdot 10^{-22}$	$2.6 \cdot 10^{-26}$	[125]
NC Pd	8–11	Ag	393	SIMS	$3.5 \cdot 10^{-17}$	$1.3 \cdot 10^{-21}$	[125]
	8–11	Au	393	SIMS	$8.1 \cdot 10^{-19}$	$1.5 \cdot 10^{-20}$	[125]
	20	^{59}Fe	473	RT	$4.1 \cdot 10^{-20}$	$1.3 \cdot 10^{-21}$	[126]
	52	^{59}Fe	523	RT	$1.0 \cdot 10^{-19}$	$1.0 \cdot 10^{-19}$	[126]
NC Ni	70	Au	483	RBS	$3.3 \cdot 10^{-20}$	$5.6 \cdot 10^{-21}$	[127]
	70	^{63}Ni	293	RT	$1.14 \cdot 10^{-20}$	$1.4 \cdot 10^{-26}$	[127]
	70	^{63}Ni	373	RT	$1.05 \cdot 10^{-18}$	$3.3 \cdot 10^{-22}$	[127]
	70	^{63}Ni	423	RT	$4.0 \cdot 10^{-18}$	$7.7 \cdot 10^{-20}$	[127]
	70	^{63}Ni	473	RT	$2.0 \cdot 10^{-17}$	$8.4 \cdot 10^{-19}$	[127]
	30	Cu	423	SIMS	$3.8 \cdot 10^{-17}$	$4.3 \cdot 10^{-19}$	[128]
NC Fe	31	^{59}Fe	452	RT	$2.4 \cdot 10^{-21}$	$1.6 \cdot 10^{-22}$	[129]
	31	^{59}Fe	472	RT	$9.4 \cdot 10^{-21}$	$1.3 \cdot 10^{-21}$	[129]
	31	^{59}Fe	499	RT	$2.9 \cdot 10^{-21}$	$2.9 \cdot 10^{-21}$	[129]
NS Pd	80–150	^{59}Fe	371	RT	$1.0 \cdot 10^{-21}$	$8.0 \cdot 10^{-25}$	[126]
	80–150	^{59}Fe	401	RT	$3.0 \cdot 10^{-20}$	$1.2 \cdot 10^{-23}$	[126]
	80–150	^{59}Fe	473	RT	$5.0 \cdot 10^{-20}$	$6.0 \cdot 10^{-21}$	[126]
	80–150	^{59}Fe	577	RT	$9.0 \cdot 10^{-20}$	$3.0 \cdot 10^{-18}$	[126]
NS Fe	150	^{59}Fe	528	RT	$4.0 \cdot 10^{-19}$	$1.0 \cdot 10^{-19}$	[126]
NS Ni	300	Cu	398	SIMS	$5.06 \cdot 10^{-15}$	$4.64 \cdot 10^{-20}$	[128]
	300	Cu	423	SIMS	$9.6 \cdot 10^{-15}$	$4.31 \cdot 10^{-19}$	[128]
	300	Cu	448	SIMS	$2.2 \cdot 10^{-14}$	$2.14 \cdot 10^{-18}$	[128]

Comment: Diffusion coefficients D_n–in NC and NS metals; D_b–in polycrystals along GBs. Experimental methods: RT–radiotracer; EPMA–electron probe microanalysis; AES–Auger electron spectroscopy; SIMS–secondary ion-mass spectroscopy.

A detailed investigation of self-diffusion in nanocrystalline Ni was carried out in[127], and it was found that the self-diffusion coefficient in nanocrystalline Ni is considerably higher than the GB diffusion coefficient in polycrystals, but their difference decreases with temperature growth. It was also found that the self-diffusion activation energy in NC Ni is by more than a factor of two lower than the activation energy of self-diffusion along grain boundaries of polycrystalline Ni, their values being 50.7 and 117 kJ/mole, respectively. In[123] it is noted that the self-diffusion activation energy in NC Ni is close to that of self-diffusion along a free surface of Ni. Based on these results a cluster model of NC materials was suggested in[127], and later it was developed in[131] and called the bimodal structure model (Fig. 3.20).

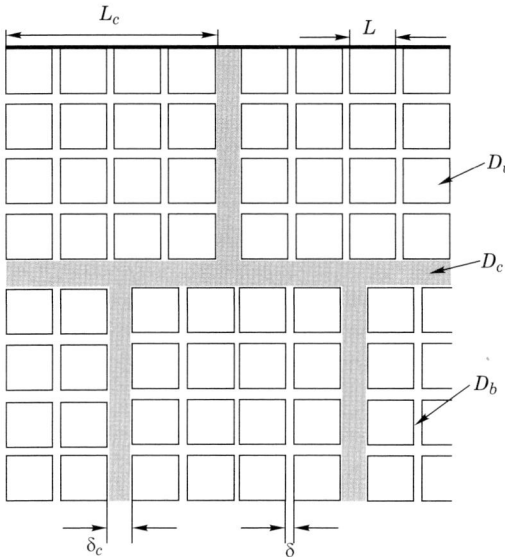

Fig. 3.20. A nanocrystalline material model[131]: The squares correspond to individual nano-grains (of size of L) with the bulk diffusivity D_v. The inter-agglomerate boundaries with a relatively larger fraction of the associated free volume (the diffusivity D_c and the width δ_c) and regular nanocrystalline (inside agglomerate) boundaries (the diffusivity D_b and the width δ are distinguished)

It was assumed that nanocrystals the size of L form clusters (agglomerates) the size of L_C, which was confirmed experimentally[131]. There are two types of boundaries in such structures: GBs inside the clusters and inter-cluster boundaries. Grain boundaries inside clusters separating nanocrystals practically do not differ from common GBs in polycrystals (with diffusivity D_b and grain width δ). Inter-cluster boundaries with a relatively large free volume associated with them have a higher diffusion coefficient, D_c, and larger grain boundary width, δ_c, compared to that of inside boundaries.

It is stated in[132] that micropores (sources and sinks of vacancies) as well as vacancies proper play an essential role in diffusion acceleration in nanocrystalline materials. The validity of such explanation of high diffusivities in NC materials is in good agreement with the cluster model of their structure described above, which found an experimental confirmation in[133] at HREM studies of Ag and Pd (the crystallite size of 8–10 nm). In triple junctions of GBs in these materials pores were observed.

In recent years a severe plastic deformation (SPD) is widely used to obtain nanostructured and submicrocrystalline materials, its mostly used techniques being equal-channel angular pressing (ECAP) and high pressure torsion (HPT). Such treatments enable to attain nanostructure with crystallite sizes of 100–300 nm. Peculiarities of NS materials obtained by SPD result from the fact that they don't contain pores and GBs in them are highly non-equilibrium (they are characterized with high elastic stresses, intrinsic and lattice dislocations, facets and steps of deformation origin). Considerable distortions are observed in near-boundary areas testifying directly the presence of elastic stresses and consequently the non-equilibrium state of GBs[9, 119].

The first studies of diffusion in such materials were carried out in[126,135] by an example of Fe isotopes diffusion in Pd. A marked increase of GB diffusion coefficients was found (see Table 3.2) as well as a decrease of activation energy of grain-boundary diffusion by a factor of 1.5. Similar results were obtained in the investigation of self-diffusion in nanostructured Fe[134].

Diffusion of Cu in nanostructured Ni was studied in[119, 128, 136–138]. It was found that GB diffusion coefficients in NS Ni are by 4–5 orders of magnitude higher and activation energy is by a factor of two lower than in coarse-grained Ni. Thus, similar to the above-discussed publications[118, 127] on diffusion studies of NC metals, activation energy in NS Ni is close to its value for a free surface in spite of the absence of pores in it[128].

Since the conditions dominating during the synthesis of "nanocrystalline" materials are far from thermodynamic equilibrium, the initial structure of interfaces in such materials may strongly depend on the time-temperature history. It is especially essential for SPD obtained specimens, but may be also observed in case of the materials prepared by the cluster condensation and compaction route. For nanocrystalline metals prepared by crystallite condensation in an inert-gas atmosphere with subsequent crystallite compaction, extensive experimental evidence exists that structural relaxation occurs at slightly elevated temperatures, and this structural relaxation affects their diffusion behavior[120].

Similar observations were made in[128] for nanostructured Ni prepared by SPD. In this case, the diffusivity of Cu at 423K was found to decrease by more than three orders of magnitude upon pre-annealing of a sample at 523K prior to the onset of grain growth. Both in NS Fe[134] and NS Ni[128] the interfacial diffusion coefficients in the relaxed state appear to be similar to or only slightly higher than the values expected for conventional GBs in coarse-grained materials from an extrapolation of high-temperature data.

The enhanced diffusivity observed in NS metals prior to relaxation may indicate a non-equilibrium structure of interfaces. The enhanced diffusivities found in such materials are ascribed to the lattice dislocations absorption at GBs[139–140].

In case of nanocrystalline metals, in which dislocation activity ceases due to small crystallite sizes[141], enhanced diffusivities may arise from local excess free volume, which in case of inert-gas-condensed samples may remain after the high-pressure compaction step. Evidence for the existence of vacancy-type interfacial free volumes and their variation upon annealing at slightly elevated temperatures could be derived from positron annihilation studies[142]. This situation resembles the well-known structural relaxation in amorphous alloys, where the excess volume reduction at annealing causes a decrease in diffusivity[143].

Unlike the case of the amorphous alloys, the microstructure of which remains stable over a relatively wide temperature range below the onset of crystallization, the relaxed structure of nanocrystalline metals is highly prone to interface migration and grain growth. This is particularly true for porosity-free nanocrystalline metals of high purity, in which pinning centers for interfaces have largely been eliminated[126]. In that case, the assessment of the diffusion behavior is affected by the concomitant GB migration.

The complications introduced by grain growth in nanocrystalline metals are even more serious when attempting to ascribe a physical interpretation to experimental values for the activation energies of diffusion, because grain growth seriously limits the temperature range of such studies.

In NC materials obtained by gas condensation and compacting the residual porosity is of great importance. When the residual porosity decreases, GB diffusion parameters in NC materials approach to their values for coarse-grained polycrystals. Thus, diffusion studies in densely-compacted (98% of theoretical density) NC Pd shows that GB diffusion coefficients of ^{59}Fe and ^{103}Pd isotopes do not differ from the appropriate calculated values for coarse-grained materials[126]. The same result was obtained in[129, 134] at self-diffusion investigation in densely-compacted NC Fe, including that prepared by explosion pressing[129].

8. EMISSION MÖSSBAUER SPECTROSCOPY OF GRAIN BOUNDARIES

Emission Mössbauer spectroscopy is a method enabling to judge directly on physical properties of GBs, and this section is devoted to its capabilities. A method of investigation of GBs and near-boundary areas in polycrystals based on the use of accelerated diffusion along grain boundaries together with the emission NGR spectroscopy was worked out by Kaigorodov and Klotsman[144].

The procedure of specimens-sources preparation in this method is as follows. A radionuclide is deposited on a polycrystalline specimen surface, and then the specimen is annealed at temperatures when volume diffusion is knowingly frozen. At the annealing the Mössbauer isotope atoms diffuse along grain boundaries. The last step consists in the removal of non-diffused isotope from the surface together with a thin surface layer in which the radioisotope atoms could penetrate not along grain boundaries, but by dislocations and low-angle boundaries. This procedure is shown in Fig. 3.21.

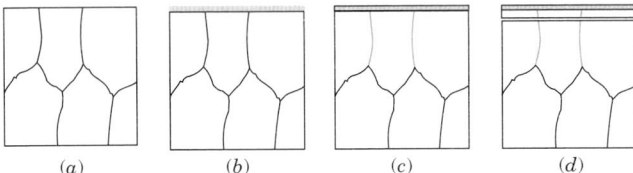

Fig. 3.21. A specimen-source preparation for GBs study by the emission Mössbauer spectroscopy: (*a*) initial polycrystalline specimen; (*b*) the specimen after the radioisotope deposition on its surface; (*c*) the specimen after the diffusion annealing; (*d*) the non-diffused isotope and a surface layer removal

A Mössbauer spectrum is taken from the as-prepared specimen. Then higher temperature annealing is carried out, and after every annealing a spectrum is taken. At this experimental procedure the Mössbauer atoms remained in the specimen after the first annealing and the surface layer removal only redistribute between a GB and the surrounding matrix, the total amount of the radionuclide being invariable.

It has been found that in all NGR spectra of various metals studied by now there are always two components (see, *e.g.*[109, 110, 117, 145]). It is illustrated by Fig. 3.22 in which the Mössbauer spectra of ^{119m}Sn in polycrystalline Nb are shown. The isomer shift of one line (component 1) appreciably differs from that of atoms located in a regular lattice of the element under study. It means that atoms forming this line are located in severely distorted areas of crystallite interfaces, *i.e.*, in GB cores. For another line (component 2) the isomer shift is close to that for the regular lattice, testifying that the atoms forming that line are in positions close to the regular lattice, *i.e.*, in near-boundary areas. With temperature growth the component 1 intensity is decreasing up to its complete vanishing, whereas that of component 2 is growing, and at highest temperatures it is the only line in a Mössbauer spectrum.

It should be emphasized that these two components are present in all the Mössbauer spectra of polycrystalline metals even at the lowest annealing temperatures when the volume diffusion is knowingly frozen, and it is difficult to explain this result from the viewpoint of Fisher's model in its present interpretation. As a rule, it's assumed that when the volume diffusion is frozen the regime *C* is realized, and atoms of a diffusing impurity are located only in GBs and don't transfer in crystallites' volume. On the other hand, the Fisher's model assumes that there is a local equilibrium at an interface of a GB and the surrounding matrix, given by Eq. (17), but to establish this equilibrium some amount of a diffusant must transfer into near-boundary areas of a matrix. An existence of equilibrium near-boundary layers is not explicitly included in the Fisher's model, but in conditions when a diffusant penetration depth in a grain bulk is commensurable with a diffusion width of a grain boundary it must be taken into account. In a model suggested in[111–112] these layers are explicitly taken into account, and this model is illustrated schematically in Fig. 3.23.

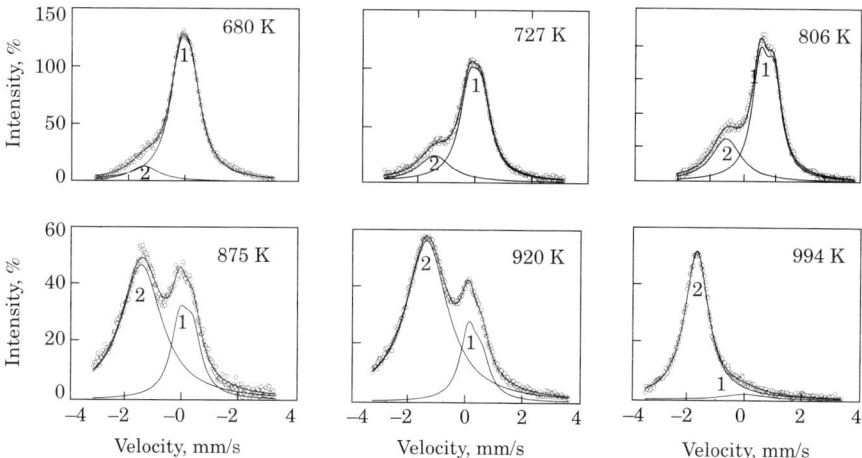

Fig. 3.22. Emission Mössbauer ^{119m}Sn spectra in polycrystalline Nb annealed at various temperatures indicated in the figures

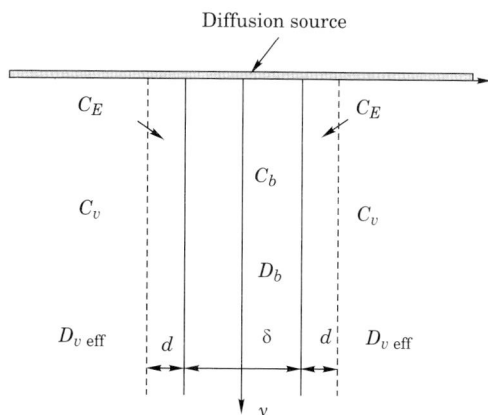

Fig. 3.23. A GB diffusion model with an account of equilibrium composition near-boundary layers

In this case it is assumed that layers the width of d are in contact with a GB, and an equilibrium concentration of a diffusing element, C_E, is established in these layers. Penetration of this element in farther areas of a matrix results from diffusion in crystallites bulk. It is assumed that mass transfer from a GB into crystallites volume is realized not only by volume diffusion but by other mechanisms as well. That's why in our model not a volume diffusion coefficient is considered, but some effective diffusivity $D_{V\,\text{eff}}$ characterizing the rate of atoms outflow from a GB.

Equilibrium concentration in near-boundary areas is established as a result of a reaction at an interface of a GB and the surrounding matrix. It is suggested that this reaction proceeds much faster than diffusion into volume. With an account of an equilibrium composition near-boundary layer the width of d Eq. (14) becomes

$$\frac{\partial C_b}{\partial t} = D_b \frac{\partial^2 C_b}{\partial y^2} + \frac{2D_{V\,\text{eff}}}{\delta}\left(\frac{\partial C_V}{\partial x}\right)_{|x|=\delta/2+d} - \frac{2d}{s\delta}\frac{\partial C_b}{\partial t} \qquad \ldots(35)$$

where the last member accounts for a substance transfer from a GB into equilibrium composition layers.

After rearrangement this equation takes a form

$$\left(1 + \frac{2d}{s\delta}\right)\frac{\partial C_b}{\partial t} = D_b \frac{\partial^2 C_b}{\partial y^2} + \frac{2D_{V\,\text{eff}}}{\delta}\left(\frac{\partial C_V}{\partial x}\right)_{|x|=\delta/2+d} \qquad \ldots(36)$$

It is obvious that it differs from Eq. (14) only by a factor of $1 + 2d/(s\delta)$ in the left part. Based on this consideration it is easy to demonstrate that solutions for regime B do not change and for regime C a small correction is added.

From the viewpoint of GB diffusion with an account of equilibrium composition near-boundary areas the presence of two components in NGR spectra is explained as follows. Component 1 is formed by Mössbauer atoms localized in GBs, whereas component 2 is associated with atoms located in near-boundary areas and crystallites volume.

If it's assumed that electron density on an isotope nucleus is primarily determined by the volume occupied by the Mössbauer atom, then judging from the isomer shift of component 1 connected with the electron density on a nucleus one can make certain suggestions on the grain-boundary diffusion mechanism. If the electron density on the Mössbauer isotope nuclei in GBs is higher than in crystallites volume, it means that the diffusing impurity atoms are in constrained positions, *i.e.*, most likely in interstitial sites. Thus, it may be suggested that inter-crystallite diffusion occurs by interstitial sites. If the electron density on nuclei in GBs is lower than in the volume, it may be concluded that the vacancy mechanism of grain-boundary diffusion is realized.

From isomer shifts of components 1 and 2 determined by the electron density on the Mössbauer isotope nuclei located respectively in GBs and near-boundary areas one can judge on composition and defectiveness of these areas. As a rule it is suggested that interstitials cause the decrease of the electron density on nuclei, while vacancies affect in opposite direction.

Let's consider what conclusions can be made from dependences of relative intensities of Mössbauer spectrum components on the annealing temperature. Figure 3.24 demonstrates examples of such dependences of ^{57}Co (^{57}Fe) spectral lines in polycrystalline Ta, W, Cr, Pd, Pt and Au on homologous temperature of annealing according to[106, 109, 110]. Two sections are clearly seen on these curves.

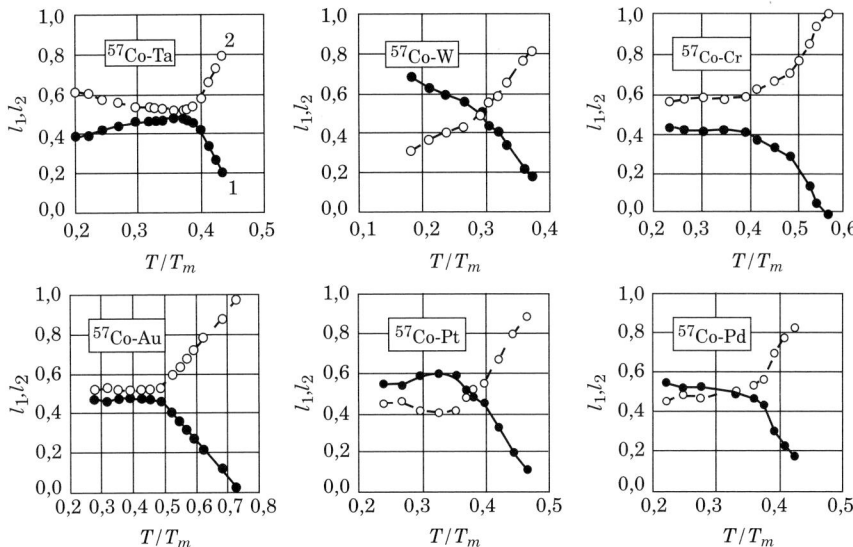

Fig. 3.24. Temperature dependences of relative intensities of spectral lines of ^{57}Co isotope in various matrixes (1 and 2 denote the spectrum components)

In the low-temperature section spectral lines relative intensities don't change or change only slightly. From the viewpoint of the model accounting for the existence of equilibrium composition near-boundary areas this section corresponds to the diffusing impurity redistribution between a GB core and the near-boundary layers due to temperature dependence of segregation factor. At this stage occupancies of sites q_i are connected with a GB diffusion width, the equilibrium composition layers width and the segregation factor by the following relationship:

$$\frac{q_1}{q_2} = \frac{s}{2(d/\delta)} \qquad \text{...(37)}$$

Relative intensities of the components are related to occupancies of states as follows[146]:

$$I_i = \varphi f_i q_i \qquad \text{...(38)}$$

where φ is a constant of NGR spectrometer and f_i is a nuclear gamma resonance probability.

To transfer from relative intensities of NGR spectrum components to occupancies of states one must know the Debye temperature θ_D for all types of states occupied in the zone of inter-crystalline diffusion. From θ_D the probabilities of the effect for spectrum components are calculated at room temperature at which the spectra were taken, and occupancies of states are determined from an expression:

$$q_i = \frac{(I_i/f_i)}{\sum_j (I_i/f_i)} \qquad \ldots(39)$$

Determination of local Debye temperatures is an extremely laborious and not always reliable procedure, and we did not use it. On the other hand, at least for the low-temperature section in temperature dependences of relative intensities of lines corresponding to different states local Debye temperatures for the states in GBs and near-boundary layers may be considered to be equal. It is confirmed both by experimental determination of local Debye temperatures in polycrystalline W[147], and by an obvious suggestion that it is hardly correct to assume different Debye temperatures in neighboring atomic layers. Thus in our calculations we assumed local Debye temperatures of atoms in GBs and near-boundary areas to be equal. Based on such consideration we used in calculations relative intensities of lines instead of occupancies of states. In this case formula (37) takes the form:

$$\frac{I_1}{I_2} = \frac{s}{2(d/\delta)} \qquad \ldots(40)$$

As seen from Fig. 3.24, in some systems (Co in Au and Pd) relative intensities of spectral lines proportional to occupancies of states almost don't change with temperature (on a low-temperature section). This means that for them grain-boundary segregation is poor or absent, i.e., $s \approx 1$. Taking this into account, based on the data of spectral line intensities ratio one can estimate the width of equilibrium-composition near-boundary layers by formula (40), and such estimation shows that their width is about $\delta/2$. Consequently, formula (40) takes the form:

$$s \approx \frac{I_1}{I_2} \qquad \ldots(41)$$

With an account of the available estimations of GB width (0.5–0.7 nm[96, 148]), it may be concluded that the near-boundary layers thickness is 0.25–0.35 nm, which approximately equals a metal atom diameter. Thus, one may say that these layers are nearly monatomic in thickness.

As mentioned above, GB segregation factor is an Ahrenius function of temperature [Eq. (18)]. Taking it into account the segregation energy may be determined from temperature dependences of spectral lines relative intensities. As follows from (18) and (41),

$$E_s = -R \frac{\partial \ln(I_1/I_2)}{\partial(1/T)} \qquad \ldots(42)$$

The high-temperature section of spectral lines relative intensities corresponds to the beginning of the Mössbauer atoms diffusion in crystallites volume (according to Fisher's model their outflow from GBs occurs by volume diffusion). With an assumption of volume diffusion mechanism Kurkin et al. in[107–108] obtained expressions enabling to determine volume diffusivity based on temperature dependences of spectral lines relative intensities. We used the results of theoretical analysis made in[107–108], but took into account the presence of equilibrium composition layers adjacent to GBs. Besides, we assumed atoms outflow from a GB not only by volume diffusion but by realization of other mechanisms of mass transfer. That is why as mentioned above in all equations not the volume diffusivity but some effective diffusion coefficient ($D_{V\,eff}$) is used characterizing the rate of atoms outflow from a GB into the volume.

It can be shown that in this case the following expression is valid

$$\frac{I_1}{I_1 + I_2} = \frac{s}{s + 2d/\delta} \exp \tau \times erfc\,\tau \qquad ...(43)$$

where

$$\tau = \frac{4 D_{V\,eff}\, t}{\delta^2 (s + 2d/\delta)^2} \qquad ...(44)$$

As the temperature dependence of GB segregation factor can be determined based on temperature dependences of spectral lines relative intensities in their low-temperature sections, and parameters δ and d are assumed known ($\delta \approx 0.5$ nm and $d \approx \delta/2$), the value of τ may be determined from expression (43). However this value can not be directly put in Eq. (44) because from (43) one determines the total value τ_N resulting from N successive annealings. The total value τ_N is an additive sum of τ_I collected at every separate annealing, i.e.,

$$\tau_N = \sum_{I=1}^{N} \tau_I \qquad ...(45)$$

Thus, the value of τ obtained at the n-th annealing may be calculated by the formula

$$\tau_n = \tau_N - \tau_{N-1} \qquad ...(46)$$

From (46) and (44) one can obtain an expression for calculation of the effective coefficient of pumping

$$D_{V\,eff} = \delta^2 \frac{(s + 2d/\delta)^2}{4 t_n} (\tau_N - \tau_{N-1}) \qquad ...(47)$$

where t_n is the n-th annealing time.

A large volume of the Mössbauer studies of GBs in polycrystalline metals has been done by now (see, e.g.[109, 110, 117, 145]). In all these studies equilibrium GBs of recrystallization origin were studied. To avoid GB migration, specimens before radionuclide deposition are subjected to a stabilizing annealing at temperatures considerably higher than that of further annealings. In this case GBs acquire the structure close to the equilibrium one.

Figure 3.25 demonstrates dependences of isomer shifts of ^{57}Co (^{57}Fe) spectral lines in polycrystalline Ta, W, Cr, Pd, Pt and Au on homologous temperature according to the data of[106, 109, 110]. The isomer shift is given for potassium ferrocyanide $K_4Fe(CN)_6 \times H_2O$ absorber relative to a specimen-source.

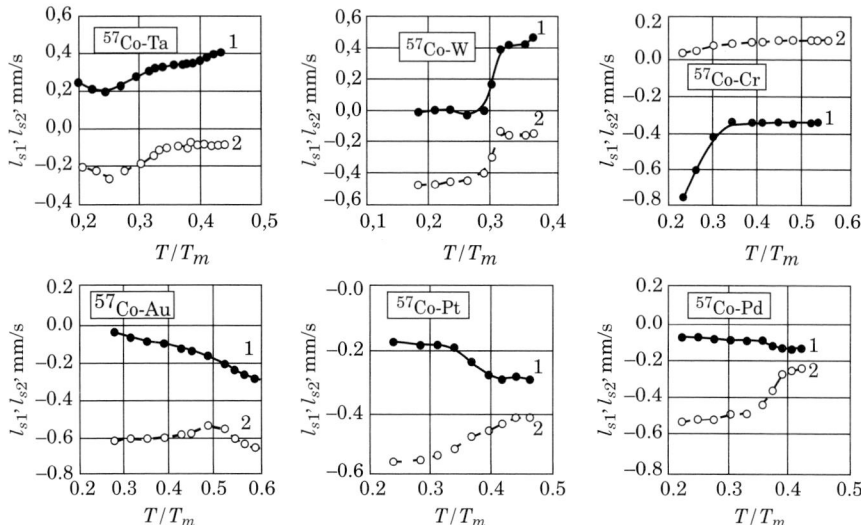

Fig. 3.25. Temperature dependences of isomer shifts of spectral lines of an absorber relative to specimens-sources ^{57}Co (^{57}Fe) in various matrixes (1 and 2 denote spectra components)

Analysis of these curves shows that for all the systems considered (except ^{57}Co-Cr) the isomer shift of component 2 is higher than that of component 1. It means that for all systems (except ^{57}Co-Cr) electron density on the Mössbauer isotope nuclei in a GB is higher than in the regular lattice, and for ^{57}Co in Cr it is lower.

Analysis of the results of Mössbauer studies carried out in[106, 109, 110, 145, 149] shows that electron density on Co nuclei located in GBs of polycrystalline Mo, W, Nb, Ta, Pt and Pd is higher than in their regular lattices. On the contrary, electron densities on nuclei of Co atoms located in GBs of Cr polycrystals and on nuclei of Sn atoms located in GBs of polycrystalline Nb are lower than in the regular lattices of corresponding matrixes, which makes possible to suggest that in the former intercrystalline diffusion occurs by interstitial sites, while in the latter it proceeds by vacancy mechanism. Realization of this or that diffusion mechanism is determined by the atomic size ratio of a basic metal and a diffusing impurity which is illustrated by Table 3.3.

If an atomic size of a diffusant is considerably smaller than that of basic metal, an interstitial mechanism of GB diffusion is realized. This case is observed for GB diffusion of Co in polycrystals of Nb, Ta, W, Mo, Au, Pt and Pd. If a diffusant atom radius is close to or larger than that of the matrix then the vacancy mechanism of GB diffusion is realized as it is found in case of Co diffusion in Cr and Sn diffusion in Nb.

Table 3.3. Atomic radii of diffusants and matrixes[150]

Diffusant/matrix	Co/Nb	Co/Ta	Co/Cr	Co/Mo	Co/W	Co/Pt	Co/Au	Co/Pd	Sn/Nb
R_{dif}, nm	0.125	0.125	0.125	0.125	0.125	0.125	0.125	0.125	0.158
R_m, nm	0.145	0.146	0.127	0.140	0.141	0.139	0.144	0.137	0.145

From the data on spectral lines intensities in low-temperature sections (before the diffusion in the volume started) one can obtain expressions of temperature dependences of GB segregation factors. It should be noted that by now GB segregation factors and especially segregation energies are determined with quite a low accuracy. Nevertheless, GB segregation factors determination from the Mössbauer data is the only way of examination the data on these parameters obtained from measurement in the diffusion regimes B and C.

We have treated the data published in[109, 110, 145] and obtained expressions for GB segregation factors in several systems (see Table 3.4).

Table 3.4. Expressions for Co and Sn GB segregation factors in various metals

Metal	Impurity	Temperature interval, K	Equations, s
Cr	Co	504–841	$(0.63 \pm 0.04)\exp\left(\dfrac{805 \pm 330}{RT}\right)$
Ta	Co	659–1114	$(1.63 \pm 0.09)\exp\left(-\dfrac{5340 \pm 490}{RT}\right)$
W	Co	673–1072	$(0.330 \pm 0.05)\exp\left(\dfrac{10700 \pm 900}{RT}\right)$
Au	Co	373–603	$(0.97 \pm 0.07)\exp\left(-\dfrac{235 \pm 250}{RT}\right)$
Pt	Co	423–623	$(2.44 \pm 0.63)\exp\left(-\dfrac{2500 \pm 1000}{RT}\right)$
Pd	Co	403–603	$(0.70 \pm 0.13)\exp\left(\dfrac{1790 \pm 670}{RT}\right)$
Nb	Sn	680–806	$(0.014 \pm 0.006)\exp\left(\dfrac{33200 \pm 2500}{RT}\right)$

As mentioned above, analysis of spectral lines isomer shifts allows making certain conclusions on composition and state of grain boundaries and near-boundary areas. Let's analyze isomer shift of component 2 characterizing the state of near-boundary areas. As seen from Fig. 3.25, for all metal matrixes shown except Au at low temperatures (lower than $0.3T_m$) the component 2 isomer shift (I_{s2LT}) is considerably lower than the volume value of ^{57}Co (^{57}Fe) isomer shift (I_{sV}) (see Table 3.5). It means that electron density on the Mössbauer nuclei in near-boundary areas is lower than in the volume. Analogous effect is observed for ^{119m}Sn diffusion in Nb polycrystals. Thus, one may conclude that positions occupied by the Mössbauer atoms in near-boundary areas are looser than in the volume. Most obvious are the following two explanations of this effect. Firstly, interstitial segregations are probable in near-boundary areas which results in an increase of lattice period and, consequently, electron density decrease on the Mössbauer atoms nuclei. Another possible explanation is that in a monatomic near-boundary layer the lattice is extended due to extension stresses.

Table 3.5. Isomer shifts of component 2 of ^{57}Co (^{57}Fe) and ^{119m}Sn isotopes in various metals at temperatures before the beginning of diffusion from GBs into the volume ($I_{s2\ LT}$) and after the pronounced development of this process ($I_{s2\ HT}$) compared to literature data for volume isomer shift (I_{sV})

Matrix	Ta	Cr	W	Au	Pd	Pt	Nb
Isotope	^{57}Fe	^{57}Fe	^{57}Fe	^{57}Fe	^{57}Fe	^{57}Fe	^{119}Sn
I_{sV}, mm/s[151, 152]	–0,07	0.10	–0,18	–0,69	–0,22	–0,39	–1,63
$I_{s2\ LT}$	– 0,22	0,05	–0,48	–0,60	–0,53	–0,55	–1,12
$I_{s2\ HT}$	–0,08	0,11	–0,16	–0,69	–0,22	–0,41	–1,63

With the annealing temperature growth isomer shift of component 2 increases, and at some temperature (higher than about $(0,35 \div 0,40)\,T_m$) it reaches some constant value of I_{s2HT} which is close to the volume isomer shift (Table 3.5). Temperature interval at which the isomer shift grows coincides with the interval in which the Mössbauer atoms diffusion in crystallites volume starts (see Fig. 3.24). That is why it's natural to associate the isomer shift growth with the Mössbauer atoms transfer from highly distorted monatomic near-boundary layers into the volume. Comparison of data on variation of spectral lines intensities and isomer shift of component 2 with temperature (Figs. 3.24 and 3.25) enables to conclude that as near as several inter-atomic spaces from a GB the state of the diffusing atoms is practically the same as in the volume.

The ^{57}Co-Au system essentially differs from the others. In this system the component 2 isomer shift at low temperatures is higher than its volume value which testifies that the Mössbauer atoms in the monatomic near-boundary layer are in more constraint positions than in the volume of crystallites. With temperature growth the isomer shift at first slightly grows and then decreases, and at $T/T_m = {\sim}0.7$ it reaches the volume value. The temperature at which the isomer shift starts to decrease coincides with the temperature at which Co diffusion into crystallite volumes begins. This temperature is about $0.5T_m$ which is appreciably higher than for other matrixes in which diffusion from GBs into the volume is possible at as low temperatures as $(0.35{-}0.4)\,T_m$. It's quite difficult to explain such behavior. In[153] the specific behavior of ^{57}Co (^{57}Fe)-Au system is associated with practically absolute absence of GB segregation of interstitials whereas in other systems it is quite possible.

Whereas temperature dependences of isomer shifts of component 2 are mainly understandable, the corresponding dependences of this parameter for component 1 raise a number of problems. As seen from Fig. 3.25, for various metals these dependences considerably differ. For polycrystalline Ta and W temperature dependence of isomer shift of component 1 to a great degree repeats the corresponding curves for component 2, namely, up to $\sim 0.3T_m$ the isomer shift remains constant and has a relatively low value. Then at $(0.35 \div 0.40)\,T_m$ it grows considerably, and at further temperature growth it only slightly increases. These results demonstrate that at temperatures of the onset of diffusion in the volume the Mössbauer atoms positions in GBs change, becoming more compact. Similar situation is observed for ^{57}Co (^{57}Fe)-Cr system, but in this case the isomer shift changes occur at lower homological temperatures ($\sim 0.2T_m$).

For ^{57}Co-Pt, ^{57}Co-Pd and ^{57}Co-Au systems temperature dependences of component 1 isomer shifts noticeably differ from that considered previously. For all these systems the isomer shift decreases with temperature growth. At present there is no way to explain unambiguously specific features of temperature dependences of component 1 isomer shifts. Presumably, calculations of GB structure are required for that.

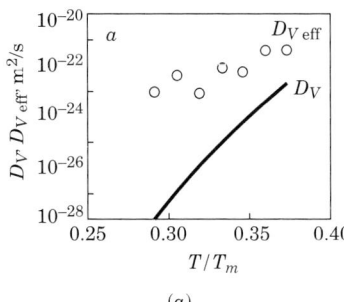
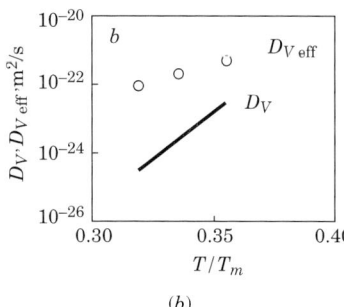

Fig. 3.26. Comparison of the effective coefficients of Co outflow from GBs of W (a) and Sn from GBs of Nb (b) with volume diffusivities of Co in W[154] and Sn in Nb[155]

As shown above, based on the analysis of temperature dependences of spectral lines relative intensities one can estimate an effective diffusion coefficient characterizing the rate of a diffusing impurity outflow from GBs.

Such estimation can be made only for a very narrow temperature interval limited from below by the temperature at which the diffusion into volume is already possible and from above by the temperature at which component 1 vanishes from a NGR spectrum. We compared effective coefficients of outflow from GBs and volume diffusivities for the Co-W and Sn-Nb systems. Figure 3.26 demonstrates the comparison of the effective coefficients of Co atoms outflow from GBs of W and Sn from GBs of Nb, calculated based on the Mössbauer studies, with the volume diffusion coefficient for Co-W and Sn-Nb systems. It is seen that at temperatures lower than ~$0.4T_m$ the pumping coefficient is considerably higher than the volume diffusivity, their difference growing with temperature decrease. It obviously results from marked differences between the properties of near-boundary areas and crystallites volume. This result is in agreement with estimations of the component 2 isomer shift which show that the state of atoms in several atomic near-boundary layers appreciably differs from their state in the volume.

The results given above demonstrate that the analysis of the Mössbauer GB spectra parameters allows to judge on physical properties of equilibrium GBs.

Relatively recently the emission Mössbauer spectroscopy came into use for non-equilibrium boundaries studies[156–160]. The main problem in such cases is that stabilizing annealing cannot be carried out, and that even at low temperature annealing the structure and properties of non-equilibrium boundaries may change. That is why analyzing the results obtained one must take into account a probability of GB migration. Besides, up to now there is no unambiguous understanding of non-equilibrium boundaries structure, and it makes difficult to interpret the results of NGR studies.

Nevertheless, certain useful information can be obtained from the Mössbauer studies of non-equilibrium boundaries. Particularly, comparing parameters of GB spectra of the materials under study with that possessing equilibrium boundaries, one can see whether the boundaries in the former are really non-equilibrium and to judge on an extent of their deviation from equilibrium.

As an example we consider the results of non-equilibrium boundaries Mössbauer investigations in Nb nanostructured by SPD[156–157]. For the comparison we use polycrystalline Nb obtained by recrystallization 800°C annealing of single-crystalline cold-worked Nb. The latter treatment results in the structure consisting of crystallites the size of ~40 μm divided with high-angle boundaries of recrystallization origin the state of which is close to an equilibrium one. Nanostructured Nb was obtained by HPT (5 turns of anvils, 4 GPa) at room temperature, this treatment resulting in a structure consisting of crystallites the size of about 100 nm separated by non-equilibrium boundaries of deformation origin. A conclusion on non-equilibrium state of the latter was made judging from specific contrast in electron-microscope images. TEM studies of thermal stability of the structure obtained showed that up to 873 K crystallite sizes do not vary but the state of boundaries changes due to recovery processes[161]. NGR studies were carried out with ^{119m}Sn Mössbauer isotope.

Figure 3.27 demonstrates temperature dependences of spectral lines relative intensities for polycrystalline and nanostructured Nb. As seen from the Figure, relative intensity of volume line (component 2) in nanostructured Nb is much higher than in polycrystalline. It means that in the former Sn atoms much easier transfer from GBs into near-boundary areas.

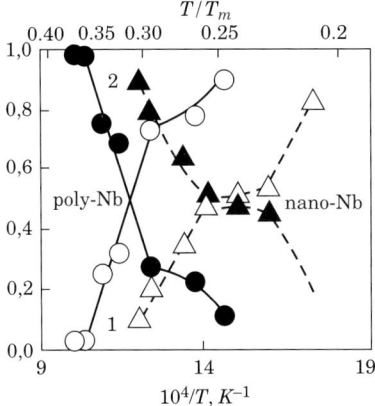

Fig. 3.27. Temperature dependences of ^{119m}Sn spectral lines relative intensities in poly- and nano-Nb (1 and 2 denote spectrum components)

If GB segregation factors in nano- and poly-Nb are assumed to be equal, one can estimate the thickness of near-boundary layers in nanostructured Nb from the relative intensities of spectral lines. This estimation gives the value of ~1.5 nm.

Figure 3.28 demonstrates temperature dependences of isomer shifts and spectral line widths of ^{119m}Sn in poly- and nano-Nb. Horizontal line shows the ^{119m}Sn isomer shift in Nb regular lattice determined on a single-crystal specimen.

It can be seen that component 1 parameters for poly- and nano-Nb are practically the same, testifying that the state of the Mössbauer atoms in equilibrium and non-equilibrium GBs differs only little. On the contrary, parameters of component 2 are markedly different. In polycrystalline Nb the isomer shift for component 2 is higher than in the regular lattice. It means that in case of equilibrium boundaries electron density on nuclei in near-boundary areas is lower than in the regular lattice, and it may be concluded that in this case near-boundary areas are in the extended condition which is most likely due to segregation of the residual interstitials in these areas.

In nanostructured Nb the isomer shift of component 2 is lower than in a regular lattice, which may be due to high concentration of non-equilibrium vacancies in near-boundary areas of crystallites. The width of component 2 in nano-Nb is much larger than in coarse-grained niobium which testifies a larger variety of positions in near-boundary areas of non-equilibrium grain boundaries. This can be obviously explained by a non-uniform field of elastic stresses induced by a non-equilibrium boundary. With temperature growth the grain-boundary spectra parameters of poly- and nano-Nb approach each other, and at 830 K the isomer shifts and widths of component 2 are almost equal, which means that grain boundaries in nanostructured Nb at this temperature are no more non-equilibrium. It should be noted that the state of grain boundaries approaches to equilibrium at lower temperatures than migration of grain boundaries is possible.

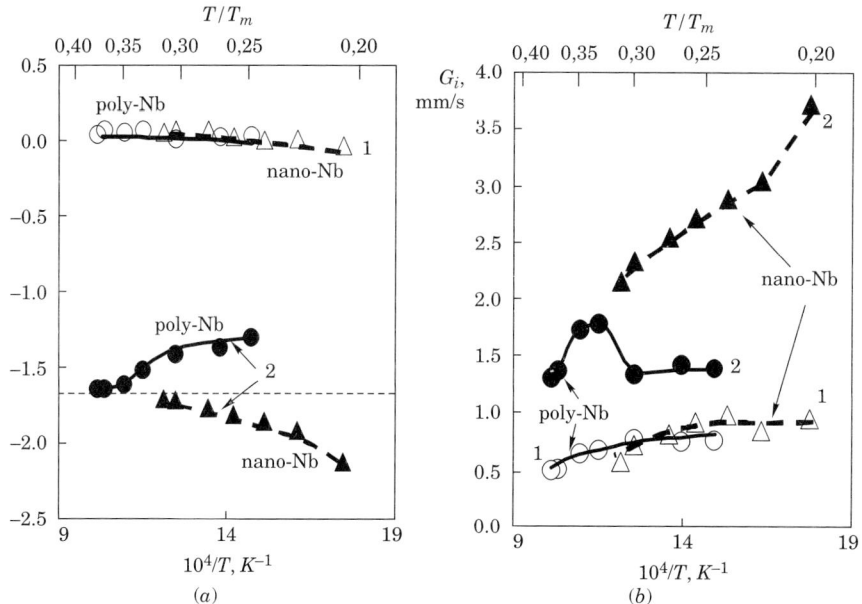

Fig. 3.28. Temperature dependences of isomer shifts (a) and widths (b) of ^{119m}Sn spectral lines in polycrystalline and nanostructured Nb (1 and 2 are spectra components). Horizontal line shows the ^{119m}Sn isomer shift in single-crystal Nb

Thus, based on the emission Mössbauer studies one can reveal the non-equilibrium state of GBs and even judge on the degree of their deviation from equilibrium.

9. CONCLUSIONS

This chapter is a brief review on the structure and properties of grain boundaries. The subject of this review being manifold, we had to restrict our consideration to several points we think to be of greatest interest. Thus, we did not touch upon mechanical properties of grain boundaries (grain-boundary sliding, GB embrittlement, Coble creep, etc.) as these problems are presented in detail in other publications (see, *e.g.*[6, 162]). The main attention in this chapter is paid to the models of GBs structure, their energy and diffusion along them. It is demonstrated that structure and properties of GBs can be understood only based on combined use of simulation and a number of various experimental techniques such as HREM, diffusion investigations, and emission Mössbauer spectroscopy.

The structure and properties of equilibrium grain boundaries are quite deeply studied and understood by now, they are successfully calculated and simulated. As for non-equilibrium boundaries, the state of affairs with them is not as favorable. At present great attention is attracted by submicrocrystalline and nanostructured materials obtained by severe plastic deformation, the unique properties of which are to a great extent determined by their non-equilibrium GBs. That is why the study and simulation of non-equilibrium boundaries are nowadays of special significance.

Some models of non-equilibrium boundaries are considered in this chapter. They are far from being perfect yet, though recently a considerable progress has been achieved in the development of this problem. It has been realized by now that structure and properties of non-equilibrium boundaries are to a great extent determined by this or that method of a material preparation. For example, non-equilibrium boundaries in nanocrystalline materials obtained by gas condensation and compacting are likely to differ essentially in their structure and properties from that of GBs in materials nanostructured by SPD.

One might expect that in the nearest future the main attention of the materials science researchers will be concentrated on simulation and investigation of non-equilibrium grain boundaries in nanocrystalline and nanostructured materials.

REFERENCES

1. Gleiter, H. and Chalmers, B. 1972. High-angle Grain Boundaries. Pergamon Press, Oxford, New York, Toronto, Sydney, Braunschweig.
2. Orlov, A.N., Pereverzentsev, V.N. and Rybin, V.V. 1980. Grain Boundaries in Metals. Metallurgiya (Metallurgy), Moscow (in Russian).
3. Sutton, A.P. 1984. Grain-boundary structure, *Int. Met. Rev.* **29(5)**: 377–402.
4. Kaibyshev, O.A. and Valiev, R.Z. 1987. Grain boundaries and Properties of Metals, Metallurgiya (Metallurgy), Moscow (in Russian).
5. Kopetsky, C.V., Orlov, A.N. and Fionava, L.K. 1987. Grain Boundaries in Pure Materials, Nauka (Science), Moscow (in Russian).
6. Sutton, A.P. and Balluffi, R.W. 1995. Interfaces in Crystalline Materials. Clarendon Press, Oxford.
7. Thomas, G.J., Siegel, R.W. and Eastman, J.A. 1990. Grain boundaries in nanophase palladium: high resolution electron microscopy and image simulation, *Scripta Metall. Mater.* **24(1)**: 201–6.
8. Gleiter, H. 2000. Nanostructured materials: basic concepts and microstructure, *Acta Mater.* **41(1)**: 1–29.

9. Valiev, R.Z. and Aleksandrov, I.V. 2007. Bulk Nanostructured Materials, «Academkniga», (in Russian) Moscow.
10. Vitek, V., Sutton, A.P., Wang, G.J. and Schwartz, D. 1983. On the multiplicity of structures and grain boundaries, *Scripta Metall.* **17(2)**: 183–9.
11. Kronberg, M.L. and Wilson, F.H. 1949. Structure of high-angle grain boundaries, *Trans. AIME.* **185**: 506–8.
12. Grimmer, H., Bollmann, W. and Warrington, D.H. 1974. Coincidence-site lattices and complete pattern-shift lattices in cubic crystals, *Acta Crystall.* **30A**: 197–207.
13. Grimmer, H. 1974. A method of determining the coincidence site lattices for cubic crystals, *Acta Crystall.* **30A(5)**: 680.
14. Joshi, A. and Ballufi, R.W. 1975. Grain boundary structure and properties, *J. Phys.* **36**: C-4-VII: 458.
15. Bonnet, R. 1976. Note on general analytical method to find a basis for DSC lattices–derivation of a basis for CSL, *Scripta Metall.* **10(9)**: 801–6.
16. Bonnet, R. and Cousineau, E. 1977. Computation of coincident and near-concident cells or any 2 lattices, *Acta Crystallogr.* **33A(5)**: 850–6.
17. Bonnet, R. and Durand, F. 1975. General analytical method to find a basis for DSC lattices, *Scripta Metall.* **9(9)**: 935–9.
18. Fortes, M.A. 1977. Coincidence site lattices in non-cubic lattices, *Phys. Stat. Sol. B* **82(1)**: 377–82.
19. Bollman, W. 1967. On the geometry of grain and phase boundaries. I. General theory, *Phil. Mag.* **16(140)**: 363–82.
20. Bollman, W. 1970. Crystal Defects and Crystalline Interfaces. Springer-Verlag. Berlin.
21. Hirth, J.P. and Lothe, J. 1967. Theory of Dislocations. McGraw-Hill, New York.
22. Brandon, D.G. 1966. The structure of high angle grain boundaries, *Acta Metall.* **14(11)**: 1479–84.
23. Balluffi, R.W., Komem, Y. and Schober, T. 1972. Electron microscope studies of grain boundary dislocation behavior, *Surf. Sci.* **31(1)**: 68–101.
24. Ashby, M.F., Spaepen, F. and Williams, S. 1978. The structure of grain boundaries described as a packing of polihedra, *Acta Metall.* **26(9)**: 1647–68.
25. Pond, R.C., Smith, D.A. and Vitec, V.V. 1979. Computer simulation of <110> tilt boundaries structure and symmetry, *Acta Met.* **27(2)**: 235–46.
26. Bernal, J.D. 1964. On polihedra model for liquid structure, *Proc. Roy. Soc. London A.* **280**: 299–312.
27. Sutton, A.P. and Vitek, V. 1983. On the structure of tilt boundaries in cubic metals. I. Symmetrical tilt boundaries, *Phil. Trans. Roy Soc. London. A* **309(1506)**: 1–36.
28. Sutton, A.P. and Vitek, V. 1983. On the structure of tilt boundaries in cubic metals. II. Asymmetrical tilt boundaries, *Phil. Trans. Roy Soc. London. A* **309(1506)**: 37–54.
29. Sutton, A.P. and Vitek, V. 1983. On the structure of tilt boundaries in cubic metals. III. Generation of the structural study and applications for the properties of grain boundaries, *Phil. Trans. Roy Soc. London. A* **309(1506)**: 55–68.
30. Sutton, A.P. 1984. Grain-boundary structure, *Int. Metals Rev.* **29(5)**: 377–402.
31. Pawaskar, D.N., Miller, R. and Phillips, R. 2001. Structure and energetics of long-period tilt grain boundaries using an effective Hamiltonian, *Phys. Rev. B* **63(21)**: 214105–14.
32. Rouviere, J.-L. and Bourret, A. 1990. Analysis of structures of symmetrical [001] tilt grain boundaries in silicon and germanium, *Coll. Phys.* **51(1)**: 329–34.
33. Thibault, J., Putaux, J.L., Jaques, A., George, A., Michaud, H.M. and Ballin, X. 1993. Structure and characterization of dislocations in tilt grain boundaries

between $\Sigma = 1$ and $\Sigma = 3$: a high resolution electron microscopy study, *Mater. Sci. Eng. A.* **164(1–2)**: 93–100.
34. Rouviere, J.L. and Bourret, A. 1989. Multiple structures of a [001] Σ = 13 tilt grain boundary in germanium, *In: Polycrystalline Semiconductors.* Springer-Verlag, Berlin. pp. 19–24.
35. Sutton, A.P. and Vitec, V. 1980. On coincidence site lattice and DSC dislocation network model of high angle grain boundary structure, *Scripta Metal.* **14(1)**: 129–32.
36. Sutton, A.P., Balluffi, R.W. and Vitec, V. 1981. On intrinsic secondary grain boundary dislocation arrays in high angle symmetric tilt grain boundaries, *Scripta Metal.* **15(9)**: 989–94.
37. Balluffi, R.W. and Bristowe, P.D. 1984. On structural unit/grain boundary dislocation model for grain boundary structure, *Surf. Sci.* **144(1)**: 28–43.
38. Hsieh, T.E. and Balluffi, R.W. 1989. Experimental study of grain boundaries melting in aluminium, *Acta Metal.* **37(6)**: 529–41.
39. Sutton, A.P. 1989. On the structural unit model of grain boundary structure, *Phil. Mag. Lett.* **59(2)**: 53–59.
40. Sutton, A.P. and Balluffi, R.W. 1990. Rules for combining structural units of grain boundaries, *Phil. Mag. Lett.* **61(3)**: 91–94.
41. Li, J.C.M. 1972. Dislocation model of high angle grain boundaries, *Surf. Sci.* **31(11)**: 1479–84.
42. Shih, K.K. and Li, J.C.M. 1975. Energy of grain boundaries between cusp misorientations, *Surf. Sci.* **50(1)**: 109–24.
43. Valiev, R.Z., Vladimirov, V.I., Gertsman, V.Y., Nazarov, A.A. and Romanov, L.E. 1990. Disclination-structural model and energy of grain boundaries in FCC metals, *Fiz. Met. Metalloved* (in Russian) **3**: 31–38.
44. Gertsman, V.Y., Nazarov, A.A., Romanov, A.E., Valiev, R.Z. and Vladimirov, V.I. 1989. Disclination-structural unit model of grain boundaries, *Phil. Mag. A.* **59(5)**: 1113–8.
45. Shenderova, O.A., Brenner, D.W., Nazarov, A.A., Romanov, A.E. and Yang, L. 1998. Multiscale modeling approach for calculating grain boundary energies from first principles, *Phys. Rev. B* **57(6)**: 3181–4.
46. Sorensen, M.R., Mishin, Y. and Voter, F. 2000. Diffusion mechanisms in Cu grain boundaries, *Phys. Rev.* **62(6)**: 3658–73.
47. Suzuki, A. and Mishin, Y. 2003. Atomistic Modeling of Point Defects and Diffusion in Copper Grain Boundaries, *Interf. Sci.* **11(4)**: 131–48.
48. Suzuki, A. and Mishin, Y. 2004. Diffusion Mechanisms in Grain Boundaries, *J. Metastable Nanocrystalline Mat.*, **19**: 1–24.
49. Ma, Q., Liu, C.L., Adams, J.B. and Balluffi, R.W. 1993. Diffusion along [001] tilt boundaries in the Au/Ag system. I. Experimental results, *Acta Metall. Mater.* **41(1)**: 133–41.
50. Ma, Q., Liu, C.L., Adams, J.B. and Balluffi, R.W. 1993. Diffusion along [001] tilt boundaries in the Au/Ag system. II. Atomistic modeling and interpretation, *Acta Metall. Mater.* **41(1)**: 143–51.
51. Nomura, M. and Adams, J.B. 1995. Interstitial Diffusion Along Twist Grain-Boundaries in Cu, *J. Mater. Res.* **10(11)**: 2916–92.
52. Grabski, M.W. and Korski, R. 1970. Grain boundaries as sink for dislocations, *Phil. Mag.* **22(1178)**: 707–15.
53. Pumphrey, P.H. and Gleiter, H. 1975. On the structure of non-equilibrium high-angle grain boundaries, *Phil. Mag.* **32(4)**: 881–5.
54. Varin, R.A. 1979. Spreading of extrinsic grain boundary dislocations in austenitic steel, *Phys. Stat. Sol. A* **52**: 347–56.

55. Valiev, R.Z., Kaibyshev, O.A. and Khananov, S.K. 1979. Grain boundaries during superplastic deformation, *Phys. Stat. Sol.* **52(2)**: 447–53.
56. Lastigue, S. and Priester, L. 1983. Stability of extrinsic grain boundary dislocations in relation with intergranular segregation and precipitation, *Acta Metal.* **31(11)**: 1809–19.
57. Kopetsky, I.V. and Fionova, L.K. 1984. Special grain boundaries in metals, *Poverhnost: Fiz., Chim, Mech., (Surface: Phys., Chem, Mech.)* (in Russian) **2**: 5.
58. Valiev, R.Z., Gertsman, V.Y. and Kaibyshev, O.A. 1986. Grain boundary structure and properties under external influences, *Phys. Stat. Sol.* **97(1)**: 11–56.
59. Nazarov, A.A., Romanov, A.E. and Valiev, R.Z. 1993. On the structure, stress fields and energy of nonequilibrium grain boundaries, *Acta Metall. Mater.* **41(4)**: 1033–40.
60. Nazarov, A.A. 1998. Non-equilibrium dislocation ensembles in grain boundaries and their role in properties of polycrystals. Phys. Math. Dr. Dissertation (in Russian), Russian Academy of Sciences, Institute of problems of super-plasticity of metals.
61. Gertsman, B.Y. and Valiev, R.Z. 1982. *Poverhnost (Surface)* (in Russian) **8**: 101.
62. Valiev, R.Z., Gertsman, V.Y., Kaibyshev, O.A. and Khannanov, S.K. 1983. Non-equilibrium state and recovery of grain boundary structure, *Phys. Stat. Sol. A* **77(1)**: 97–105.
63. Gleiter, H. 1972. The interaction of lattice defects and grain boundaries, *J. Less-Common Metals.* **28(2)**: 237–324.
64. Balluffi, R.W., Komem, Y. and Schober, T. 1972. Electron microscope studies of grain boundary dislocation behaviour, *Surf. Sci.* **31(1)**: 68–103.
65. Hovell, P.R., Jones, A.R., Horswell, A. and Rakph, B. 1976. The creation and accommodation of extrinsic dislocations at grain boundaries, *Phil. Mag.* **33(1)**: 21–31.
66. Dingley, D.J. and Pond, R.C. 1979. On the interaction of lattice dislocations with grain boundaries, *Acta Metal.* **27(4)**: 667–82.
67. Gertsman, V.Y., Bonus, V.Y., Valiev, R.Z. and Kaibyshev, O.A. 1984. On the role of grain boundaries in deformation strengthening of a fine-grained polycrystal, *Physica Tverdogo Tela (Solid State Physics*, in Russian) **26(6)**: 1712–8.
68. Jenecek, M. and Tangi, K. 1991. Structure evolution and flow behavior of AISI 316L stainless steel polycrystals at room temperature, *Mater. Sci. Eng. A* **138**: 237–45.
69. Rybin, V.V., Zisman, A.A. and Zolotorevsky, N.Y. 1985. Junction disclinations in plastically deformed crystals, *Physica Tverdogo Tela (Solid State Physics*, in Russian) **27(1)**: 181–6.
70. Rybin, V.V. 1987. Dislocation-disclination structures forming at developed plastic deformation. *In*: Problems of theory of defects in crystals. Leningrad.: Nauka (Science) (in Russian). C. 68–84.
71. Rybin, V.V., Zisman, A.A. and Zolotorevsky, N.Y. 1993. Junction dislocations in plastically deformed crystals, *Acta Metall. Mater.* **41(7)**: 2211–7.
72. Rybin, V.V. 1986. Large Plastic Deformations and Fracture of Metals. *Metallurgiya (Metallurgy)* (in Russian), Moscow.
73. Valiev, R.Z., Korznikov, A.V. and Mulyukov, R.R. 1993. Structure and properties of ultrafine-grained materials produced by severe plastic-deformation, *Mater. Sci. Eng.* **168(2)**: 141–8.
74. Islamgaliev, R.K. and Valiev, R.Z. 1995. Electron-microscopic study of grain-boundaries in ultrafine germanium, *Fizika Tverdogo Tela (Solid State Physics*, in Russian) **37(12)**: 3597–3606.

75. Islamgaliev, R.K. and Valiev, R.Z. 1999. Formation of solidification textures upon quenching of molten pure metals on a rotating heat absorber, *Phys. Met. Metallogr.* **87(3)**: 215–20.
76. Mullins, W.W. 1957. Theory of thermal grooving, *J. Appl. Phys.* **28(3)**: 333–9.
77. Schober, T. and Balluffi, R.W. 1970. Quantative observation of misfit dislocation array in low and high angle twist grain boundaries, *Phil. Mag.* **21(169)**: 109–23.
78. Herrman, G., Gleiter, H. and Baro, G. 1976. Investigation of low energy grain boundaries in metals by a sintering technique, *Acta Metall.* **24(4)**: 353–9.
79. Glaudhari, P. and Matthews, J.W. 1970. Coincidence twist boundaries between crystals of mgo smoke, *Appl. Phys. Letter.* **17(3)**: 115–8.
80. Matthews, J.W. and Glaudhari, P.J. 1971. Coincidence twist boundaries between crystalline smoke particles, *Appl. Phys.* **8**: 3063–71.
81. Chuveldiev, V.N. 1996. Micromechanism of deformation–induced grain-boundary self-diffusion. I. Effect of excess free volume on free energy and diffusion parameters of grain boundaries. *Phys. Met. Metallogr.* **81(5)**: 463–8.
82. Chuveldiev, V.N. 1996. Micromechanism of deformation-induced grain-boundary self-diffusion. II. Effect of lattice dislocations introduced in boundaries on the diffusion properties of grain boundaries, *Phys. Met. Metallogr.* **81(6)**: 583–8.
83. Mott, N.F. 1948. Slip at grain boundaries and grain growth in metals, *Proc. Phys. Soc.* **60(4)**: 391–4.
84. Nazarov, A.A., Romanov, A.E. and Valiev, R.Z. 1994. On the nature of high internal stresses in ultrafine-grained materials, *Nanostructured Materials*. **4(1)**: 93–101.
85. Nazarov, A.A. 1997. Ensembles of gliding grain boundary dislocations in ultrafine grained materials produced by severe plastic deformation, *Scripta Mater.* **37(8)**: 1155–61.
86. Nazarov, A.A., Romanov, A.E. and Valiev, R.Z. 1996. Random disclination ensembles in ultrafine materials produced by severe plastic deformation, *Scripta Mater.* **34(5)**: 729–34.
87. Nazarov, A.A. 1995. Stress fields of disordered dislocation arrays: a double wall consisting of dislocation dipoles, *Phil. Mag. Lett.* **72(1)**: 49–53.
88. Fisher, J.C. 1951. Calculation of diffusion penetration curves of surface and grain boundary diffusion, *J. Appl. Phys.* **22**: 74–80.
89. Whipple, R.T.P. 1954. Conception contours in grain boundary diffusion, *Phil. Mag.* **45(351)**: 1225–36.
90. Suzuoka, T. 1961. Lattice and grain boundary diffusion in polycrystals, *Trans. Jap. Inst. Met.* **2(1)**: 25–32.
91. Suzuoka, T. 1964. Exact solution of two ideal cases in grain boundary diffusion problem and application to sectioning method, *J. Phys. Soc. Jpn.* **19**: 839–51.
92. Kaur, I., Mishin, Y. and Gust, W. 1995. Fundamentals of Grain and Interphase Boundary Diffusion. John Wiley & Sons Ltd.
93. Mishin, Y., Herzig, C., Bernardini, J. and Gust, W. 1997. Grain boundary diffusion: fundamentals to resent developments, *Int. Mater. Rev.* **42(4)**: 155–78.
94. Mishin, Y. and Herzig, C. 1999. Grain boundary diffusion: recent progress and future research, *Mater. Sci. Eng. A* **260**: 55–71.
95. Harrison, L.G. 1961. Influence of dislocations on diffusion kinetics in solids with particular reference to the alkali halides, *Trans. Faraday Soc.* **57**: 1191–9.
96. Sommer, J. and Herzig, C. 1992. Direct determination of grain-boundary and dislocation self-diffusion coefficients in silver from experiments in type-C kinetics, *J. Appl. Phys.* **72(7)**: 2758–66.
97. Herzig, C., Geise, J. and Mishin, Y. 1993. Grain-boundary diffusion and grain-boundary segregation of tellurium in silver, *Acta Mater.* **41(6)**: 1683–91.

98. Surholt, T., Mishin, Y.M. and Herzig, C. 1994. Grain boundary diffusion and segregation of gold in copper–Investigation of type B and type C kinetic regimes, *Phys. Rev.* **50(6)**: 3577–87.
99. Surholt, T., Minkwitz, C. and Herzig, C. 1998. Solute diffusion and segregation in grain boundaries of silver and copper, *Defect Diff. Forum* **156**: 59.
100. Surholt, T., Minkwitz, C. and Herzig, C. 1998. Nickel and selenium grain boundary solute diffusion and segregation in silver, *Acta Mater.* **46(6)**: 1849–59.
101. Kaur, I., Gust, W. and Kozma, L. 1989. Handbook of Grain and Interphase Boundary Diffusion Data. Vols. 1–2, Ziegler press, Stutgart.
102. Kondratev, V.V. and Trachtenberg, I.S. 1992. Intergranular diffusion in real polycrystals, *Phys. Stat. Sol. B* **171**: 303–15.
103. Mishin, Y.M. and Yurovitskii, I.V. 1991. A generated model of grain boundary diffusion, *Phil. Mag.* **64(6)**: 1239–49.
104. Gapontcev, V.L. and Koloskov, V.M. 1996. The model of heterogeneous grain boundary with nonuniform diffusivity: I. The C regime of grain-boundary diffusion in the case of short- and medium-time diffusion annealing, *Phys. Met. Metallogr.* **81(1)**: 1–8.
105. Klinger, L. and Rabkin, E. 1999. Beyond the Fisher model of grain boundary diffusion: effect of structural inhomogenity in the bulk, *Acta Mater.* **47(3)**: 725–34.
106. Dudarev, M.S., Dyakin, V.Y., Kaigorodov, V.N., Klotsman, S.M. and Kurkin, M.I. 1995. Experimental determination of diffusion pumping coefficients from grain boundaries at diffusion in the regime of isochronous annealing of polycrystals, *Fiz. Met. Metalloved (Phys. Met. Metallogr.*, in Russian) **79(5)**: 136–51.
107. Kurkin, M.I., Klotsman, S.M. and Dyakin, V.V. 1996. Effect of saturation of the "pumping-out" zone upon ultrafast diffusion along grain boundaries in polycrystals, *Phys. Met. Metallogr.* **81(4)**: 366–73.
108. Kurkin, M.I., Klotsman, S.M., Kaigorodov, V.N. and Dyakin, V.V. On the nature of states occupied by atomic probes in the zone of intercrystallite diffusion: the case of extremely small bulk-diffusion length, *Phys. Met. Metallogr.* **82(4)**: 419–26.
109. Kaigorodov, V.N. and Klotsman, S.M. 1994. Impurity states in the grain boundaries and adjacent to them crystalline regions, *Phys. Rev.* **49(14)**: 9374–99.
110. Klotsman, S.M., Kaigorodov, V.N., Kurkin, M.I. and Dyakin, V.V. 2000. Segregation of ^{57}Co Atomic Probes in the Cores of Grain Boundaries in d-Transition Metals, *Interf. Sci.* **8(4)**: 323–34.
111. Popov, V.V. 2006. Model of grain-boundary diffusion with allowance for near-boundary layers of equilibrium composition, *Phys. Met. Metallogr.* **102(5)**: 453–61.
112. Popov, V.V. 2008. Analysis of Possibilities of Fisher's Model Development, *Solid State Phen.* **138**: 133–44.
113. Bokshtein, B.S., Kopetcky, C.V. and Shvindlerman, L.S. 1986. Thermodynamics and Kinetics of Grain Boundaries in Metals, *Metallurgiya (Metallurgy)* (in Russian), Moscow.
114. Philibert, J. 1991. Atom Movements–Diffusion and Mass Transport in Solids. Les Editions de Physique, France.
115. Bokshtein, B.S. and Yaroslavtsev, Y.B. 2005. Diffusion of Atoms and Ions in Solids, *"MISIS" (Moskow Institute of Steels and Alloys)* (in Russian), Moscow.
116. Peterson, N.L. 1983. *Int. Met. Rev.* **28**: 85.
117. Popov, V.V. 2006. Mössbauer investigations of grain-boundary diffusion and segregation. *Defect Diff. Forum* **258–260**: 497–508.
118. Gleiter, H. 1992. Diffusion in nanostructured metals, *Phys. Stat. Sol. B* **172(1)**: 41–51.

119. Kolobov, Y.R., Valiev, R.Z., Grabovetskaya, G.P. et al. 2001. Grain-boundary diffusion and properties of nanostructured materials. Nauka (Science) (in Russian), Novosibirsk, Russia.
120. Würschum, R., Herth, S. and Brossmann, U. 2002 Diffusion in nanocrystalline metals and alloys–A status report, *Proc. Conf. "Nanomaterials by Severe Plastic Deformation–NANOSPD2"*, December 9–13, Vienna, Austria. pp. 755–66.
121. Thomas, G.J., Siegel, R.W. and Eastman, J.A., 1989. High resolution electron microscopy of grain boundaries in nanophase palladium, *Mat. Res. Soc. Symp. Proc.* **153**: 13–20.
122. Siegel, R.W. and Thomas, G.J. 1991. On the nature of grain boundary structures in nanophase materials, *Mat. Res. Soc. Symp. Proc.* **209**: 18–26.
123. Horvath, J., Biringer, R. and Gleiter, H. 1987. Diffusion in Nanocrystalline Materials, *Sol. State Comm.* **62(5)**: 319–22.
124. Schumaher, S., Biringer, R., Straub, S. and Gleiter, H. 1989. Diffusion of silver in nanocrystalline copper between 303 and 373 K, *Acta Metal.* **37(9)**: 2485–8.
125. Hoffer, H.J., Averback, R.S., Hahn, H. and Gleiter, H. 1993. Diffusion of bismuth and gold in nanocrystalline copper, *J. Appl. Phys.* **74(6)**: 3832–9.
126. Würschum, R., Reimann, K., Gruss, S., Kubler, A., Scharwaechter, P., Frank, W., Kruse, O., Carstanjen, H.D. and Schaefer, H.E. 1997. Structure and diffusional properties of nanocrystalline Pd, *Phil. Mag. B* **76(4)**: 407–17.
127. Bokstein, B.S., Brose, H.D., Trusov, L.L. and Khvostantseva, T.P. 1995. Diffusion in nanocrystalline nickel, Nanostructured materials. *Proceeding of second international conference on nanostructured materials*. Germany: Stuttgart University. pp. 873–6.
128. Kolobov, Y.R., Grabovetskaya, G.P., Ivanov, M.B., Zhilyaev, A.P. and Vasliev, R.Z. 2001. Grain boundary diffusion characteristics of nanostructured nickel, *Scripta Met.* **44**: 873–8.
129. Tanimoto, H., Pasquini, L., Prümmer, R., Kronmüller, H. and Schaefer, H.-E. 2000. Self-diffusion and magnetic properties in explosion densifield nanocrystalline Fe, *Scripta Mater.* **42(10)**: 961–6.
130. Klotsman, S.M. 1993. Diffusion in Nanocrystalline Materials, *Defect Diff. Forum* **99–100**: 25–52.
131. Divinski, S., Hisker, F., Hang, Y.-S., Lee, J.S. and Herzig, C. 2002. Tracer diffusion of ^{63}Ni in nano-γ-FeNi produced by powder metallurgical method: ^{59}Fe grain boundary diffusion in nanostructured γ-Fe-Ni. Part II. Effect of bimodal microstructure on diffusion behavior in type-c kinetic regime, *Z. Metallkund.* **93(4)**: 265 72.
132. Larikov, L.N. 1995. Diffusion processes in nanocrystalline materials, *Metallophysica I Noveishie Tehnologii* (*Metal Physics and the Newest Technologies*) (in Russian) **17(1)**: 3–29.
133. Ishida, Y., Ichinose, H., Kizuka, T. and Suenaga, K. 1995. High-resolution electron microscopy of interfaces in nanocrystalline materials. *Proceedings of the second international conference on nanostructured materials*. Germany, Stuttgart University. pp. 115–24.
134. Tanimoto, H., Farber, P., Würschum, R., Valiev, R.Z. and Schaefer, H.E. 1999. Self-diffusion in high-density nanocrystalline Fe, *Nanostructured Mater.* **12 (5–8)**: 681–4.
135. Schafer, H.E., Würschum, R., Gessmann, T., Stöckl, G., Scharwaecchter, P., Valiev, R.Z., Fetch, H.J. and Moelle 1995. Diffusion and free volumes in nanocrystalline Pd. *Proc. 2nd Int. conf. on nanostructured materials*. Germany, Stuttgart University. pp. 869–72.
136. Kolobov, Y.R., Grabovetskaya, G.P., Ratochka, I.V., Kabanova, E.V., Naidenkin, E.V. and Lowe, T.C. 1996. Effect of grain-boundary diffusion fluxes of copper on

the acceleration of creep in submicrocrystalline nickel, *Annales de Chimie* **21 (6–7)**: 483–91.
137. Grabovetskaya, G.P., Ratochka, I.V., Kolobov, Y.R. and Puchkareva, L.N. 1997. Comparative-studies of grain-boundary copper diffusion in submicro-crystalline and coarse-crystalline nickel, *Phys. Met. Metallog.* **83(3)**: 310–3.
138. Kolobov, Y.R., Grabovetskaya, G.P., Ratochka, I.V. and Ivanov, K.V. 1998. Peculiarities of creep and diffusion parameters of submicrocrystalline materials, *Izv. VUZ. Fizika* (in Russian) **42(3)**: 77–82.
139. Kornelyuk, L.G., Lozovoi, A.Y. and Razumovskii, I.M. 1997. Diffusion along non-equilibrium grain boundaries: investigation by radiotracer technique, *Defect Diff. Forum* **143–147**: 1481–6.
140. Valiev, R.Z., Razumovskii, I.M. and Sergeev, V.I. 1993. Diffusion Along Grain-Boundaries with Nonequilibrium Structure, *Phys. Stat. Sol. A* **139(2)**: 321–35.
141. Yamakov, V., Wolf, D., Phillpot, S.R. and Gleiter, H. 2002. Grain-boundary diffusion creep in nanocrystalline palladium by molecular-dynamics simulation, *Acta Mater.* **50(1)**: 61–73.
142. Würschum, R. and Scheafer, H.E. 1996. In Nanomaterials: Synthesis, Properties and Applications (*Eds.*: Edelstein, A.S. and Camarata, R.C.), Institute of Physics, Bristol. pp. 277–301.
143. Frank, W. 1997. Diffusion in amorphous solids–Metallic alloys and elemental semiconductors, *Defect Diff. Forum* pp. 143–147; 695–710.
144. Kaigorodov, V.N. and Klotsman, S.M. 1978. Nuclear gamma-resonance on nuclei of Fe-57 located in grain boundaries of copper, *Pisma v JETF (Letters to Journal of Experimental Physics)* (in Russian). **28(6)**: 386–9.
145. Kaigorodov, V.N., Popov, V.V., Popova, E.N., Pavlov, T.N. and Efremova, S.V. 2005. Mossbauer Investigation of Sn diffusion and segregation in grain boundaries of polycrystalline, *J. Phase Equilibr. Diff.* **26(5)**: 510–5.
146. Belozersky, G.N. 1990. Mossbauer spectroscopy as a method of surface investigation. *Energoatomizdat* (in Russian), Moskow.
147. Kaigorodov, V.N., Klotsman, S.M., Kurkin, M.I., Dyakin, V.V. and Zherebtsov, D.V. 1998. Intercrystalline diffusion of cobalt in polycrystalline tungsten: II. Experimental study of diffusion in the core of crystallite-conjugation regions and adjacent zones, *Phys. Met. Metallogr.* **85(2)**: 212–7.
148. Atkinson, A. and Taylor, R.I. 1981. The diffusion of ^{63}Ni along grain boundaries in nickel oxide, *Phil. Mag.* **43(4)**: 979–98.
149. Kaigorodav, S.M., Klotsman, S.M., Koloskov, V.M. and Tatarinova, G.N. 1988. Investigation of grain boundaries in Nb and Mo by NGR method, *Fiz. Met. Metalloved* (in Russian) **66(5)**: 958–65.
150. Physical parameters: Reference book. Babichev, A.P., Babushkina, N.A., Bratkovskaya, A.M. *et al.* 1991. Ed. by Grigoriev, I.S. and Meilihov, E.Z.M.. *Energoatomizdat* (in Russian), Moskow, pp. 1232c.
151. Mössbauer Effect Data Center (Web: www.unca.edu/medc).
152. Kaigorodov, V.N., Klotsman, S.M., Koloskov, V.M. and Shlyapnikov, S.N. 1987. Mossbauer study of elertronic and dynamic properties of high-angle boundaries in tungsten. *Poverhnost: Physica, Chimiya, Mechanica (Surface: Physics, Chemistry and Mechanics)* (in Russian) **(3)**: 124–9.
153. Kaigorodov, V.N., Klotsman, S.M. and Shlyapnikov, S.N. 1993. Structure and properties of interstitials localized in grain boundaries and adjacent areas of crystallites of Au, *Fiz. Met. Metalloved* (in Russian) **75(3)**: 5–18.
154. Klotsman, S.M., Osetrov, S.V. and Timofeev, A.N. 1992. Volume diffusion of cobalt in tungsten single crystals, *Phys. Rev. B* **46(5)**: 2831–7.
155. Askill, O. 1965. Tracer diffusion of 113Sn in niobium, *Phys. Stat. Sol.* **9**: K167–K168.

156. Popov, V.V., Kaigorodov, V.N., Popova, E.N. and Stolbovsky, A.V. 2007. Mössbauer Emission Spectroscopy of Grain Boundaries in Poly- and Nanocrystalline Niobium, *Bull. Russian Academy Sci.: Physics.* **71(9)**: 1244–8.
157. Popov, V.V., Kaigorodov, V.N., Popova, E.N. and Stolbovsky, A.V. 2007. NGR Investigation of Grain-Boundary Diffusion in Poly- and Nanocrystalline Nb, *Defect Diff. Forum* **263**: 69–74.
158. Popov, V.V. 2008. Mössbauer Spectroscopy Studies of Grain Boundaries in Nanostructured Metals, *Defect Diff. Forum* pp. 273–276; 506–13.
159. Popov, V.V., Grabovetskaya, G.P., Sergeev, A.V. and Mishin, I.P. 2008. Mössbauer spectroscopy of grain boundaries in submicrocrystalline molybdenum obtained by severe plastic deformation, *Phys. Metals Metallogr.* **106(5)**: 490–4.
160. Popov, V.V. 2009. Emission Mössbauer Spectroscopy of Grain Boundaries in Poly- and Nanocrystalline Metals, *Defect Diff. Forum* pp. 289–292; 633–40.
161. Popova, E.N., Popov, V.V., Romanov, E.P. and Pilyugin, V.P. 2006. Thermal Stability of Nanocrystalline Nb Produced by Severe Plastic Deformation, *Phys. Met. Metallogr.* **101(1)**: 52–57.
162. Physical Metallurgy. *Edited* by Cahn, R.W. and Haansen, P. 1983. North Holland Physics Publishing, Amsterdam.

4

A Review on Thermal Lattice Monte Carlo Analysis

T. Fiedler[1,*], I.V. Belova[1], A. Öchsner[2] and G.E. Murch[1]

ABSTRACT

In this review chapter, we introduce the recently developed Lattice Monte Carlo method for addressing and solving phenomenologically-based thermal diffusion problems especially for composite and porous materials. We describe in detail the application of this method to calculate effective thermal diffusivities and to determine temperature profiles including conditions of temperature-dependent material properties and materials exhibiting a phase change. Where possible, results of the method are compared with results of exact or finite element methods. Excellent agreement is demonstrated.

NOTATION

C	specific heat
d	dimensions
E_p	thermal particle energy
g	area fraction of inclusion
L	latent heat
m	mass
n	number of particles at a lattice site
N_n	number of particles in the source plane
N_{tot}	total number of particles
p_j	jump probability
p_s	selection probability
Q	thermal energy
$\dot{Q}$	heat flux

---

[1] University Centre for Mass and Thermal Transport in Engineering Materials, Priority Research Centre for Geotechnical and Materials Modelling, School of Engineering, The University of Newcastle, Callaghan, NSW 2308, AUSTRALIA.
[2] Department of Applied Mechanics, Faculty of Mechanical Engineering, Technical University of Malaysia, 81310 UTM Skudai, Johor, MALAYSIA.
[*] *Corresponding author* : E-mail : thomas.fiedler@newcastle.edu.au

s	lattice spacing, distance
T_c	constant temperature
T_D	phase change temperature
T_{init}	initial temperature
T_{ref}	minimum possible temperature in transient simulation
V	volume
X	coordinate of solid–liquid phase boundary

GREEK LETTERS

κ	thermal diffusivity
λ	thermal conductivity
ρ	density
υ_s	temperature distribution inside solid phase
υ_l	temperature distribution inside liquid phase

SUBSCRIPTS

0	inclusion
1	matrix
Al	aluminium
eff	effective property
i, j	phase indices

1. INTRODUCTION

The Monte Carlo method was first developed at Los Alamos during the WWII Manhattan Project for the purposes of modelling neutron trajectories during fission. Since that time, the Monte Carlo method has undergone numerous developments and has enjoyed applications in virtually every area of science and engineering. Intrinsically a computationally very demanding method, the Monte Carlo method has naturally become far more popular as computers have become faster, less expensive and more accessible. The Monte Carlo method has been a popular method to address both mass and thermal diffusion problems in materials. For mass diffusion, the Monte Carlo method has been used for many years for addressing atomistic problems in crystalline solids in such problems it is now usually called the Kinetic Monte Carlo (KMC) method; see Murch (1984) for an early review and Mishin *et al.* (2005) for a typical recent KMC calculation. More recently, a related Monte Carlo method has been used for addressing phenomenological problems, where it is called the Lattice Monte Carlo (LMC) method (Belova and Murch, 2007; Belova *et al.*, 2007a). For thermal diffusion, the Monte Carlo method has also been used for many years for addressing transient thermal conduction problems in homogeneous materials, where a continuous random walk method has been used (Haji-

Sheikh and Sparrow, 1967). Recently, the LMC method has been adapted to address steady state and transient thermal problems in inhomogeneous materials (Fiedler et al., 2007; Belova and Murch, 2008).

The LMC method, the focus of this contribution, is basically a type of finite-difference method that is embedded in a quasi-simulation of the physical process. It has shown itself to be very useful in addressing and solving complex phenomenological mass and thermal diffusion problems. In Sections two and three of this Chapter, we provide a review at an introductory level of the LMC method, with an emphasis on how the method is actually employed to address and solve phenomenological thermal diffusion problems.

In the fourth Section of this Chapter, thermal analysis accounting for temperature-dependent material properties in composite materials is addressed. A theoretical understanding of thermal transport in composite materials, including cellular metals, has generally been achieved using fairly primitive geometric models that are then addressed with Finite Element (FE) or analytical analysis (Fiedler et al., 2008a; Fiedler and Öchsner, 2007; Fiedler et al., 2008b; Zhao et al., 2008). However, the complex structures of real composites are not captured in this way. But meshing from images from computed tomography (CT) of real materials with FE is problematic due to a possible mesh dependence of the solution and restrictions of the model size (Maire et al., 2003). A valuable alternative to FE analysis is now the Lattice Monte Carlo method. In Section 4, the LMC calculation strategy is enhanced in order to incorporate temperature dependencies of thermal conductivities $\lambda(T)$ and specific heats $C(T)$ in transient thermal analyses. The computational LMC models are based directly on obtained CT data (cf. Fig. 4.1), allowing for a detailed representation of the *real* geometry. Initially, simple model geometries are addressed. This enables LMC results to be compared with results from FE analysis of the same models for purposes of validation. The second part of this Section addresses transient heat transfer into cellular metals. The information gained can be used to assess their potential to enhance thermal transfer in compact heat sinks or as fire retardant materials (Lu and Chen, 1999).

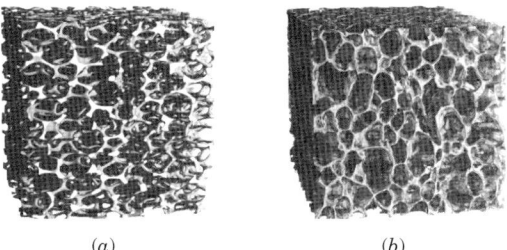

(a) (b)

Fig. 4.1. Three dimensional Computed tomography reconstructions of analysed cellular metals: (a) M-Pore®, (b) Alporas®

As far as porous materials are addressed, we should first consider that heat transfer is brought about by three different mechanisms: thermal radiation, thermal convection and thermal conduction. It has been demonstrated (Lu and Chen, 1999; Öchsner and Gracio, 2005) that the influence of thermal radiation on the effective thermal conductivity of cellular metals, especially at temperatures below 700 K, is low and its contribution to the effective thermal conductivity can be disregarded within these studies. On the other hand, the movement of gas through the interconnected porosity of the M-Pore® structure (*i.e.*, natural convection) could partially contribute to its effective thermal conductivity, but the contribution in the case of open-cell aluminium foams is estimated to be lower than 10% (Zhao *et al.*, 2005). The effect of thermal convection within the closed-cell Alporas® structures is small and can be neglected. Accordingly, energy transfer due to the bulk movement of gas through interconnected porosity can be disregarded and the validity of results obtained, *i.e.*, temperature profiles, is restricted to cases where heat transfer occurs predominantly by thermal conduction.

Section 5 addresses transient thermal analysis including materials exhibiting a phase change. As an example, thermal conductivity enhancement in compact heat sinks is presented. The effective thermal conductivity of phase change materials (PCM), such as paraffin, is distinctly smaller than for example aluminium. Accordingly, following a sudden temperature increase, thermal energy is first stored within the metallic matrix before it is slowly conducted within the PCM. For short time intervals, the thermal conductivity of the PCM can therefore be neglected.

2. EFFECTIVE THERMAL DIFFUSIVITIES/CONDUCTIVITIES

In thermal diffusion problems, the long-time limit effective thermal transport quantity that is best known is probably the effective thermal conductivity of a composite material. In Section 2.1 we review the LMC method for calculating the effective thermal diffusivity/conductivity and in Section 2.2 we highlight the various published LMC calculations of the effective thermal diffusivity/conductivity.

2.1. The LMC Method for Determining the Effective Thermal Diffusivity/Conductivity

Thermal diffusion, like mass diffusion, is a random process that can be represented by random walks of particles. In the case of thermal diffusion, the particles are virtual thermal energy quantities. The Einstein-Smoluchowski (ES) Equation describes the thermal diffusivity κ in d dimensions ($d = 1, 2, 3$):

$$\kappa = \frac{\langle R^2 \rangle}{2dt} \quad \ldots(1)$$

It should be noted that the thermal conductivity λ_i in a phase i is directly related to the thermal diffusivity κ_i in that phase by the well-known expression $\kappa_i = \lambda_i/(\rho_i\, C_i)$ where ρ_i is the density of phase i and C_i is its specific heat. In a model composite, by simply requiring that the densities and the specific heats take values equal to unity everywhere in the calculation, the effective thermal conductivity λ_{eff} then simply equals the effective thermal diffusivity κ_{eff}. Before each jump attempt, the calculation time t is incremented. For the dependence on the material properties of the origin and destination lattice node, a jump probability p_j is calculated. A jump probability is basically a scaled thermal conductivity and can possess only values between zero and one. The jump probability is then compared to a random number (taken out of a uniformly distributed population between zero and one) and, if successful, the coordinates of the virtual energy particle are updated. In order to calculate the effective thermal diffusivity in Eq. (1), simply the mean square displacement $\langle R^2 \rangle$ of a large population of thermal energy particles in a lattice model needs to be calculated. To the authors' knowledge, in a thermal diffusion context the ES Equation has no physical meaning in the sense that it can be made use of experimentally. It purely provides a very useful and simple means for calculating the effective thermal diffusivity (conductivity) from random walks of virtual particles using the LMC method.

2.2. LMC Calculations of the Effective Thermal Diffusivity/Conductivity

Several calculations have been specifically directed to the LMC calculation of the effective thermal conductivity (Fiedler et al., 2007; Belova et al., 2007b; Belova and Murch, 2004; Fiedler et al., 2008c; Fiedler et al., 2008d). These include calculations of the effective thermal conductivity of models of syntactic metallic hollow sphere structures (Fiedler et al., 2008c) and compact heat sinks based on cellular metals (Fiedler et al., 2008d). An example of the results of a LMC calculation of the effective thermal conductivity is shown in Fig. 4.2 for the case of circular inclusions in a matrix (Fiedler et al., 2007). In the same Figure are also the results for the effective thermal conductivity using finite element analysis. The agreement is seen to be excellent. Four different ratios of the thermal conductivities of the inclusion λ_0 and matrix λ_1 are considered and the maximum thermal conductivity is always scaled to unity. If matrix and inclusions exhibit identical thermal properties (i.e., $\lambda_0 = \lambda_1 = 1$), the thermal conductivity is independent of the area fraction of the inclusions g. In the case that the thermal conductivity of the matrix exceeds the corresponding value of the inclusions ($\lambda_0/\lambda_1 < 1$), the thermal conductivity plotted as a function of the area fraction g shows an approximately linear characteristic. In contrast, for highly thermal-conducting inclusions (i.e., $\lambda_0/\lambda_1 = 10$) an increase of g coincides with an increase of the effective thermal conductivity.

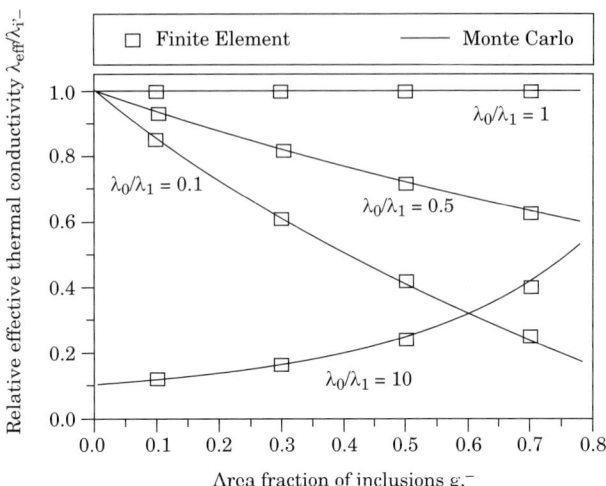

Fig. 4.2. Comparison of results from Lattice Monte Carlo and Finite Element Method calculations of the relative effective thermal conductivity of a composite with circular inclusions (0) in a matrix (1) in a square planar arrangement as a function of area fraction of inclusions for several values of the matrix and dispersed phase thermal conductivities

There have been no direct LMC calculations of the effective thermal diffusivity. However, this quantity can be readily formed from the effective thermal conductivity using the following equation:

$$\kappa_{\text{eff}} = \frac{\lambda_{\text{eff}}}{(\rho C)_{\text{eff}}} \qquad ...(2)$$

where,
$$(\rho C)_{\text{eff}} = \sum_{\text{all phases } i} \rho_i C_i g_i.$$

3. TEMPERATURE PROFILES

In Section 3.1, we review the LMC method for determining temperature profiles and in Section 3.2 we highlight several LMC calculations of temperature profiles.

3.1. The LMC Method for Determining Temperature Profiles

Because of the role of the density and specific heat in transient thermal transport problems, the determination of temperature profiles requires quite a different procedure from the (steady state) calculation of the thermal diffusivity described above. As an example frequently found experimentally, we will focus on the situation where the temperature at the surface or source is held constant at T_c. The number of virtual thermal particles in the source plane is N_n. The amount of thermal energy E_p corresponding to a virtual thermal energy particle is then given by the relation:

$$E_p = T_c \cdot s^3 \cdot \rho \cdot C \cdot \frac{1}{N_n} \qquad ...(3)$$

where s is the distance between two neighbouring lattice sites.

We organize the random walks of the particles in such a way that they are now directed by *two* parameters: the jump probability p_j (the scaled thermal conductivity) and the selection probability p_s (the scaled inverse product $\rho_i \cdot C_i$). We treat the selection probability as an 'amount of inertia' assigned to a virtual thermal energy particle in the particular phase: i.e., the higher the specific heat in the phase the slower the virtual thermal particle. Both jump and selection probabilities depend on the material parameters of the phase i. Analogous to thermal diffusivity calculations, the jump and selection probabilities p_j and p_s are defined to have values between zero (the event never occurs) and unity (the event always occurs). Then the jump probability in a phase must be scaled with respect to the highest thermal conductivity so that:

$$p_{j,i} = \lambda_i / \lambda_{max} \qquad ...(4)$$

and the selection probability $p_{s,i}$ is scaled with respect to the lowest value of the product $\rho_i \cdot C_i$:

$$p_{s,i} = \frac{(\rho_j C_j)_{min}}{\rho_i C_i} \qquad ...(5)$$

According to Eq. (5), materials with a high specific heat and density possess a low selection probability. Different selection probabilities between sites from different material regions results in the modified Eq. (3) definition of the energy E_p corresponding now to a virtual thermal particle:

$$E_p = T_c \cdot s^3 \cdot (\rho \cdot C_p)_{min} \cdot \frac{1}{N_n} \qquad ...(6)$$

Let us assume that phase 1 has a lower selection probability than phase 2 i.e.,: $p_{s,1} < p_{s,2}$. The overall probability of a jump of a probing thermal particle in phase 1 is then equal to $p_{s,1} \cdot p_{j,1}$ and similarly for a particle in phase 2. (This value is, in fact, a scaled thermal diffusivity.) The increased number of unsuccessful jump attempts in phase 1 simulates an accumulation of virtual thermal particles in that phase. It should be mentioned here that the selection of p_s and jump probabilities p_j of virtual thermal particles inside the source plane ($x = 0$) are equal to unity.

At the beginning of each time-step, a particle is randomly selected. Next, a random number between 0 and 1 is generated and compared to the selection probability corresponding to the phase at that lattice site. If this random number is higher than the selection probability, the attempt is unsuccessful; the LMC time is increased and another particle is randomly chosen. Otherwise, a jump direction for the particle is randomly chosen and, depending on the phase(s) of the starting and target lattice sites, the jump probability p_j is now determined. In the case that the jump attempt is successful, the coordinates of the particle are updated before the LMC time is increased and a new particle is selected. The incremental increase to the Monte Carlo time t_{MC} depends on the total number N_{ges} of virtual thermal particles in the system. At the end of the specified time t_{MC} the final positions of all of the particles are recorded.

The results of the LMC analyses are virtual thermal energy (particle) distributions that correspond to particular Monte Carlo times t_{MC}. In order to obtain temperature profiles, the thermal particles are translated into site temperatures T according to:

$$T = \frac{n \cdot E_p}{s^3 \cdot (\rho_i \cdot C_{p,i})_{min}} \qquad \ldots(7)$$

where n is the number of virtual thermal particles currently located at the site.

Next, the Monte Carlo time t_{MC} needs to be converted into real time t. To do this, we will use a standard parametric analysis approach. Consider the Heat Equation in its dimensionless form:

$$\frac{dT}{dt'} = \frac{\kappa t^*}{(x^*)^2} \, \text{div}\,(\nabla T) \qquad \ldots(8)$$

where x^* is a characteristic length for which we will use the jump distance s: $x^* = s$, t^* is characteristic time that is most naturally a jump attempt per virtual thermal particle and t^* should be determined in real time units; t' and x' are the dimensionless time and space coordinates. $\kappa t^*/(x^*)^2$ is the dimensionless parameter of the thermal diffusion processes. It is obvious that the value of this parameter used in LMC simulations should be equal to its value in real units.

For the case of multiphase material, it is clear that we need to consider this correspondence only in one phase, provided that the other phases are modelled correctly. Let us choose phase i where the thermal conductivity is the highest λ_{max}. Then the LMC value of the thermal diffusion parameter is:

$$\left(\frac{\kappa t^*}{(x^*)^2}\right)^{MC} = \frac{1}{6} \frac{(\rho C)_{min}}{\rho_i C_i} \frac{1}{1^2} \qquad \ldots(9)$$

Equating this value to the thermal diffusion parameter in the real units we have that:

$$\frac{\lambda_{max} t^*}{\rho_i C_i s^2} = \frac{\kappa t^*}{(x^*)^2} = \left(\frac{\kappa t^*}{(x^*)^2}\right)^{MC} = \frac{1}{6} \frac{(\rho C)_{min}}{\rho_i C_i} \qquad \ldots(10)$$

Solving the equation between the leftmost term and the rightmost term we soon arrive at:

$$t^* = \frac{1}{6} \cdot \frac{s^2 (\rho C)_{\min}}{\lambda_{\max}} \qquad \ldots(11)$$

Therefore, for the total time in real units we have the following connection to the LMC simulation time as follows:

$$t = t^* t_{\mathrm{MC}} = \frac{t_{\mathrm{MC}}}{6} \cdot \frac{s^2 (\rho C)_{\min}}{\lambda_{\max}} \qquad \ldots(12)$$

It would be possible in principle to determine the effective thermal diffusivity of a model composite by analyzing simulated temperature profiles. This would assume that there has been averaging of the temperature profile over all possible locations of the second phase. The temperature profile could then be processed to give the effective diffusivity in much the same way as an experimental temperature profile would be processed by using the appropriate solution to the Heat Equation. In practice, it is much easier to calculate the effective thermal diffusivity of the model composite in a separate calculation using the ES Equation along the lines of what has been described above.

3.2. LMC Calculations of Temperature Profiles

Figure 4.3 shows three example temperature profiles in aluminium calculated by the LMC method for the condition of a constant surface temperature. In the same figure, we include the exact analytical error function solution (dashed lines) and also a determination of the temperature profiles using finite element analysis (solid lines).

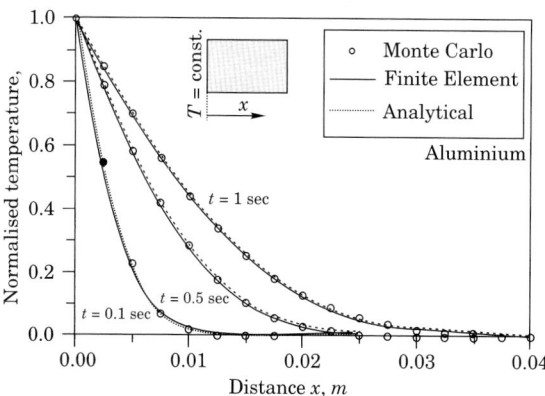

Fig. 4.3. Temperature profiles in homogeneous aluminium

It can be seen that there is excellent agreement between all three results. Figure 4.4 shows an example temperature profile obtained by LMC in a layered composite of aluminium and paraffin with the layers arranged normal to the heat flow. Here, the thermal parameters of the two phases are of course very different. In the same figure are the results of a determination of the temperature profile using finite element analysis. It can be seen that there is excellent agreement between the two methods.

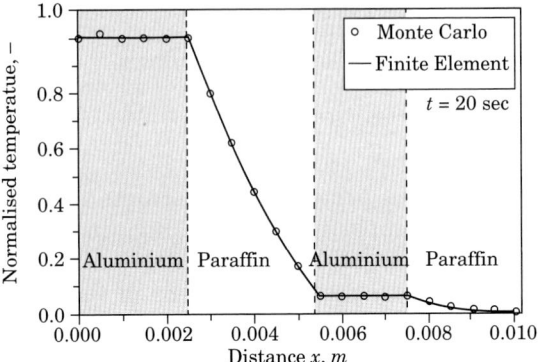

Fig. 4.4. Temperature profile in a layered two phase aluminium-paraffin composite

The analysis is repeated for an reversed arrangement of the aluminium and paraffin phases. Again, an excellent agreement between the LMC and Finite Element solutions can be found (see Fig. 4.5).

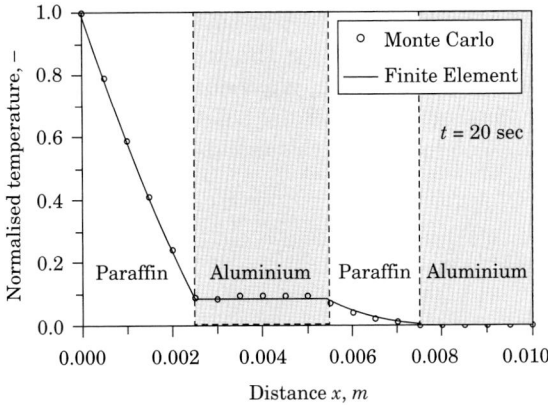

Fig. 4.5. Temperature profile in a layered two-phase paraffin-aluminium composite

4. TEMPERATURE-DEPENDENT MATERIAL PROPERTIES

In the following, the Lattice Monte Carlo method is enhanced in order to account for temperature-dependent material properties.

4.1. Extension of the LMC Method

As already mentioned in Section 2.1, the LMC method is based on the fact that thermal diffusion is a random process that can be represented by random walks of particles on a discrete lattice. In thermal LMC analyses, the random walks are performed by virtual thermal energy particles.

Accounting for temperature-dependent material properties, the constant energy E_p [cf. Eq. (6)] assigned to particle is defined by:

$$E_p = s^3 \, (\rho \cdot C(T))_{\min} \frac{T_c}{n} \qquad \ldots(13)$$

It should be mentioned here that the change of density (and therefore also of the step length s) with temperature is small in the considered temperature range and, accordingly, constant values are presumed for the step-length s and the density ρ. If multiple phases exist, the particular combination of phase and temperature resulting in the *minimum* product $\rho \cdot C(T)$ must be identified and used for the calculation of the particle energy. The number of particles n can be selected freely: higher numbers of particles yield more accurate results, but of course proportionately longer computation times.

In the case of temperature dependencies, the jump probability $p_{j,i}(T)$ (the scaled thermal conductivity) and the selection probability $p_{s,i}(T)$ [the scaled inverse product $\rho_i \cdot C_i(T)$] are also functions of temperature:

$$p_{j,i}(T) = \lambda_i(T)/\lambda(T)_{\max} \qquad \ldots(14)$$

The selection probability $p_{s,i}(T)$ is scaled with respect to the lowest $(\rho \cdot C(T))_{\min}$ already used in Eq. (13):

$$p_{s,i}(T) = \frac{(\rho \cdot C(T))_{\min}}{\rho_i \cdot C_i(T)} \qquad \ldots(15)$$

The temperature-dependent material properties considered in subsequent calculations are shown in Fig. 4.6.

Prior to LMC analyses, initial and boundary conditions need to be implemented [cf. Fig. 4.7 (a)]. Therefore, a global reference temperature T_{ref} is introduced. This temperature corresponds to zero thermal energy particles present at a lattice node and is therefore the minimum temperature any node can exhibit. Next, virtual thermal particles are added to each node until the initial temperature T_{init} is reached. For obvious reasons $T_{\text{ref}} \leq T_{\text{init}}$ must be valid. The number of virtual particles required depends on the local density ρ_i and specific heat $C_i(T_{\text{init}})$ and can be obtained using Eq. (16). In addition, at one (or more) of the six free surfaces of the lattice model, a constant temperature boundary condition T_c ($\geq T_{\text{ref}}$) is prescribed. To this end, additional particles are added (or removed) until the condition $T = T_c$ is fulfilled. The temperature T at a node with coordinates (x, y, z) and time t can be obtained for temperature-dependency from the relation:

$$T(x, y, z, t) = \frac{n(x, y, z, t) \cdot E_p}{s^3 \cdot \rho_i \cdot C_{p,i}(T)} + T_{\text{ref}} \qquad \ldots(16)$$

where $n(x, y, z, t)$ is the number of energy particles present at the node and the constant particle energy E_p is defined according to Eq. (13).

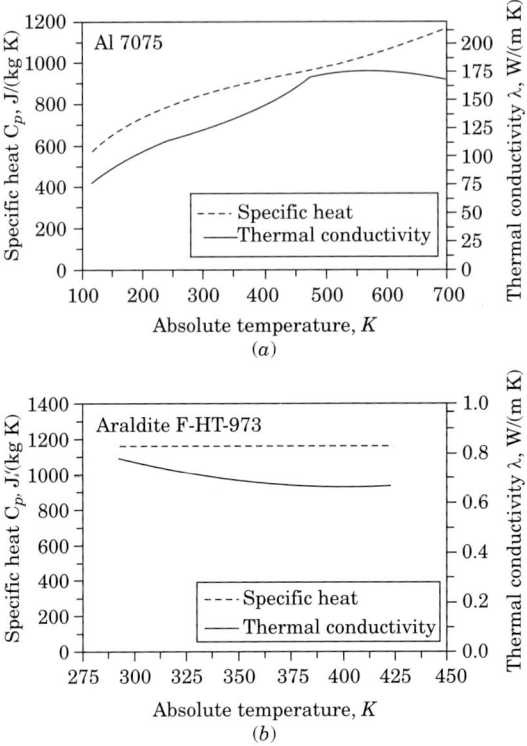

Fig. 4.6. Temperature-dependent thermal properties of the base materials: (a) Al 7075, (b) Araldite® F-HT-973 (Burghartz and Schulz, 1991; Archer, 1993; Garcia et al., 1991)

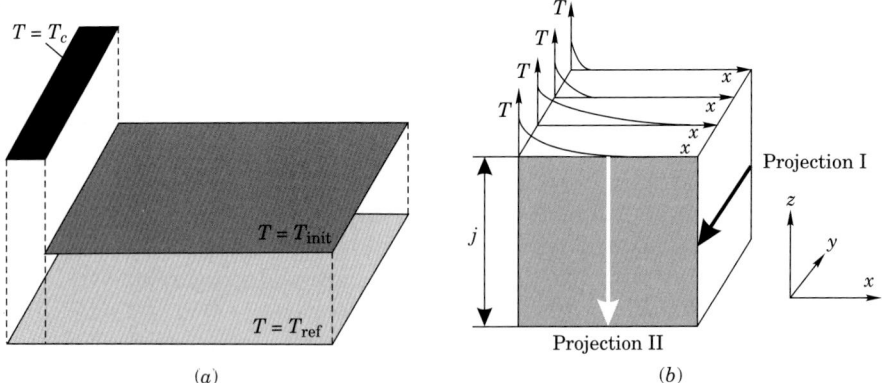

Fig. 4.7. (a) Initial conditions and boundary conditions; (b) Projection algorithm of temperature profiles

In order to sustain the constant temperature boundary condition T_c, as soon as a particle leaves the boundary plane a new one is generated. Accordingly, the number of particles will increase over time. This increase

n^* of energy particles can be related to the total amount of thermal energy $Q(t) = \sum n^*(t) \cdot E_p$ conducted into the structure. The derivation of the conversion of t_{MC} to real time t has already been shown in Section 3.1 [cf. Eq. (12)]. Incorporating temperature dependencies of the specific heat and thermal conductivity, the following equation is valid:

$$t = \frac{t_{MC}}{6} \cdot \frac{s^2 \, (\rho \cdot C(T))_{min}}{\lambda(T)_{max}} \quad \ldots(17)$$

In order to visualise 3D temperature distributions as temperature profiles in composite materials, a two-step projection algorithm schematically drawn in Fig. 4.7 (b) is applied. Starting in each node within the plane where the constant temperature boundary condition is prescribed, j^2 temperature profiles in the x-direction are obtained. Next, j provisional temperature profiles are obtained by averaging the j^2 temperature profiles along the y-direction (Projection I). The resulting two dimensional temperature distribution (gray plane) is then averaged in the z-direction to obtain the final temperature profile. For the case of cellular metals it is important to mention that only nodes with non-zero thermal conductivity are included in this averaging process.

The material properties of all considered materials are summarized in Table 4.1.

Table 4.1. Physical material properties (Burghartz and Schulz, 1991; Archer, 1993; Garcia et al., 1991; Mettawee and Assassa, 2007)

Material	Thermal conductivity [W/(mK)]	Specific heat [J/(kg K)]	Density [kg/m³]
Aluminium 7075	cf. Fig. 4.6	cf. Fig. 4.6	2700
Araldite F-HT-973	cf. Fig. 4.6	cf. Fig. 4.6	1580
Paraffin wax	0.21	2500	900

4.2. Results of Temperature-Dependent LMC Analysis

4.2.1. *Homogeneous material*

Figure 4.8 (a) displays temperature profiles within a 7075 aluminium cube with a side length of 10 mm. The LMC results are represented by drawn lines and finite element results are shown by circular markers at appropriate intervals for comparison. Excellent agreement between the two numerical methods is found. For the time $t = 0.5$ s, dotted lines also give the standard error function solutions for constant values of $\lambda = \lambda$ (300 K) and $\lambda = \lambda$ (650 K) which straddle the range of the input $\lambda = \lambda(T)$. The importance of accounting for the temperature dependence of the material properties is clear.

Figure 4.8 (b) shows the heat transfer of energy Q versus time t. After a fast initial increase of stored energy (linear slope), the energy slowly approximates to an asymptotic maximum E_{max}. The numerical value of the maximum amount of transferable energy (under the prescribed initial and boundary conditions) can be calculated according to

$$E_{max} = \rho_{Al} \cdot C_{Al}(T_c) \cdot V \cdot (T_c - T_{init}),$$

where the subscript Al indicates the material properties of the aluminium alloy Al 7075 [cf. Fig. 4.6 (a)] and V is the volume of the considered geometry.

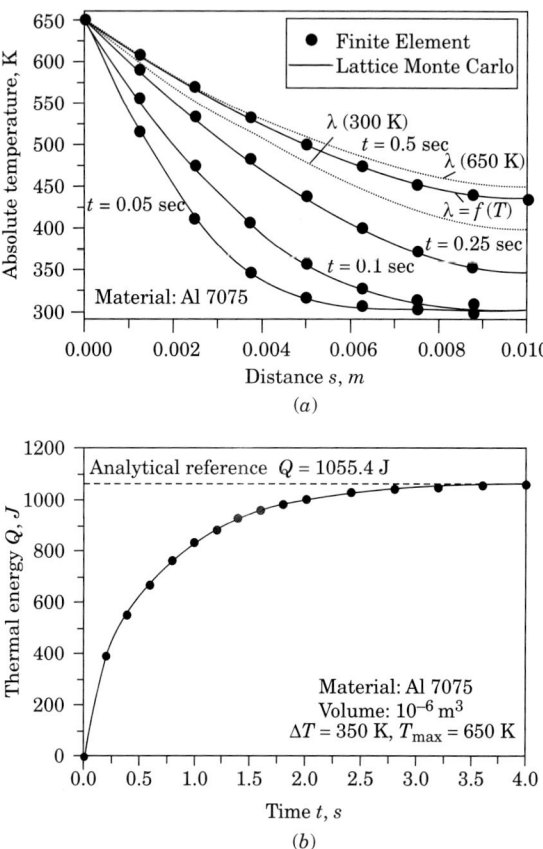

Fig. 4.8. Temperature profiles and energy transport inside 7075 aluminium sheet

4.2.2. *Layered two-phase material*

In order to decrease the heat transfer of energy, the aluminium can be coated with a thin layer (w = 1 mm) of Araldite® F-HT-973. The thermal conductivity of the polymer is two orders of magnitude lower than aluminium and it therefore acts as a thermal insulator. Temperature profiles at different times t are plotted in Fig. 4.9. A high temperature gradient forms within the thermal insulator, whereas the temperature within the higher thermally conducting aluminium alloy is approximately constant. Comparing Figs. 4.8 and 4.9, it becomes obvious that the thermally insulating layer significantly decreases the thermal transfer of energy. In addition to the LMC results, FE data are shown by circular markers. Analogous to the homogeneous material, excellent agreement between the two numerical methods is found.

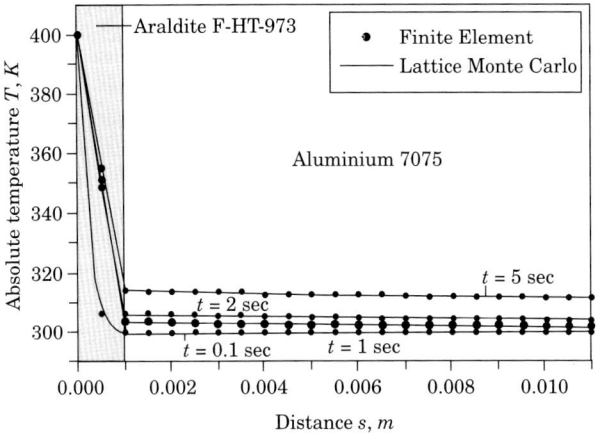

Fig. 4.9. Temperature profile inside an 7075 aluminium sheet coated by Araldite® F-HT-973

4.2.3. Cellular metals

In the following, the results of transient heat transfer into cellular metals are presented. In this case, the lattice models for these analyses are generated directly using the computed tomography data shown in Fig. 4.1, where each lattice node corresponds to a voxel (resolution 800^3 voxels). Figure 4.10 shows the temperature profiles of cellular metals at times $t = 0.5$ sec and 2.0 sec. The transformation of a three-dimensional temperature distribution to a one dimensional temperature profile is schematically shown in Fig. 4.7 (b). The cell wall material of both cellular metals is Al 7075 and both structures exhibit a relative density (density of the cellular metal divided by the density of the cell wall material) of 8.4%.

The faster temperature increase is found inside the Alporas® material. The difference from the M-Pore® sponge can be explained by a small difference in the effective thermal conductivity (Fiedler et al., 2008e). Comparing Figs. 4.8 and 4.10 we note also a distinct decrease of thermal energy transfer in the case of the cellular aluminium. The explanation for this deviation is the high volume fraction (91.6%) of non-conducting pores in the M-Pore® and Aluminium® structures. In addition to the cellular metals, approximate temperature profiles of the PCM paraffin (assumed temperature-independent material properties $\lambda_{\text{Paraffin}}(T) = 0.21$ W/(mK), $C_{\text{Paraffin}} = 2500$ J/kg) (Mettawee and Assassa, 2007) are shown. The comparison of the temperature distributions inside the cellular metals and paraffin at particular times t (i.e., 0.5 and 2 s, cf. Fig. 4.10) reveals that the temperature increases faster inside the M-Pore® and Aluminium® structures. This clearly indicates the possibility of improving the performance of this type of heat sink by using cellular metals as a matrix within low thermally conducting PCMs.

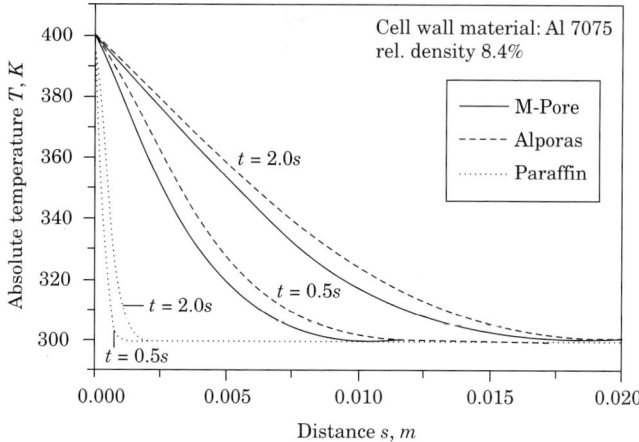

Fig. 4.10. Temperature distribution inside the matrix material of Empore® and Alporas® structures and homogeneous paraffin

4.2.4. *Two-dimensional heat transfer*

Up to this point, only one-dimensional heat transfer has been considered, *i.e.*, the temperature boundary condition T_c was only defined on one free surface. Although this load case is appropriate for many applications, many engineering problems require the definition of more complex boundary conditions. The LMC method usefully allows for the definition of an arbitrary temperature T_c for each single node. As an example of application to a more complicated situation, LMC results for a two-dimensional heat transfer problem are now presented. In these calculations, a constant temperature boundary condition is prescribed at two neighbouring surfaces of the geometry. The visualisation of these results is best achieved with two-dimensional temperature maps which can be obtained by using the projection I shown in Fig. 4.7 (*b*). Figure 4.11 (*a*) shows the temperature distribution for a homogeneous aluminium cube with a side length of 0.011 m. As expected, a symmetric temperature distribution is obtained. A temperature profile extracted far enough from the second temperature boundary condition would be identical to Fig. 4.8 (*a*). A different picture emerges for a composite structure which is composed of alternating layers of Araldite® F-HT-973 and Aluminium 7075 [cf. Fig. 4.11 (*b*)]. As mentioned above, the epoxy acts as a thermal insulator and energy transfer or change of temperature only slowly occurs inside the Araldite®. In contrast, the temperature increases rapidly inside the highly thermally conductive aluminium.

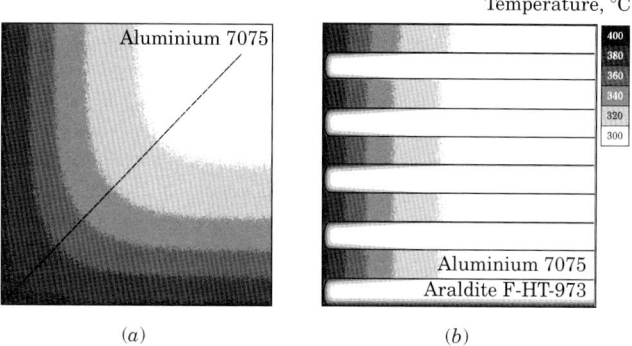

Fig. 4.11. Two-dimensional heat transfer problems ($t = 0.5$ s, $T_{\text{init}} = 300$ K, $T_c = 400$ K). (*a*) homogeneous aluminium, (*b*) layered structure

5. PHASE CHANGE MATERIALS

In the final section of this chapter, transient thermal transport including phase change of materials is addressed. The methodology is presented for the example of thermal conductivity enhancement in compact heat sinks. As passive thermal management systems for cyclic thermal loading heat sinks are continuously gaining importance, *e.g.* due to rapid developments in the electronic industry (Kulkarni and Das, 2005; Walsh and Grimes, 2007; Wang *et al.*, 2008). Alternative applications comprise solar energy storage where heat sinks decrease maximum temperatures during the daytime and provide heat at low temperatures (Kenisarin and Mahkamov, 2007; Najjar and Hasan, 2008) or are used for the heating of water (Talmatsky and Kribus, 2008; Mazman *et al.*, 2008). In principle, every material can be used as a heat sink. However, engineering problems often put constraints on weight or volume and compact lightweight solutions are required. To this end, PCMs allow additional storage of thermal energy in the latent heat of the material (Regin *et al.*, 2008). Figure 4.12 schematically shows the interdependence of temperature T and stored thermal energy per mass unit Q/m. At the phase transition temperature T_Δ the latent heat L must be transferred before the temperature further increases. Essential requirements for PCMs are a high value of the heat of fusion L and specific heat C, a suitable phase change temperature T_Δ with respect to operation temperatures, chemical stability and non-corrosiveness, a small volume change during the phase change and low cost. In addition, a high thermal conductivity is required.

An excellent material with respect to most of these requirements is paraffin. However, paraffin exhibits a low thermal conductivity. Recent approaches aim for the increase of its effective thermal conductivity by creating paraffin-metal (Fiedler *et al.*, 2008d) or paraffin-graphite (Zhang and Fang, 2006) composite structures. Hong and Herling conducted experiments on open-cell aluminium foams filled with paraffin (Hong and

Herling, 2006). A distinct increase of cooling and heating times could be observed, potentially decreasing the thermal stress of a device under cyclic thermal loading. Analytical solutions for the transient heat transfer including phase change are limited to a few special cases (Carslaw and Jaeger, 1959) and are merely useful for validation purposes. A finite-difference method using an enthalpy-based model was presented in (Liu and Majumdar, 2002). However, this interesting approach is restricted to relatively simple geometries and boundary conditions. As already shown in Section 4.2.3. a major advantage of the LMC method is the ability to address very complex geometries, *e.g.* obtained by computed tomography (Fiedler *et al.*, 2008e). As shown in Section 4, further non-linearities such as temperature-dependent material properties can readily be included in the calculation.

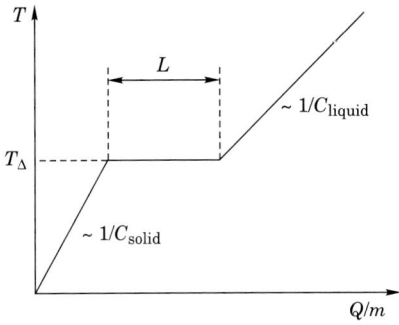

Fig. 4.12. Change of temperature due to thermal energy transfer

Analogous to Section 4, a simple load case is considered where a constant temperature boundary condition T_c is defined at the edge of the geometry while the structure is initially at temperature T_{init} [cf. Fig. 4.7 (a)]. Two cases can be distinguished: for $T_{init} > T_\Delta$ the initial phase is liquid and $T_c < T_\Delta$ must be valid in order to observe solidification. Analogously, $T_{init} < T_\Delta$ imposes a solid PCM that melts starting from the boundary for $T_c > T_\Delta$.

5.1. Methodology

5.1.1. *Analytical solution*

Assuming incompressible material behaviour, the maximum storable/available energy per unit mass of each material Q_j/m can be calculated by integration of

$$\frac{Q_j}{m} = \int_{T_{init}}^{T_c} C(T)\,dT \qquad \ldots(18)$$

In the case of composite materials, the arithmetic mean needs to be determined in order to obtain the effective thermal energy Q_{eff}

$$\frac{Q_{\text{eff}}}{m} = \frac{1}{m} \frac{\sum_j Q_j \cdot m_j}{\sum_j m_j} = \frac{\sum_j Q_j \cdot m_j}{m^2} \qquad \ldots(19)$$

where m is the total mass of the heat sink and m_j the mass of each material j. This energy maximum is reached for steady state $T \to T_c$.

For special cases, an analytical solution of the temperature distribution inside a PCM exists. If the PCM is initially at temperature $T_{\text{init}} > T_\Delta$ and $T_c = 0$ K then the following relations are valid (Carslaw and Jaeger, 1959)

$$\upsilon_s(x,t) = \frac{T_\Delta}{\text{erf}(\alpha)} \text{erf}\left(\frac{x}{2\sqrt{\kappa_s t}}\right)$$

$$\upsilon_l(x,t) = T_i - \frac{T_i - T_\Delta}{\text{erfc}(\alpha\sqrt{\kappa_s/\kappa_l})} \text{erfc}\left(\frac{x}{2\sqrt{\kappa_s t}}\right) \qquad \ldots(20)$$

where υ_s is the temperature inside the solid, υ_l is the temperature inside the liquid phase and coordinate x describes the distance from the 0 K boundary condition. The variable t is the lapsed time and $\kappa_k = \lambda_k/(\rho \cdot C_k)$ is the thermal diffusivity of phase k. For each material, a constant density ρ needs to be presumed whereas the thermal conductivity λ_k may vary between the liquid and solid state. Finally, α is a numerical parameter which can be determined by the solution of (Carslaw and Jaeger, 1959):

$$\frac{e^{-\alpha^2}}{\text{erf}(\alpha)} - \frac{\lambda_l \sqrt{\kappa_s} \, (T_{\text{init}} - T_\Delta) \, e^{\frac{-\kappa_s \alpha^2}{\kappa_l}}}{\lambda_s \sqrt{\kappa_l} \, T_\Delta \text{erfc}\left(\alpha\sqrt{\frac{\kappa_s}{\kappa_l}}\right)} = \frac{\alpha L \sqrt{\pi}}{C_s T_\Delta}$$

5.1.2. Lattice monte carlo method for transient thermal analysis of a PCM

The principle of transient LMC analysis has already been explained in Sections 3.1 and 4.1. However, some modifications are required to account for the latent heat related to phase change phenomena. In the following analysis, the temperature boundary conditions as shown in Fig. 4.7 (a) are defined.

The temperature T at a node with coordinates (x, y, z) and time t can be obtained from the relation (16): $T(x, y, z, t) = \frac{n(x, y, z, t) \cdot E_p}{s^3 \cdot \rho_i \cdot C_{p,i}(T)} + T_{\text{ref}}$.

However, it is important to note that Eq. (16) is only valid for temperatures $T \neq T_\Delta$. For example, a solid that exceeds the phase change temperature enters the two-phase area represented by the constant temperature line in Fig. 4.12. During the phase change, additional particles that enter the site are 'destroyed' and do not increase $n(x, y, z, t)$. Instead, the latent heat L_n of the node is incrementally increased by E_p until the phase change is completed for $L_n > L$.

Within the LMC analysis, two different geometries are considered: homogeneous paraffin and a paraffin-copper composite [cf. Figs. 12 (a) and 12 (b) respectively]. The composite is assembled by alternating layers of paraffin and copper where the copper occupies 10% of the total volume. At the boundaries in y- and z-direction, periodic symmetry conditions are defined. According to the thermal boundary conditions shown in Fig. 4.7 (a) a constant temperature T_c is prescribed at the left hand side of the structures whereas the right hand side is thermally insulated ($\dot{Q} = 0$). The width of all structures in x-direction is 0.002 m.

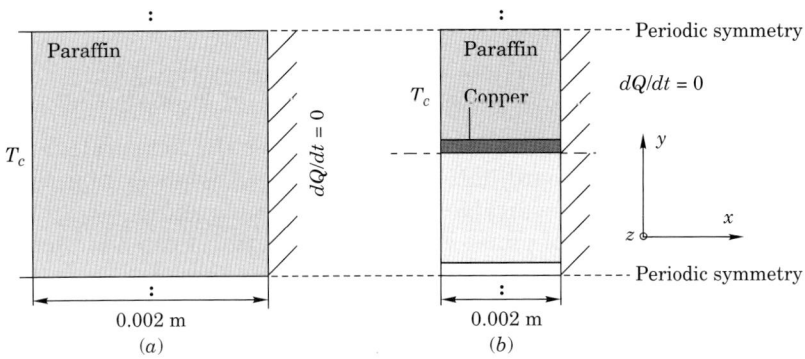

Fig. 4.13. Geometries and boundary conditions of the LMC analysis

5.2. Results of LMC Analysis on Phase Change Materials

In the following, the results of LMC analysis accounting for phase change are presented. First, the temperature distributions are validated by comparison with the analytical solution given by Eq. (20). Next, the heating and cooling of the PCM paraffin with and without copper ribs are considered.

5.2.1. Verification

Figure 4.14 shows the results of the analytical solution Eq. (20) and LMC simulations. The chosen temperature boundary conditions are $T_{init} = 350$ K and $T_c = 0$ K, the material properties of the solid and liquid phases of paraffin are: $\lambda_s = 0.24$ W/(m · K), $C_s = 2200$ J/(kg · K) and $\lambda_l = 0.15$ W/(m · K), $C_l = 2790$ J/(kg · K), respectively. The analytical solution (20) requires a constant density of both phases which is assumed to be $\rho = 800$ kg/m³. In addition to the temperature profiles at times $t = 0.1, 0.2$ and 0.4 seconds the position of the phase boundary $X(t)$ is indicated in the graph. An excellent agreement between the analytical and numerical solutions is found.

5.2.2. Heating of paraffin

The material parameters of the PCM paraffin are defined in Section 5.2.1. However, in order to improve the accuracy of the numerical model the density of the solid phase is now defined as 891.2 kg/m³. Using Eqs. (18) and (19) the maximum amounts of storable energy per unit mass Q_{eff}/m for $T_{init} = 290$ K and $T_c = 500$ K can be calculated. The results are 717 kJ/kg for pure paraffin and 364 kJ/kg for a composite material where a copper matrix occupies 10% of the total volume. The material properties of copper are defined as $\lambda_c = 386$ W/(m · K), $C_c = 385$ J/(kg · K) and $\rho_c = 8960$ kg/m³ (ASTM, 1990). The values indicate that the copper distinctly decreases the maximum amount of Q_{eff}/m since the metal has no phase transition within the considered temperature range and has a low specific heat capacity in comparison with paraffin.

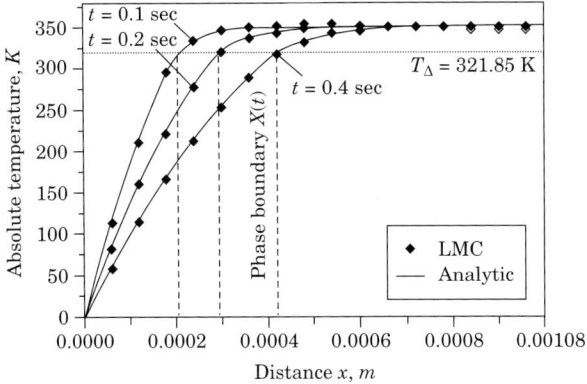

Fig. 4.14. Comparison of analytical and LMC solutions

Figure 4.15 shows the transferred thermal energy per unit mass plotted versus time. In the case of pure paraffin, energy is only slowly transferred. In the composite structure, periodic copper ribs increase the surface area of the PCM [cf. Fig. 4.13 (b)] and, due to their high thermal conductivity, accelerate energy transport: at $t = 1$ sec, approximately 1/3 more energy is

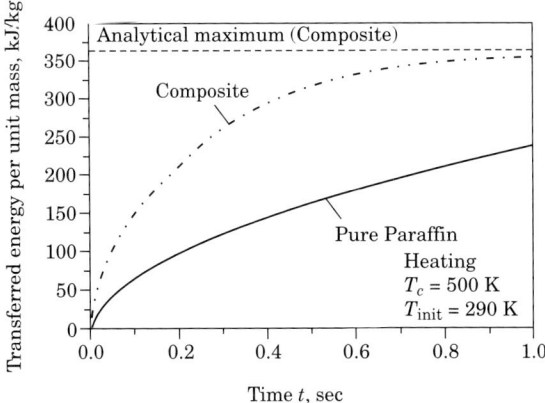

Fig 4.15. Energy storage

transferred into the composite structure in comparison to pure paraffin. The results indicate that by varying the composition of paraffin-metal composites, heat sinks can in fact be optimized between the requirements of maximum energy storage capacity and a fast dynamic response. Increase of the fraction of metal accelerates energy transport but simultaneously decreases the maximum energy storage capacity.

Two dimensional temperature maps are shown in Fig. 4.16 for the time $t = 1$ sec. Figure 4.16 (a) represents the case of pure paraffin. The temperature increases in proximity of the temperature boundary condition and the distribution in x-direction can be approximated by a modified error function (cf. Eq. (20) or Fig. 4.14). A different picture emerges in Fig. 4.16 (b) where high thermally conducting copper ribs protrude into the PCM. The temperature in the copper ribs rises rapidly, thereby storing a fraction of the thermal energy. Originating from the copper ribs, thermal energy is now transferred into the heat sinks resulting into a two dimensional temperature field. The temperature distribution is symmetrical with regard to the centreline of the copper rib.

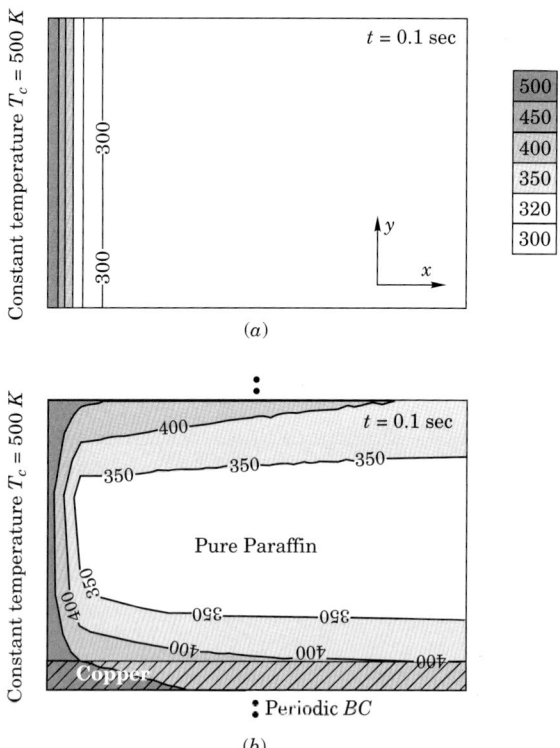

Fig. 4.16. Two dimensional temperature maps: (a) pure paraffin, (b) copper-paraffin composite

5.2.3. Cooling of paraffin

In the following, the inverse process of the heating process considered in Section 5.2.2 is addressed. The initial temperature of the liquid paraffin is now $T_{init} = 500$ K and the constant temperature boundary condition is prescribed as $T_c = 290$ K. The specific internal energy is plotted versus time in Fig. 4.17. The composite structure releases energy faster than pure paraffin. Again, the explanation is the high thermal conductivity of the copper ribs. At $t = 1$ sec the composite structure reaches 90% of the analytical maximum. In comparison, the pure paraffin only releases 30% of the maximum amount of energy possible.

For pure paraffin, the released energy after 1 sec is approximately 8% lower than the energy stored in an inverse heating process (cf. Section 5.2.2). The reason is that in the case of heating, initially only solid paraffin exists whereas the cooling process starts with a superheated fluid. In addition, the thermal conductivity of solid paraffin exceeds the one of paraffin oil: $\lambda_s = 0.24$ W/(m · K) $> \lambda_l = 0.15$ W/(m · K). The deviation between stored and released energy of the composite is smaller (5%), since the initial material properties of the copper ribs are independent of the load case. Accordingly, for both heat sinks, energy is stored slightly faster than it can be released.

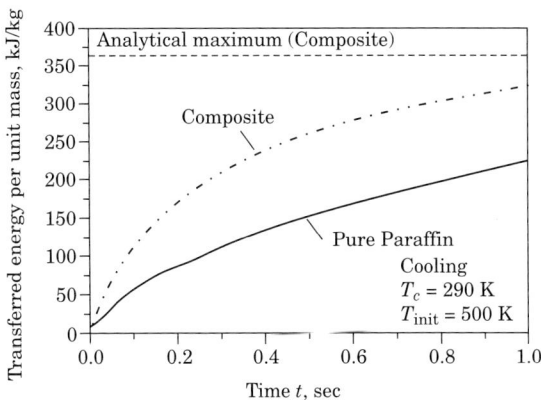

Fig. 4.17. Energy release

6. CONCLUSIONS

In this review chapter, we have introduced the recently developed LMC method for addressing and solving phenomenologically-based thermal diffusion problems especially for composite and porous materials. We described in detail the application of this method to calculate effective thermal conductivities/diffusivities (Section 2) and to determine temperature profiles (Section 3). Results of the LMC method were compared with results of exact or finite element methods. Excellent agreement was always shown. In Section 4, temperature-dependent material behaviour, *i.e.*, variation of thermal conductivity and specific heat with temperature, were addressed. The first part of this Section addressed the determination of temperature profiles for homogeneous and layered materials. Again, the comparison of the LMC results obtained with FE results showed excellent agreement. The second part of Section 4 comprised the calculation of temperature profiles in two different types of cellular metals. In this case, due to the complex geometries, only LMC analyses were performed. The results indicate a slightly faster temperature increase in the Alporas® aluminium foam. However, comparison of the energy transfer over time shows an increase for both materials in comparison to the common PCM paraffin. Chapter 5 deals with LMC analysis of transient heat transfer in composite structures containing PCMs. The validity of the numerical method was shown by comparison with an analytical solution for a simple case where excellent agreement was obtained. In addition, pure paraffin and a paraffin-copper composite structure were subjected to heating and cooling processes with a phase change of the PCM. The composite structure allowed for faster energy transfer. In contrast, the pure PCM exhibits a higher maximum thermal energy storage capacity. Both heat sinks stored energy slightly faster than could be released to the environment. These applications demonstrate the great power, flexibility and reliability of the LMC method for addressing phenomenological thermal diffusion problems.

REFERENCES

1. ASM, 1990. Metals Handbook, Vol. 2-Properties and Selection: Nonferrous Alloys and Special-Purpose Materials, Proceedings of the 10th Edition, ASM International.
2. Archer, D.A. 1993. Thermodynamic Properties of Synthetic Sapphire (α-Al2O3), Standard Reference Material 720 and the Effect of Temperature-Scale Differences on Thermodynamic Properties, *J. Phys. Chem. Ref. Data* **22**: 1441–53.
3. Belova, I.V. and Murch, G.E. 2008. Cellular and Porous Materials. Thermal Properties, Simulation and Prediction. In: Öchsner, A., Murch, G.E., de Lemos, J.S. *Eds.*, Wiley VCH Weinheim.
4. Belova, I.V. and Murch, G.E. 2007. Bridging Different Length and Time Scales in Diffusion Problems Using a Lattice Monte Carlo Methods, *Solid State Phenomena* **129**: 1–10.
5. Belova, I.V., Murch, G.E., Muthubandara, N. and Öchsner, A. 2007a. Analysis of Oxygen Segregation at Metal-Oxide Interfaces Using a New Lattice Monte Carlo Method, *Solid State Phenomena* **129**: 111–8.
6. Belova, I.V., Öchsner, A. and Murch, G.E. 2007b. Modelling of Oxygen Diffusion and Segregation at Interfaces in Ag-MgO Composites, *Def. Diff. Forum.* **266**: 29–38.
7. Belova, I.V. and Murch, G.E. 2004. Monte Carlo Simulation of the Effective Thermal Conductivity in Two-Phase Material, *J. Materials Proces. Tech.* **153–154**: 741–5.
8. Burghartz, S. and Schulz, B. 1994. Thermophysical properties of sapphire, AlN and MgAl2O4 down to 70 K, *J. Nuclear Mater.* **212**: 1065–8.
9. Carslaw, H.S. and Jaeger, J.C. 1959. Conduction of Heat in Solids, Second Edition, Clarendon Press, Oxford.

10. Fiedler, T., Solórzano, E. and Öchsner, A. 2008a. Numerical and experimental analysis of the thermal conductivity of metallic hollow sphere structures, *Mater. Lett.* **62**: 1204–7.
11. Fiedler, T., Öchsner, A., Belova, I.V. and Murch, G.E. 2008b. Recent Advances in the Prediction of the Thermal Properties of Syntactic Metallic Hollow Sphere Structures, *Adv. Eng. Mater.* **10**: 269–73.
12. Fiedler, T., Öchsner, A., Belova, I.V. and Murch, G.E. 2008c. Recent Advances in the Prediction of the Thermal Properties of Syntactic Metallic Hollow Sphere Structures, *Adv. Eng. Mater.* **10**: 269–73.
13. Fiedler, T., Öchsner, A., Belova, I.V. and Murch, G.E. 2008d. Thermal conductivity enhancement of compact heat sinks using cellular metals, *Def. Diff. Forum* **273-276**: 222–6.
14. Fiedler, T., Solórzano, E., Garcia-Moreno, F., Öchsner, A., Belova, I.V. and Murch, G.E. 2008e. Lattice Monte Carlo and experimental analyses of the thermal conductivity of random shaped cellular aluminium, *Adv. Eng. Mater.* **11**: 843–7.
15. Fiedler, T. and Öchsner, A. 2007. Influence of the morphology of joining on the heat transfer properties of periodic metal hollow-sphere-structure, *Mater. Sci. Forum* **553**: 45–50.
16. Fiedler, T., Öchsner, A., Muthubandara, N., Belova, I.V. and Murch, G.E. 2007a. Calculation of the Effective Thermal Conductivity in Composites Using Finite Element and Monte Carlo Methods, *Mater. Sci. Forum* **553**: 51–56.
17. Garcia, A., Rojas, F. and Nava, R. 1991. Determination of thermal properties of thermal casting resin systems for encapsulated transformers, *High Pressures-High Temp.* **23**: 651–8.
18. Haji-Sheikh, A. and Sparrow, E.M. 1967. The Solution of Heat Transfer Problems by Probability Methods, *J. Heat Transfer* **89**: 121–31.
19. Hong, S.T. and Herling, D.R. 2006. Open-cell aluminium foams filled with phase change materials as compact heat sinks, *Scripta Materialia* **55**: 887–90.
20. Kenisarin, M. and Mahkamov, K. 2007. Solar energy storage using phase change materials, *Sust. Energ. Rev.* **11**: 1913–65.
21. Kulkarni, D.P. and Das, D.K. 2005. Analytical and numerical studies on microscale heat sinks for electronic applications, *Appl. Thermal Eng.* **25**: 2432–49.
22. Liu, B. and Majumdar, P. 2002. Numerical Simulation of Phase Change Heat Transfer in PCM-Encapsulated Heat Sinks, 18th IEEE Semi-Therm Symposium **2002**: 88–91.
23. Lu, T.J. and Chen, C. 1999. Thermal transport and fire retardance properties of cellular aluminium alloys, *Acta Mater.* **47**: 1469–85.
24. Maire, E., Fazekas, A., Salvo, L., Dendievel, R., Youssef, S., Cloetens, P. and Michel, Letang J. 2003. X-ray tomography applied to the characterization of cellular materials. Related finite element modeling problems, *Compos. Sci. Tech.* **63**: 2431–43.
25. Mazman, M., Cabeza, L.F., Mehling, H., Nogues, M., Evliya, H. and Paksoy, H.Ö. 2008. Utilization of phase change materials in solar domestic hot water systems, *Renew. Energ.* (in press).
26. Mettawee, E.B.S. and Assassa, G.M.R. 2007. Thermal conductivity enhancement in a latent heat storage system. *Sol. Energy* **81**: 839–45.
27. Mishin, Y., Belova, I.V. and Murch, G.E. 2005. Atomistic Modelling of Diffusion in the TiAl Compound, *Def. Diff. Forum* **237-240**: 271–6.
28. Murch, G.E. 1984. Diffusion in Crystalline Solids, Academic Press, Orlando, 1984. pp. 379.

29. Najjar, A. and Hasan, A. 2008. Modelling of greenhouse with PCM energy storage, *Energ. Convers. Manage.* **49**: 3338–42.
30. Öchsner, A. and Gracio, J. 2005. On the macroscopic thermal properties of syntactic metal foams. Multidiscipl, *Model. Mater. Struct.* **1**: 95–105.
31. Regin, A.F., Solanki, S.C. and Saini, J.S. 2008. Heat transfer characteristics of thermal energy storage system using PCM capsules: a review, *Renew. Sust. Energ. Rev.* **12**: 2438–58.
32. Talmatsky, E. and Kribus, A. 2008. PCM storage for solar DHW: An unfulfilled promise?, *Sol. Energy* **82**: 861–9.
33. Walsh, E. and Grimes, R. 2007. Low profile fan and heat sink thermal management solution for portable applications, *Int. J. Therm. Sci.* **46**: 1182–90.
34. Wang, X.Q., Yap, C. and Mujumdar, A.S. 2008. A parametric study of phase change material (PCM)–based heat sinks, *Int. J. Therm. Sci.* **47**: 1055–2068.
35. Zhang, Z. and Fang, X. 2006. Study on paraffin/expanded graphite composite phase change thermal energy storage material, *Energ. Convers. Manage.* **47**: 303–10.
36. Zhao, C.Y., Tassou, S.A. and Lu, T.J. 2008. Analytical considerations of thermal radiation in cellular metal foams with open cells, *Int. J. Heat Mass Trans.* **51**: 929–40.
37. Zhao, C.Y., Lu, T.J. and Hodson, H.P. 2005. Natural convection in metal foams with open cells, *Int. J. Heat Mass Trans.* **48**: 2452–63.

5

Predicting the Effective Thermal Conductivity of Perforated Hollow Sphere Structures (PHSS)

S.M.H. Hosseini[1,A], A. Öchsner[1,2,B], M. Merkel[3,C] and T. Fiedler[2,D]

ABSTRACT

This chapter investigates the thermal properties of a new type of hollow sphere structures. For this new type, the sphere shell is perforated by several holes in order to open the inner sphere volume and surface. The effective thermal conductivity of perforated sphere structures in several kinds of arrangements is numerically evaluated for different geometrical parameters such as: hole diameter of the perforated hollow spheres, hollow spheres' wall thickness and joining element dimension. The results are compared to classical configurations without perforation. In addition the influence of different joining techniques, i.e. sintering and adhesively bonding, on the thermal conductivity has been compared. Three dimensional finite element analysis was used in order to investigate the heat conductivity of simple cubic, body-centered cubic, face-centered cubic and hexagonal unit cell models. A linear dependency behavior was found for the heat conductivity of different hole diameters for several kinds of arrangements when the results were plotted over the average density for homogeneous models.

NOTATION

a	thickness of linking element
A	area for heat flux calculation
Al	aluminium

---

[1] Department of Applied Mechanics, Technical University of Malaysia, 81310 Skudai, MALAYSIA.
[2] Centre for Mass and Thermal Transport in Engineering Materials, The University of Newcastle, Callaghan, New South Wales 2308, AUSTRALIA.
[3] Department of Mechanical Engineering, University of Applied Sciences Aalen, 73430 Aalen, GERMANY.
* *Corresponding authors* : E-mail : [A]rsg.931@gmail.com,
[B]Andreas.Oechsner@gmail.com, [C]Markus.Merkel@htw-aalen.de,
[D]Thomas.Fiedler@newcastle.edu.au

ave	average
b_1	diameter of linking element
b_s	diameter of sintering area
BCC	body-centered cubic arrangement
CFRP	carbon fibre reinforced plastic
d_l	largest hole diameter between links in PHSS
d_s	largest hole diameter between sintered areas in PHSS
D_l	hollow sphere's total length with linking elements
D_s	hollow sphere's total length without hole (sintered)
FCC	face-centered arrangement
HC	hexagonal closest
Hex	hexagonal arrangement
k	effective thermal conductivity
k_s	thermal conductivity of the hollow sphere base material
PC	primitive cubic arrangement
$\dot{Q}$	heat flux
R	hollow sphere outer radius
$<R$	reduced radius of the sintered hollow sphere
RHS	right-hand side
sp	hollow sphere
st	steel
t	hollow sphere's wall thickness
T	temperature
UC	unit cell
V	volume

GREEK LETTERS

ρ	density
Δy	spatial difference between reference planes
ΔT	temperature difference

1. INTRODUCTION: HOLLOW SPHERE STRUCTURES IN THE CONTEXT OF CELLULAR METALS

Hollow sphere structures (HSS) are novel lightweight materials within the group of cellular metals (such as metal foams) which are characterised by high specific stiffness, the ability to absorb high amounts of energy at relatively low stress levels, potential for noise control, vibration damping and thermal insulation. A combination of these different properties opens a wide field of potential multifunctional applications *e.g.* in automotive or aerospace industry. Typical functional applications of cellular metals in the scope of heat transfer are heat exchangers (Lu et al., 2006; Lu and Stone et al., 1998; Boomsma et al., 2003), fire retardance systems (Lu and Chen, 1999) or thermal insulation. The high thermal insulation capability of hollow sphere structure has been addressed in a US Patent by Schneider and co-workers (2002). Baumeister and colleagues (2004) investigated the linear thermal expansion coefficient of corundum based hollow sphere composites (HSC) using thermomechanical analysis. They found that the thermal behaviour of HSC is mainly governed by the epoxy resin.

Lu and Kou (1993) conducted a comprehensive numerical and experimental study based on unit cells of homogeneous hollow sphere structures. However, only automatically generated finite element meshes were used and the applied approach does not allow the consideration of different material combinations. In a recent paper (Fiedler and Öchsner, 2007a), the thermal conductivity of adhesively bonded and sintered HSS was numerically investigated. In paper (Fiedler and Öchsner, 2007b), the influence of material parameters and geometrical properties of syntactic (hollow spheres completely embedded in a matrix) and partially bonded HSS (spheres joined in localised contact points) was analysed. A comparison between analytical, numerical and experimental approaches is given in (Fiedler et al., 2008a). Fiedler et al. reported in (2008b) recent advances in the prediction of the thermal properties of syntactic metallic hollow sphere structures. They described the application of the Finite Element and Lattice Monte Carlo Method in the case of syntactic periodic and random hollow sphere structures.

Manufacturing techniques for perforated hollow sphere structures are actually under development (Glatt GmbH, 2007) and we use an idealised model structure to clarify basic effects of the perforation on the thermal conductivity.

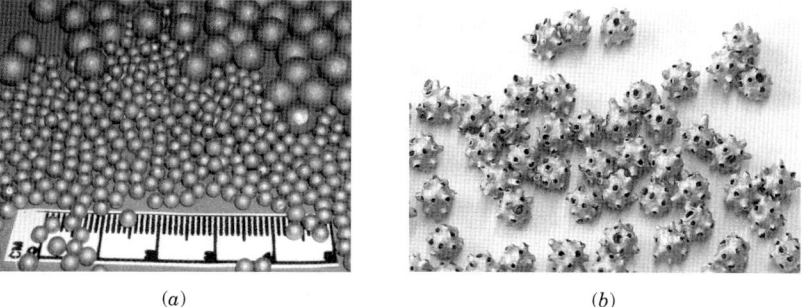

(a) (b)

Fig. 5.1. Single hollow spheres: (a) closed surface (common configuration); (b) with perforated surface (new development); (© by Glatt GmbH, Dresden, Germany)

Figure 5.1 (a) shows classical hollow sphere structures and Fig. 5.1 (b) shows an early production stage of perforated spheres. In the scope of this study, three dimensional finite element analysis is used in order to investigate the effective thermal conductivity of simple cubic, body-centered cubic, face-centered cubic and hexagonal unit cell models and the obtained values are compared to similar structures without holes. The influences of the geometrical properties and the arrangement types on the effective conductivity are analysed within this parametric computational study.

2. MANUFACTURING

A powder metallurgy based manufacturing process enables the production of metallic hollow spheres of defined geometry (Jäckel, 1987). This technology brings a significant reduction in costs in comparison to earlier applied galvanic methods and all materials suitable for sintering can be applied. EPS (expanded polystyrol) spheres are coated with a metal powder-binder suspension by turbulence coating. The green spheres produced can either be sintered separately to manufacture single hollow spheres or be pre-compacted and sintered in bulk (cf. Fig. 5.2) thus creating sintering necks between adjacent spheres (Jäckel, 1983).

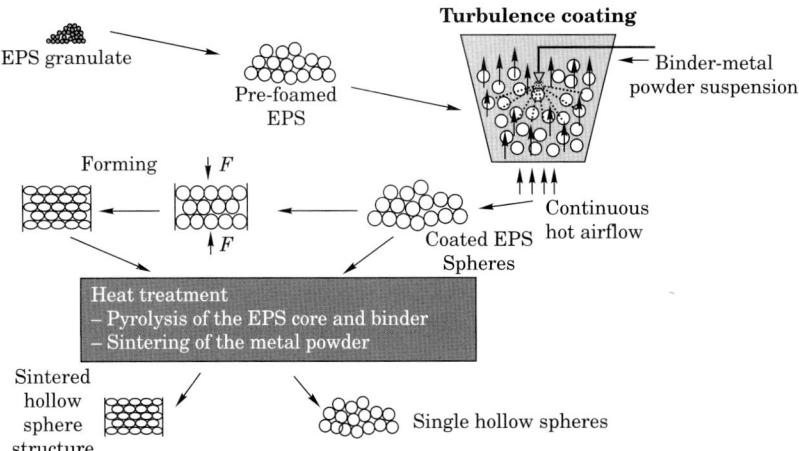

Fig. 5.2. Manufacturing processes of single hollow spheres and hollow sphere structures (Glatt GmbH, 2007)

Depending on the parameters of the sintering process the micro-porosity of the sintered cell wall can be adjusted. In a subsequent debindering process, the EPS spheres are pyrolised. The increase of the carbon content of the sintered metal by the diffusion of the incinerated binder and polymer causes degradation of mechanical properties and corrosion resistance. Special reduction processes are required to reduce this effect (Studnitzky and Andersen, 2005). Various joining technologies such as sintering, soldering and adhering can be used to assemble the single hollow spheres to interdependent structures (Rousset, Bonino, Blottiere and Rossignol, 1987; Degischer and Kriszt, 2002).

3. PERFORATED HOLLOW SPHERE STRUCTURES (PHSS)

The major idea of introducing a perforation, *i.e.*, holes of circular cross section in the sphere shells, is to make the inner sphere surface and volume usable. In this work, the holes are defined in such a way that the largest possible hole can be located between the sintering areas in a primitive cubic arrangement. In subsequent steps the size of this initial hole was gradually reduced for the models. Our modelling approach is restricted to the simple case where the holes do not intersect with the sintering and link area. The additional inner surface may be used for chemical reactions in the case of catalysts or the additional inner volume may have a positive influence on the dissipation of *e.g.* acoustic waves. A simplified arrangement of perforated hollow sphere structures in a primitive cubic pattern is shown in Fig. 5.3. The darker grey volumes represent the connecting elements ('links') which are formed in liquid phase processes such as soldering or adhesively bonding, [cf. Figs. 5.3 (*a*), (*b*)]. In the case of sintered structures, the sphere are slightly deformed and a flat sintering area connects different spheres, [cf. Figs. 5.3 (*c*), (*d*)].

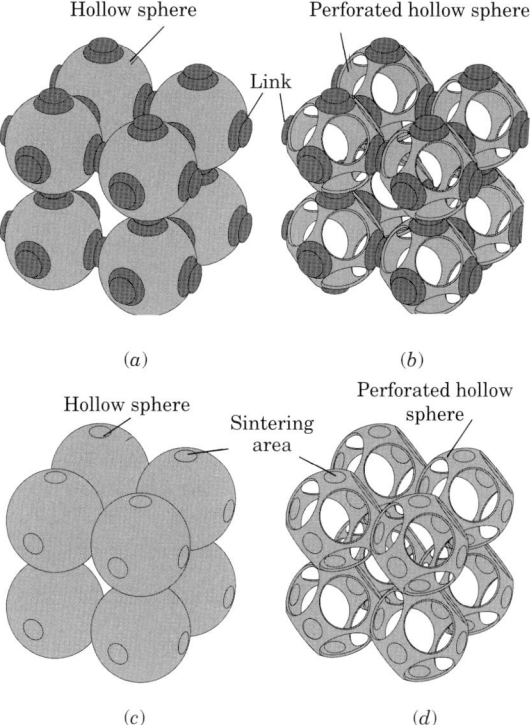

Fig. 5.3. Schematic representation of primitive cubic sphere arrangements: (*a*) classical hollow sphere structure with links; (*b*) perforated hollow spheres structure (PHSS) with links; (*c*) sintered classical hollow sphere structure; (*d*) sintered perforated hollow spheres structure

Looking at a specific example where perforated spheres with links (outer diameter 2.7 mm and shell thickness of 0.1 mm, outer hole radius 0.68 mm) are considered, the available free surface increases by 57% and the free volume increases by 259% while the bulk volume and the weight is reduced by 33%. In addition, we can state that the porosity is only increased by 3.9%.

4. MODELLING OF HOLLOW SPHERE STRUCTURES

Different sets of finite element models are used to investigate the difference between perforated and entire spheres (cf. Figs. 5.3, 5.4) and to investigate the influence of two types of joining techniques on the heat conductivity. To this end, a geometric characterization of the materials is required. By means of an image processing software, a series of micrographs was analysed and geometrical values for different model structures were derived (Veyhl et al., 2008), (cf. Figs. 5.4, 5.5, 5.10 and Table 5.1).

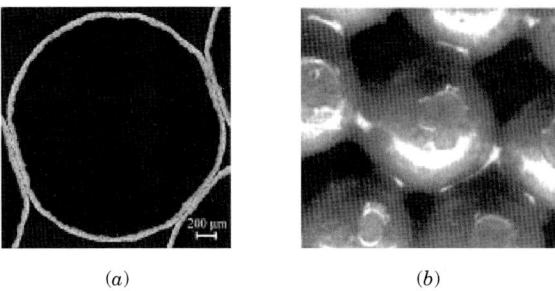

Fig. 5.4. Photos of hollow sphere structures: (a) sintered (Veyhl, 2008); (b) adhesively bonded (Fiedler, 2007)

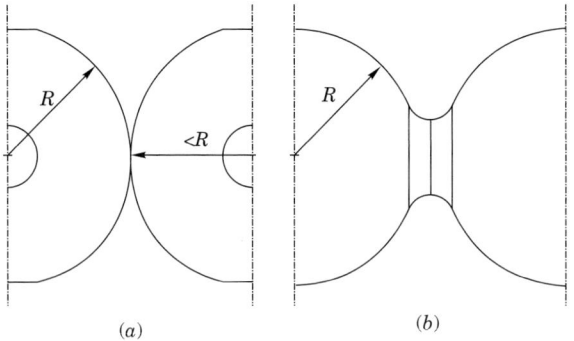

Fig. 5.5. Schematic sketch of different joining approaches: (a) sintering; (b) neck region, e.g. epoxy

In order to decrease the complexity of the calculation model and to minimise computational hardware requirements, the so-called unit-cell approach is chosen where the computational structure is reduced from a larger or infinite amount of randomly arranged spheres to a single unit-cell which is commonly based on typical space groups as in crystallography (De Graef and McHenry, 1997). This modeling approach is commonly used in finite element analysis and can give reasonable results.

The unit cell may be based on primitive cubic, hexagonal, face-centered or body-centered arrangements (Sanders and Gibson, 2003), (cf. Figs. 5.6, 5.7).

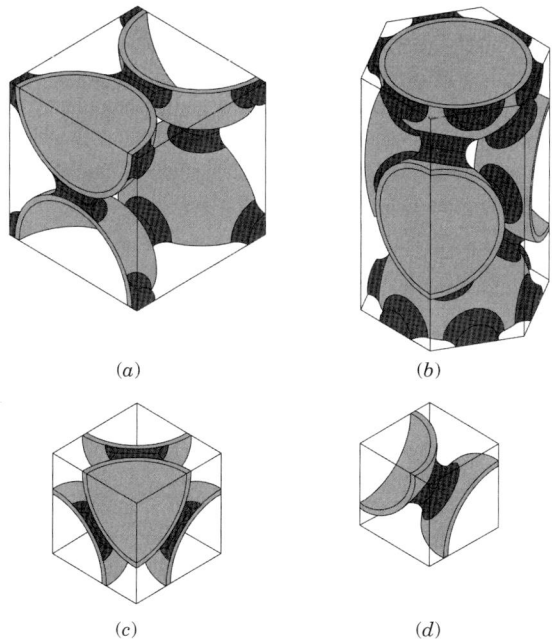

Fig. 5.6. Unit cell models for adhesively bonded spheres: (a) primitive cubic (entire UC); (b) hexagonal (entire UC); (c) face-centered cubic (one-eighth of UC); (d) body-centered (one-eighth of UC)

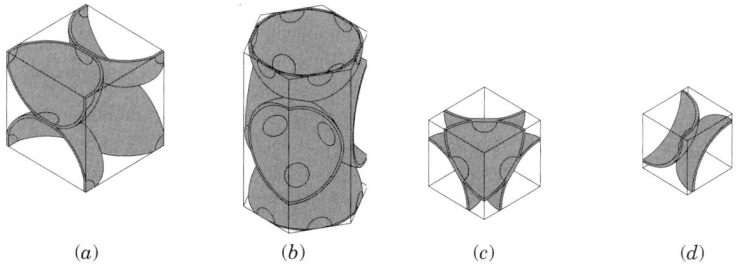

Fig. 5.7. Unit cell models for sintered spheres: (a) primitive cubic (entire UC); (b) hexagonal (entire UC); (c) face-centered cubic (one-eighth of UC); (d) body-centered (one-eighth of UC)

5. FINITE ELEMENT APPROACH: GEOMETRY, MESH, BOUNDARY CONDITIONS AND MATERIALS

Finite element analysis demands the discretisation of the geometry by subdivision into geometrically simple finite elements. Within this work, only three-dimensional geometries are considered, since such an approach gives the highest accuracy compared to models with reduced dimension (*e.g.*, 2D). For the three dimensional meshing, both tetrahedral or hexahedral elements can be employed. However, investigations (Benzley *et al.*, 1995; Fiedler *et al.*, 2006) have shown that hexahedral elements yield superior performance. Therefore, the geometry of the structures is discretised based on *regular* hexahedral elements. This approach is much more time-consuming, but it is important in order to achieve a more accurate simulation.

The *regular* meshing of the micro-structure of MHSS requires the decomposition of these complex geometries into simple sub-geometries. Based on the obtained fragments, meshing algorithms can be applied in order to obtain regular hexahedral meshes (Fiedler, 2007). These sub-geometries have to exhibit a cubical shape. Figure 5.8 demonstrates the decomposition of a syntactic face centered cubic geometry.

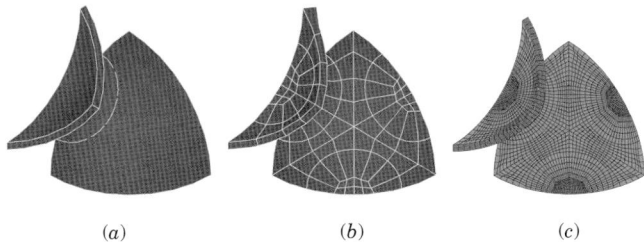

(a) (b) (c)

Fig. 5.8. Generation of the computational FCC model (one twenty fourth of UC): (*a*) CAD geometry; (*b*) subdivision of the solid geometry in simple parts; (*c*) entire mesh

In order to minimise the number of fragments, symmetries are exploited. In the example shown, only 1/24 of the final model sub-geometries needs to be meshed and, by mirroring the rest of the fragments are obtained which accumulate to the target geometry (cf. Fig. 5.9).

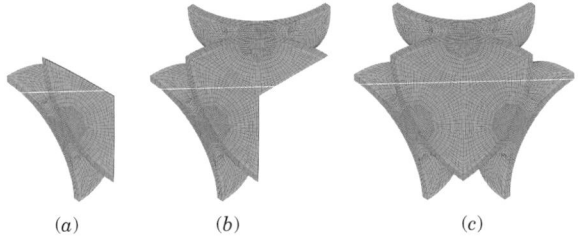

(a) (b) (c)

Fig. 5.9. Generation of the computational FCC model: (*a*) one twenty fourth of UC; (*b*) one-twelfth of UC; (*c*) one-eighth of UC

Two typical finite element meshes including boundary conditions are shown in Fig. 5.10. Constant temperature boundary conditions (T_i = const.) are prescribed at the faces of the connective elements on the left and right side of the model. In order to generate a heat flux through the structure, only $T_1 \neq T_2$ must be fulfilled.

Table 5.1. Dimensions of the cubic unit cells, (cf. Fig. 5.10)

Dimension	Value [mm]
R	1.35
D_1	2.94
D_s	2.66
a	0.12
b_1	0.8
b_s	0.6
d_1	1.36
d_s	1.47
t	0.1

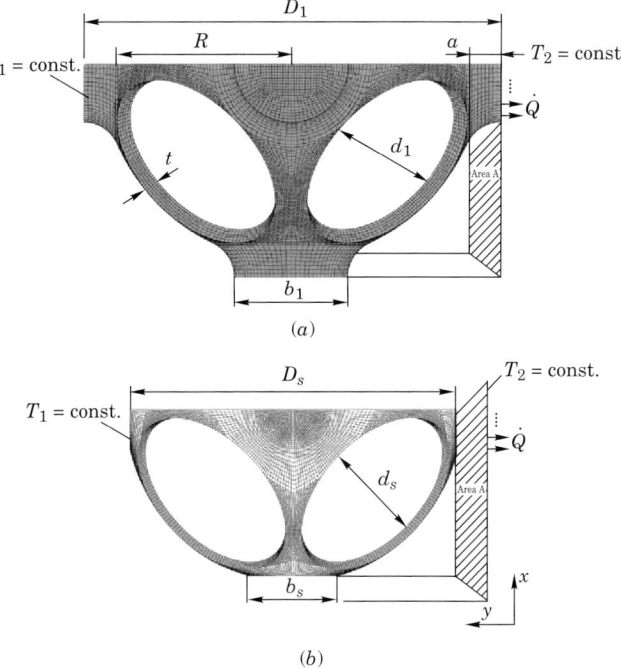

Fig. 5.10. Finite element meshes and boundary conditions of perforated primitive cubic unit cell: (*a*) adhesively bonded; (*b*) sintered

According to the assumed symmetry and simplifications (no radiative or convective heat transfer), the ratio of the heat flux perpendicular to all remaining surfaces is zero. The thermal properties of the considered base materials are shown in Table 5.2.

Table 5.2. Material properties (Fiedler, 2007; Habenicht, 2002; The Japan Carbon Fiber Association Website, 2009)

Material	Conductivity [W/(mm · K)]	Density [kg/dm³]
Steel	0.05	6.95
Epoxy resin	0.000214	1.13
Aluminium	0.232	2.67
CFRP	0.007	—

The evaluation of the effective thermal conductivities within the finite element approach is based on Fourier's law Eq. (1) where the area-related conductivity is defined by

$$k = -\frac{\dot{Q}}{A}\frac{\Delta y}{\Delta T} = -\dot{q}\frac{\Delta y}{\Delta T} \qquad ...(1)$$

The area A of the unit cell and the spatial distance Δy are given by the geometry (cf. Fig. 5.10 and Table 5.1), respectively the temperature gradient $\Delta T = T_2 - T_1$ by the boundary conditions. Only the heat flux $\dot{Q}$ remains to be determined. This is done in the scope of the finite element method by summing up all nodal values of the reaction heat flux with a user subroutine at the left or right face where a temperature boundary condition is prescribed. This approch determines the effective thermal conductivity in direction of a principal axis of the cubic arrangment. Within the relevant temperature range, the contribution of thermal radiation to the heat transfer is low (Lu and Chen, 1999). Furthermore, contributions from gaseous conduction and convection are neglected within this initial study.

The whole model consists of 27292 elements in the case of the model without hole in a primitive cubic arrangment with links. Subsequent models were generated where the hole diameter was increased by 25, 50, 75 and 100% for both, i.e., sintered and with link, models. Figure 5.11 summarises the influence of mesh density on the calculated results for the thermal conductivity. One can observe that by increasing the mesh density, the results for the normalised conductivity (related to the thermal conductivity of the base material k_s) are converging to a stable value. On the other hand the calculation time increases with the number of elements and according some kind of compromise between accuracy and calculation time must be done. Thus, as indicated in Fig. 5.11, a choice of 27292 elements is reasonable to calculate the thermal conductivity.

The geometrical parameters of the joining element (cf. parameters a and b_1 in Fig. 5.10) where modified to investigate the influence of a changing link geometry. All the simulations were done with the commercial finite element package MSC. Marc (MSC Software Corporation, Santa Ana, CA, USA).

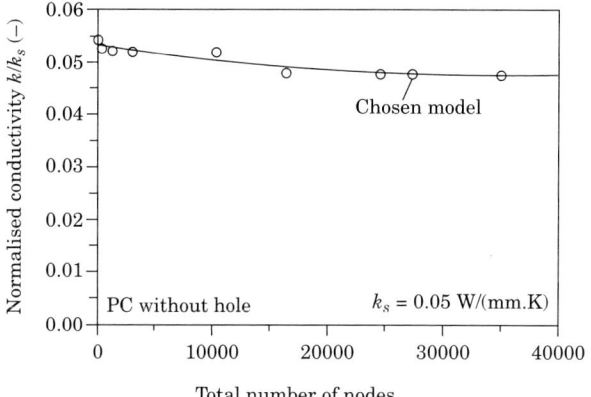

Fig. 5.11. Results of mesh refinement analysis

6. AVERAGE VALUES FOR CELLULAR MATERIALS

In the case of cellular materials, physical properties are commonly described as a function of their relative density (Gibson and Ashby, 1997)

$$\rho_{rel} = \frac{\rho}{\rho_{So}} \qquad ...(2)$$

where ρ is the density of the cellular material and ρ_{So} is the density of the solid material from which the cells are made. Scaling relations or power-laws for different physical properties which account for the density dependency can be found e.g. in Ashby et al. (2000). Since the mass of the solid material and the mass of the cellular material itself is identical, the RHS of Eq. (2) can be divided by the mass to express the relative density in terms of volumes as:

$$\rho_{rel} = \frac{V_{So}}{V_{UC}} \qquad ...(3)$$

The value of Eq. (3) is also named the relative solid volume where V_{So} is the volume of the solid material (e.g., material of the cell walls, spheres, bond material etc.) and V_{UC} is the volume of the unit cell.

It should be noted that the volume of the unit cell (UC) can be expressed as the sum of solid and free volume as:

$$V_{UC} = V_{So} + V_{free} \qquad ...(4)$$

Looking at hollow sphere structures as shown in [Figs. 5.3 (a), (b)] one can see that the solid volume is composed of a contribution from the spherical shells (index 'Sp') and the matrix (index 'link'). Thus, the total volume of a unit cell in the case of hollow sphere structures can be further distinguished into:

$$V_{UC} = V_{Sp} + V_{link} + V_{free} \qquad ...(5)$$

Assuming that the densities of the matrix (bond) material and the sphere wall material are identical, i.e., $\rho_{link} = \rho_{Sp} = \rho_{So}$, the relative density is equal to:

$$\rho_{rel} = \frac{\rho}{\rho_{So}} = \frac{V_{link}}{V_{UC}} + \frac{V_{Sp}}{V_{UC}} \qquad \ldots(6)$$

In the case that the sphere wall material and the matrix (or bond) material are different (*e.g.* aluminium and epoxy), the indication of a *relative* density is no longer possible and an *average* density ρ can be used instead:

$$\rho = \frac{V_{link}}{V_{UC}} \cdot \rho_{link} + \frac{V_{Sp}}{V_{UC}} \cdot \rho_{Sp} \qquad \ldots(7)$$

A further important field is the comparison between experimental tests with a rather random arrangement of the spheres and the simulation results with a periodic arrangement of the spheres (Veyhl, 2008). Real hollow sphere structures show a rather random arrangement of the spheres. On the other hand, most modelling (*e.g.*, within this chapter) approaches are based on periodic geometries and structures which possess different densities such as primitive cubic (PC), body-centered cubic (BCC), face-centered cubic (FCC) and hexagonal closest (HC) (cf. Figs. 5.6, 5.7). Knowing the density of the real structure, one may assign the proper topology for modelling the structure or use the result to interpolate model calculations based on the simplified periodic structures (Gibson and Ashby, 1997). In the following, results are plotted over the average density to be comparable with the experimental results which are currently under development.

7. RESULTS

The influence of the wall thickness and perforation radius d_s on the thermal conductivity of HSS and PHSS is shown in the Fig. 5.12. A distinct decrease

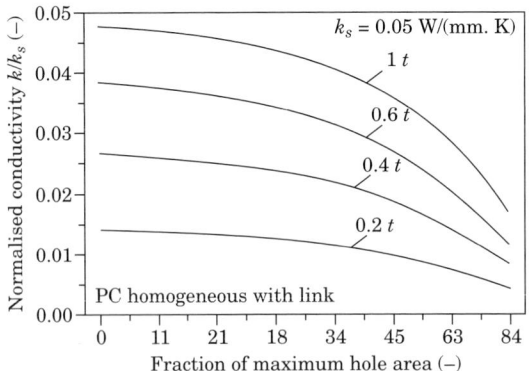

Fig. 5.12. Normalised effective thermal conductivity as a function of hole area fraction (the maximum area is $A = 9.07$ mm^2; [cf. Fig. 5.10 (a)]) and sphere shell thickness t (primitive cubic arrangement of sintered sphere with joining element). A fraction of the hole area equal to zero corresponds to the configuration without perforation

of the thermal conductivity with increasing area of the perforated hole can be observed. In a similar trend, decrease of the wall thickness also reduces the thermal conductivity and shifts the perforation diagram downwards.

In addition, the influence of changing hollow spheres' wall thicknesses has been studied for models without link (sintered). When the wall thickness changes from 0.1 mm to 0.07 mm, also the thermal conductivity is decreased. Table 5.3 summarises this reduction for different configurations.

Table 5.3. Reduction of thermal conductivity when the wall thickness of the hollow sphere is reduced from 0.1 mm to 0.07 mm for different hole diameters ($1d$ is the case where the largest possible hole is located between the sintering areas with $d_s = 1.47$ mm, cf. Fig. 5.10)

Arrangement Hole size	FCC	PC	BCC	Hex
$0.0d_s$	31.96%	31.67%	31.8%	28.74%
$0.25d_s$	31.93%	31.76%	35.13%	37.5%
$0.5d_s$	31.86%	32.12%	33.16%	38.10%
$0.75d_s$	40.66%	32.38%	40.64%	32.49%
$1.0d_s$	29.55%	33.10%	53.16%	40.36%

Figures 5.13, 5.14 and 5.15 show the normalised conductivity over the average density for different configurations. In Fig. 5.13 the results are plotted for sintered structures (without connective elements, [cf. Fig. 5.10 (b)]. Figure 5.14 shows the effective thermal conductivities for structures connected by linking elements and homogeneous material properties. Finally, Fig. 5.15 represents the case of adhesively bonded structures with linking elements and non-homogeneous material properties.

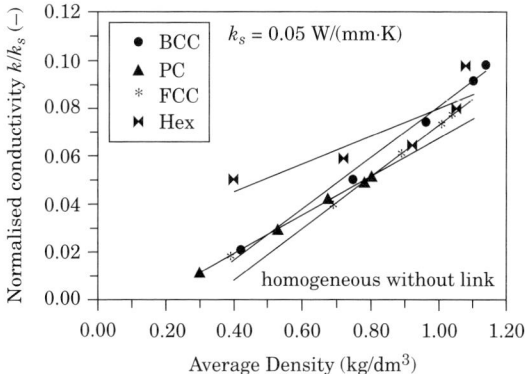

Fig. 5.13. Normalised effective thermal conductivity as a function of relative density for different hole arrangements in the case of sintered spheres without link

In all these figures, it can be seen that the heat conductivity increases with increasing average density (the hole diameter is decreasing). This trend is stronger in the case of homogenous structures. The dependency of the thermal conductivity on the average density is shown in Fig. 5.13 and Fig. 5.14, which are redundant, the homogenous models. In addition, the dependency on the different types of arrangements can be observed. It becomes clear that in all cases (PC, BCC and FCC) the thermal conductivity can be approximately expressed as a linear function of the relative density. That means that the thermal conductivity is much more influenced by the average density than by different arrangements in the case of homogenous structures with and without linking elements.

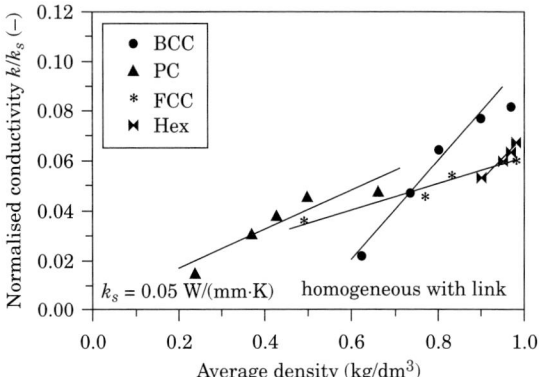

Fig. 5.14. Normalised effective thermal conductivity as a function of relative density for different hole arrangements in the case of homogeneous adhesively bonded spheres

In a different way it is clear from Fig. 5.15 that the influence of the shape arrangement on the thermal conductivity in the case of non-homogenous configurations is of great importance. In comparison, only a weak dependency on the average density can be observed. The maximum values are observed for the hexagonal arrangement. Body centred and face centred cubic arrangements yield similar values.

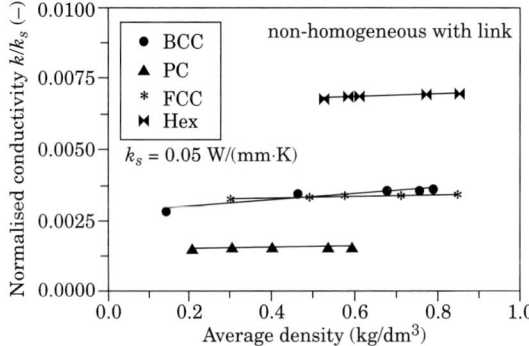

Fig. 5.15. Normalised effective thermal conductivity as a function of relative density for different hole arrangements in the case of non-homogeneous adhesively bonded spheres

Figure 5.16 compares the influence of the joining element on the effective thermal conductivity for homogeneous and non-homogeneous sphere structures. To this end, the radius of the cross sectional area b_1 and the thickness of the connective element a are systematically varied. Compared to the dependence on the average density, a change of the geometry of the connective element has only minor impact on the thermal properties of the structure in the case of homogeneous spheres. However in the case of the non-homogeneous structures the behaviour is contrary: The shape of the curve [cf. Fig. 5.16 (b)] is dominated by the geometry of the joining element while the influence of the hole diameter is marginal. It should be noted here that the values of the conductivity were normalised in all cases, i.e., homogeneous and non-homogeneous, with the sphere shell material (steel).

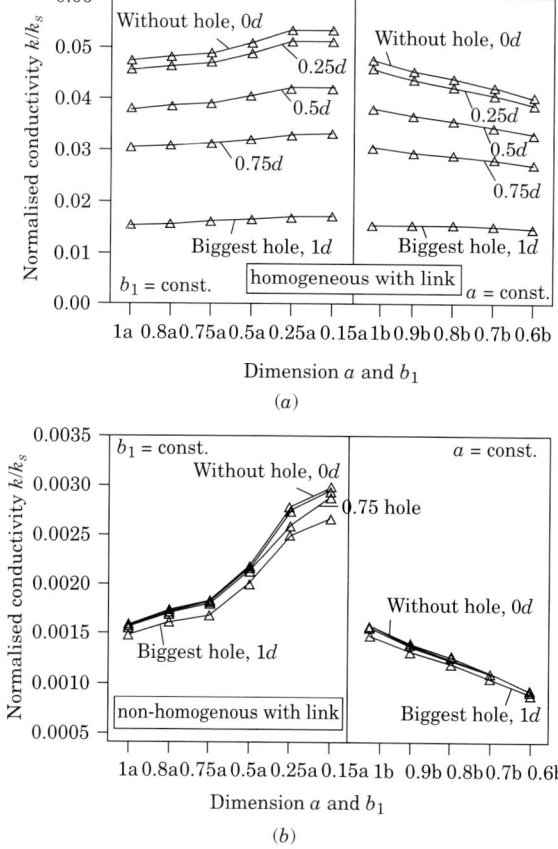

Fig. 5.16. Normalised conductivity of HSS as a function of link geometry ($a = 0.12$ mm, $b_1 = 0.4$ mm) for several hole diameter ranging from $1d = 1.36$ mm (biggest hole) till $0d$ (without hole), (cf. Fig. 5.10)

8. APPLICATION EXAMPLE: SANDWICH STRUCTURE

8.1. Sandwich Structures

Sandwich structures are essential machine elements in lightweight construction. In particular, aluminium sandwich construction has been recognised as a promising concept for structural design of lightweight transportation systems such as aircraft (Paik et al., 1999; Seibert, 2006), high-speed trains (Seibert, 2006; Belingardi et al., 2003) and fast ships (Kim et al., 2007; Knox et al., 1998). The definition of a sandwich structure is a composite where a core material is enclosed by two or more layers. The basic principle of sandwich structures in structural mechanics is for two strong face sheets to bear the applied loads whilst the core acts as a spacer which retains the face sheets in position by carrying shear stresses. Therefore, the core must be stiff enough in the direction perpendicular to the faces to ensure that they remain the correct distance apart and exhibit sufficient shear stiffness to ensure that when the panel is bent the faces do not slide over each other. The core material does not need to reach the mechanical performance of the face sheets and typically low density materials are applied (Reuterläov, 2002). Nowadays, the industrial standard for sandwich cores are honeycomb structures which exhibit excellent stiffness at very low densities. However, the processing of honeycomb structures, especially as a core between curved face sheets, is complex and therefore increases the manufacturing costs. Furthermore, honeycomb structures possess poor resistance to contact and impact loads (Allen, 1969; Silva and Gibson, 1997; Wu and Jiang, 1995) and exhibit anisotropic properties. Thus, there is a necessity for new innovative core materials. An experimental investigation of thermal contact resistance of an aluminium honeycomb (made of A13104-H19 and A13003-H16) sandwiched by two aluminium blocks was conducted by Yeh (2003) with the honeycomb specimens aligned in either axial or lateral orientations. Results show that due to the anisotropic nature in heat conduction and the close contact provided by bolted joints, the total thermal conductance of the axial honeycomb is greater than that of the honeycomb in the lateral orientation under the condition with the same specimen height. An increase of either the cell diameter or specimen height of honeycombs leads to a decrease of the axial total conductance.

In [Fig. 5.17 (a)] a M-Pore® aluminium sponge is used as core material. This commercially available material is characterised by very high porosity and therefore low density ($\rho = 0.27$ g/cm³). Figure 5.17 (b) shows the Alporas® aluminium foam with closed cells and a slightly higher density ($\rho_c = 0.34$ g/cm³) but also improved mechanical properties. In [Figs. 5.17 (c), (d)] partial ($\rho_c = 0.3$, 0.6 g/cm³) and syntactic MHSS–cores ($\rho_c = 0.75$, 1.2 g/cm³) are shown. Preliminary tests have shown that these material are able to compete with classical honeycomb structures ($\rho_c = 0.9$ g/cm³, [cf. Fig. 5.17 (e)]. The benefits of cellular metal as sandwich cores, relative to competing concepts, arise primarily in curved configurations where the isotropy of the material is advantageous (Wu and Jiang, 1997; Evans et al., 1999; Ashby et al., 2000). Due to the multi-functional properties of cellular metals the selection of the optimum core material requires the consideration of *all* characteristics with relevance to the intended application. Furthermore, the potential utilisation in safety relevant fields puts great demands on the predictability and stability of their properties.

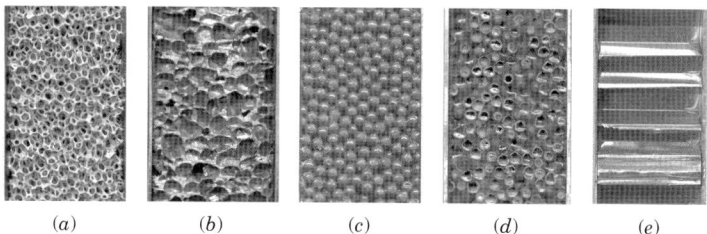

Fig. 5.17. Sandwich structures containing different core materials: (a) M-Pore®, (b) Alporas®, (c) Partially adhesively bonded MHSS, (d) Syntactic adhesively bonded MHSS, (e) Honeycomb structure (Fiedler, 2007)

In the scope of this work, MHSS and PHSS are analysed under the viewpoint of their applicability as core materials in sandwich panels for thermal insulation. Special focus is put on adhesively bonded HSS and PHSS. This joining technology allows flexible processing and the adjustment of the properties of the composite by the selection of different adhesives. Numerical and experimental analyses on their mechanical and thermal properties (non-perforated HSS) are performed in Fiedler (2007).

8.2. Heat Transfer in a Sandwich Structure with MHSS and PHSS Cores

In the following, sandwich panels with MHSS and PHSS cores in a bonded structure with a body-centered arrangement, acting as a thermal insulating layer are considered. In the case of perforated hollow spheres the hole diameter is $0.73d_s$. The effective thermal conductivity is determined in the direction of the normal vector of the face sheets. The temperatures prescribed at the upper and lower surfaces are 100 K and 200 K, respectively. The microstructure of the MHSS and PHSS are homogenised and therefore represented by plane rectangular elements in order to reduce the required calculation time. The deviation introduced by this simplification is small. A typical realisation of advanced sandwich panels is shown in Fig. 5.18 where aluminium and carbon fibber reinforced plastic are adhesively bonded to a HSS core material. Figure 5.19 summarises the results of this investigation for sandwich structures with such aluminium panels and CFRP panels (cf. Fig. 5.18).

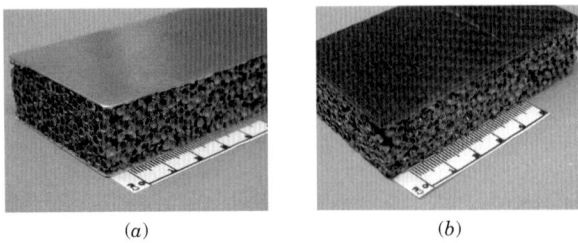

Fig. 5.18. Advanced sandwich structures with cellular core material (HSS): (a) Aluminium plates as face sheets; (b) carbon fiber reinforced plastics as face sheets; (© by Glatt GmbH, Dresden, Germany)

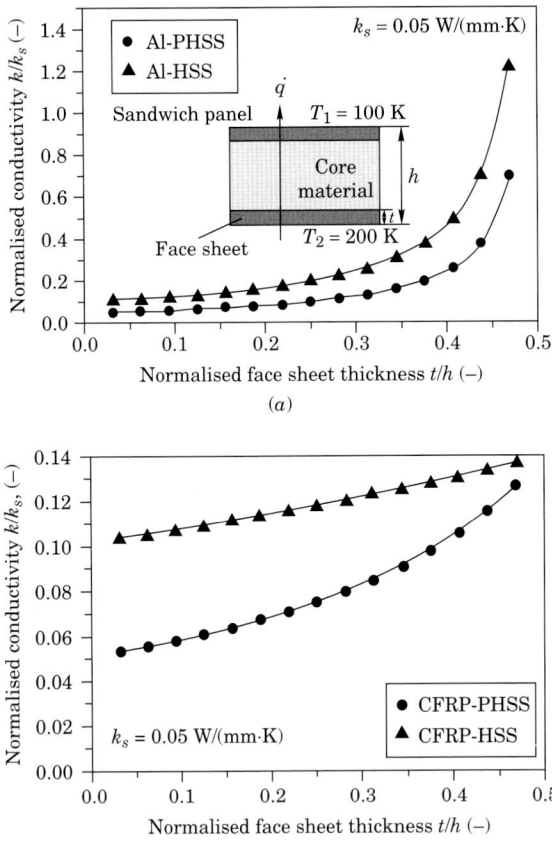

Fig. 5.19. Effective normalised thermal conductivity of sandwich panels with hollow sphere cores in dependence on the normalised face sheet thickness: (a) Aluminium face sheets; (b) carbon fiber reinforced plastic sheets. The homogenised values for the core are based on a sintered FCC configuration (configurations without hole (HSS) and with holes (PHSS) have been investigated; the average density in case of PHSS is equal to 0.00141)

The effective thermal conductivity is plotted versus the normalised face sheet thickness. This ratio is equal to the varying thickness t of the face sheets divided by the constant total height $h = 32$ mm of the structure and is defined for values between 0 (pure core material) and 0.5 (no core material, face sheets merge) (Fiedler, 2007).

The thermal conductivity of the sandwich structure increases with increasing relative thickness of the face sheets for both HSS and PHSS. This phenomenon can be explained with the high thermal conductivity of the aluminium alloy (cf. Table 5.2) in comparison to the insulating MHSS and PHSS core material. Even in the case of very thin insulating layers of PHSS ($t/h = 0.4$), the thermal conductivity of the sandwich panel only reaches approx. 5% of the values of the face sheet material. However, for further decrease of the thickness of the core, the thermal conductivity of the structure grows exponentially. In the case of the carbon fiber reinforced plastic sheets, much lower effective values are obtained since the CFRP plates have an isolating character, (cf. Table 5.2).

9. CONCLUSIONS

In the scope of this chapter, the thermal conductivity of classical and perforated hollow sphere structures was investigated. Effective values were derived based on the finite element method. Strong simplifications with respect to the geometry (only periodic model structures) and the physical effects (neglecting of any fluid flow, no radiation) were accepted to get a first estimate for the influence of the perforation on the thermal conductivity.

Future investigations will include more effects, such as random pore arrangements, or fluid flow through the open structure to derive more realistic properties which may lead to new applications of PHSS.

REFERENCES

1. Allen, H.G. 1969. Analysis and Design of Structural Sandwich Panels. Pergamon Press.
2. Ashby, M.F., Evans, A., Fleck, N.A., Gibson, L.J., Hutchinson, J.W. and Wadley, H.N.G. 2000. Metal Foams: A Design Guide, Butterworh-Heinemann.
3. Baumeister, E., Klaeger, S. and Kaldos, A. 2004. Lightweight, Hollow-Sphere-Composite (HSC) Materials for Mechanical Engineering Applications, *J. Mater. Process. Tech.* **155**: 1839–46.
4. Belingardi, G., Cavatorta, M.P. and Duella, R. 2003. Material Characterization of a Composite-Foam Sandwich for the Front Structure of a High Speed Train, *Compos. Struct.* **61**: 13–25.
5. Benzley, S.E., Perry, E., Merkle, K., Clark, B. and Sjaardema, G.F. 1995. A Comparison of All-Hexagonal and All-Tetrahedral Finite Element Meshes for Elastic and Elastic Plastic Analysis, In Fourth International Meshing Roundtable, Al-Buquerque, New Mexico. pp. 179–91.
6. Boomsma, K., Poulikakos, D. and Zwick, F. 2003. Metal Foams as Compact High Performance Heat Exchangers, *Mech. Mater.* **35**: 1161–76.
7. De Graef, M. and McHenry, M.E. 1997. Structure of Materials. An Introduction to Crystallography, Diffraction and Symmetry, Cambridge University Press.

8. Degischer, H.P. and Kriszt, B. (*Eds.*) 2002. Handbook of Cellular Metals, Wiley-VCH, Weinheim.
9. Evans, A.G., Hutchinson, J.W. and Ashby, M.F. 1999. Multifunctionality of Cellular Metal Systems, *Progress in Mater. Sci.* **43**: 171–221.
10. Fiedler, T. 2007. Numerical and Experimental Investigation of Hollow Sphere Structures in Sandwich Panels, PhD Dissertation, University of Aveiro, Portugal.
11. Fiedler, T. and Öchsner, A. 2007a. Influence of the Morphology of Joining on the Heat Transfer Properties of Periodic Metal Hollow Sphere Structures, *Mat. Sci. Forum* **553**: 45–50.
12. Fiedler, T. and Öchsner, A. 2007b. On the Thermal Conductivity of Adhesively Bonded and Sintered Hollow Sphere Structures (HSS), *Mat. Sci. Forum* **553**: 39–44.
13. Fiedler, T., Solórzano, E. and Öchsner, A. 2008a. Numerical and Experimental Analysis of the Thermal Conductivity of Metallic Hollow Sphere Structures, *Mater. Lett.* **62**: 1204–7.
14. Fiedler, T., Öchsner, A., Belova, I.V. and Murch, G.E. 2008b. Recent Advances in the Prediction of the Thermal Properties of Syntactic Metallic Hollow Sphere Structures, *Adv. Eng. Mater* **10**: 361–5.
15. Fiedler, T., Sturm, B., Öchsner, A., Gracio, J. and Kuhn, G. 2006. Modelling the Mechanical Behaviour of Adhesively Bonded and Sintered Hollow Sphere Structures, *Mech. Compos. Mater.* **42**: 559–70.
16. Gibson, L.J. and Ashby, M.F. 1997. Cellular Solids: Structures & Properties, Cambridge University Press, Cambridge.
17. Glatt GmbH, 2007. Binzen, Germany, Private Communication.
18. Habenicht, G. 2002. Kleben. Grundlagen, Technologien, Anwendungen. Springer-Verlag
19. Jäckel, M. 1983. *German Patent*, **210(3)**: 770.
20. Jäckel, M. 1987. *German Patent*, **724(3)**: 156.
21. Kim, J.S., Lee, S.J. and Shin, K.B. 2007. Manufacturing and Structural Safety Evaluation of a Composite Train Carbody, *Compos. Struct.* **78**: 468–76.
22. Knox, E.M., Cowling, M.J. and Winkle, I.E. 1998. Adhesively Bonded Steel Corrugated Core Sandwich Construction for Marine Applications, *Mar. Struct.* **11**: 185–204.
23. Lu, K.T. and Kou, H.S. 1993. The Analytical Solution of Mixed Convection in a Vertical Channel Embedded in Porous-Media with Asymmetric Wall Heat Fluxes, *Int. Commun. Heat Mass.* **20**: 489–500.
24. Lu, T.J. and Chen, C. 1999. Thermal Transport and Fire Retardance Properties of Cellular Aluminium Alloys, *Acta Mater.* **47**: 1469–85.
25. Lu, T.J., Stone, H.A. and Ashby, M.F. 1998. Heat Transfer in Open-Cell Metal Foams, *Acta Mater.* **46**: 3619–35.
26. Lu, W., Zhao, C.Y. and Tassou, S.A. 2006. Thermal Analysis on Metal-Foam Filled Heat Exchangers. Part I: Metal-Foam Filled Pipes, *Int. J. Heat Mass. Transf.* **49**: 2751–61.
27. Paik, J.K., Thayamballi, A.K. and Kim, G.S. 1999. The Strength Characteristics of Aluminum Honeycomb Sandwich Panels, *Thin Wall Struct.* **35**: 205–31.
28. Reuterläov, S. 2002. Cost Effective Infusion of Sandwich Composites for Marine Applications, *Reinforced Plast.* **46**: 30–34.
29. Rousset, A., Bonino, J.P., Blottiere, Y. and Rossignol, C. 1987. *French Patent*, **707(8)**: 440.
30. Sanders, W.S. and Gibson, L.J. 2003. Mechanics of BCC and FCC Hollow-Sphere Foams, *Mat. Sci. Eng. A-Struct.* **352**: 150–61.
31. Schneider, L., Boehm, A., Korhammer, C., Scholl, R., Voigtsberger, B. and Stephani, G. 2002. US Patent 6501784.

32. Seibert, H. 2006. Applications for PMI Foams in Aerospace Sandwich Structures, *Reinforced Plst.* **50**: 44–48.
33. Silva, M.J. and Gibson, L.J. 1997. The Effects of Non-Periodic Microstructure and Defects on the Compressive Strength of the Two-Dimensional Cellular Solids, *Int. J. Mech. Sci.* **39**: 549–63.
34. Studnitzky, T. and Andersen, O. 2005. Cellular Metals and Polymers, Trans Tech Publications.
35. The Japan Carbon Fiber Association Website 2009. http://Www.Carbonfiber.Gr.Jp/English.
36. Veyhl, C. 2008. Master Dissertation, University of Applied Science Aalen, Germany
37. Veyhl, C., Winkler, R., Merkel, M. and Öchsner, A. 2008. Structural Characterisation of Diffusion-Bonded Hollow Sphere Structure, *Defect Diffus. Forum* **280–281**: 85–96.
38. Wu, E. and Jiang, S. 1997. Axial Crush of Metallic Honeycombs, *Int. J. Impact Eng.* **19**: 439–56.
39. Wu, E. and Jiang, W.S. 1995. Crush of Honeycombs Contact and Impact Loads, *Proc. of the 10th Internat. Conf. on Composite Materials* **5**: 567–74.
40. Yeh, C.L., Chen, Y.F., Wen, C.Y. and Li, K.T. 2003. Measurement of thermal contact resistance of aluminum honeycombs, *Exp. Therm. Fluid Sci.* **27**: 271–81.

6

Food Dehydration under Forced Convection Conditions

A. Mulet[*], J.A. Carcel, N. Sanjuan and J.V. Garcia-Perez

ABSTRACT

Dehydration constitutes a common and high energy demanding operation on food processing. Reduction of water concentration contributes to preserve the food, preventing the degradation during storage and transportation. In addition, dehydration reduces the transport cost and is a common previous stage in extraction processes, avoiding the interference of water and decreasing the need for solvent. Dehydration by using hot air has been and still is the most common way at industrial scale. This work reviews the main features of this drying method, as well as it also presents new trends and modern technologies. The high importance of the relationship between water activity and moisture content is shown, not only to identify optimal storage conditions or final moisture contents on the dried product but also as a tool to be used on modeling. The identification of significant resistances to mass transfer is mandatory in order to design adequately the drying process and address accurately modeling. Drying modeling from mechanistic theories involves considering some assumptions, which will affect not only to the accuracy of the model but also to the solution method in terms of complexity. Modeling results may be used in optimization. Actually, drying optimization objectives are based on quality as well as energy aspects, although always must be kept in mind that the main goal of drying is preservation. Despite drying by forced air has been widely addressed in literature, it still present some limitations, specially a low drying rate affecting the energy efficiency. The application of new technologies as additional energy sources may be considered, in this sense, a new drying technology is introduced, the ultrasonic assisted drying.

---

[*] Grupo de Analisis y Simulacion de Procesos Agroalimentarios, Departamento de Tecnologia de Alimentos, Universidad Politecnica de Valencia, Cami de Vera s/n, 46022, Valencia-SPAIN.
[*] *Corresponding author* : E-mail : amulet@tal.upv.es

NOTATION

A	area
a_w	water activity
C	water concentration
C_G	parameter of GAB equation
D_e	effective moisture diffusivity
E_0	pre-exponential Arrhenius factor
E_a	activation energy
h	external heat transfer coefficient
K_G	parameter of GAB equation
L	half thickness
N_A	mass flux
Q	heat flux
Q_{sn}	net isosteric heat of sorption
R	universal gas constant
T	temperature
t	time
W	average moisture content
x	characteristic direction of heat and mass transfer
Ψ	dimensionless moisture content

SUBSCRIPTS

m	monolayer
s	product's surface
p	local
∞	equilibrium

1. INTRODUCTION

Dehydration is one of the oldest preservation processes available to the mankind, one that we can track since prehistoric times. The main feature of this process consists on lowering the water content in order to avoid or slow down food spoilage by microorganism. Derived from the vocabulary employed, common words found are drying or dehydration, or even dewatering. Dewatering is usually employed for the process of lowering the water content without phase change by using physical means, we can think about dewatering by centrifugation. Drying and dehydration are used commonly in the literature as synonymous, nevertheless as pointed out by Vega-Mercado et al. (2001) in some legislations those words are used to distinguish the level of water content. For instance, the US Department of Agriculture considers dehydrated foods those with water content lower than 2.5% dry basis, and dried those with a higher content. Usually the water content of foods currently consumed will fall in the category of dried.

Lowering the water content can be achieved by different ways and means. When considering taking out water from a solid food one can consider transferring water from the solid to a liquid or to a gas. This gives to raise different methodologies. We will concentrate on a midfield considering the drying medium as being gaseous (air or vapor stream) and considering material being dried (food) as particulate solids. Of course one can dry pulps and purees; the drying of slurries is nevertheless not very common, those being mostly concentrated by evaporation, membrane separation or spray drying.

During drying, there is water transfer from a dense phase (food) to a light phase (air or vapor). Under these circumstances, the transport processes may be considered in both phases, the dense phase being the food and the light phase the drying medium, usually referred as internal and external medium. The associated resistances are commonly known as internal and external, for both heat and mass transfer.

When considering transferring water from a solid to a gas, phase change occurs; this may be considered a high energy demand process unit (Lewicki and Michaluk, 2004). The food industry does not use drying intensive processes like the paper or ceramic industries, however, energy consumption being one of the major concerns in actual society. The impact associated to energy consumption may be tempered by using renewable resources or to improve the energy efficiency by applying new technologies (Cohen and Yang, 1995; Chou and Chua, 2001). These trends are considered by the food industries due to the so called environmentally friendly production as a part of a global concern on quality.

Although the primary objective of drying is preservation, quality aspects are more and more considered (Mayor and Sereno, 2004; Lewicki, 2006), in fact according to the process carried out very different products may be obtained. As it is well known food nutrient degradation, like any other chemical/biochemical reactions depends on temperature. According to the food composition the material is more or less prone to nutrient degradation, the use of different drying technologies for nutrient and in general quality preservation is a must (Chou and Chua, 2001; Achanta and Okos, 2000).

The industrial development of drying dates back to the XVIII century, the equilibrium between the solid water content and the air water content was soon recognized as being of paramount importance for the drying process. Also the long shelf life of foods was linked to their water content. In the last century it was recognized that the important factor was not the water content but the water availability for degradation reactions. Living beings reactions take place on an aqueous medium, thus if water content is lowered those reactions, mainly ageing or microbial spoilage, are disturbed or even interrupted. The water availability was shown to be linked to the water activity concept, this being the equilibrium relative moisture in the gas phase. As a consequence the knowledge sorption isotherm has been and remains an important field of research. Compilations of food isotherms exist (Chirife and Iglesias, 1982; Zhang et al., 1996; Maroulis et al., 2001). A very important step in drying knowledge was the linking of the water activity to the growing of some microorganisms. Some of the aspects so far mentioned will be addressed in this chapter in more detail.

2. WATER ACTIVITY

The water activity concept constitutes one of the main theoretical advancements in the last century. Its importance has been highlighted in the literature for the design and operation of driers as well as for dried food preservation. A main effort was undertaken in Europe for clarifying different aspects on physical properties of foods, among them water activity (Spiess and Wolf, 1983). One of the conclusions of the Cost project was that although many models are available for depicting food isotherms, the GAB (Gugghenheim, 1966; Anderson, 1946; De Boer, 1953) was the one providing the best results. This model has a physical meaning for its parameters, the model can be written in different equivalent forms like the one shown in Eq. (1)

$$W = W_m \frac{C_G K_G a_w}{(1 - K_G a_w)(1 + (C_G - 1) K_G a_w)} \qquad ...(1)$$

where W is the average moisture content, a_w the water activity and W_m represents the monolayer moisture content and C_G and K_G are GAB parameters linked to the heat of sorption in mono and multilayer, respectively. Temperature effect on isotherms may be considered in GAB model from C and K parameters (Garcia-Perez et al., 2008; Mulet et al., 2002).

Water in a fresh tissue is bound to different compounds or hold in different ways, according to its physiology, although once the water molecules are taken out during drying irreversible processes occurs. On rehydration the tissues do not hold water in the same way, this irreversibility is depicted through adsorption/desorption equilibrium curves. It should be noticed that on rehydration (sorption) at the same water content, water activity is higher as consequence water availability is higher and the food is more prone to deterioration. This phenomenon known as hysteresis has not only physiological/chemical/biochemical but also physical grounds (Fig. 6.1). Hysteresis is very important because it states that overdrying not only constitute an energy waste, brings more deterioration to the product because longer drying time, and food stability is also lowered after rehydration when in contact with ambient air.

As previously stated, the water sorption isotherms could be considered as a measurement of water availability at different moisture contents. The binding energy of water to food can be calculated by considering, among other procedures, isotherms measured at different temperatures by using the Clausins-Clapeyron equation (Eq. 2) (Mulet et al., 2002; Simal et al., 2007; Clemente et al., 2009). For that purpose the natural log a_w from

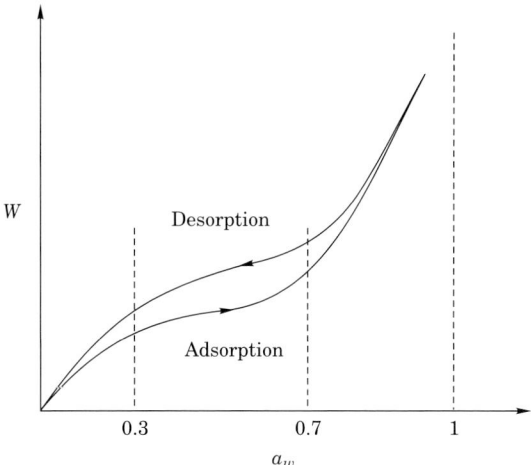

Fig. 6.1. Sorption isotherms

desorption isotherms is plotted versus $1/T$ and from the slope is obtained the net isosteric heat of sorption (q_{sn}). Those plots are carried out at different moisture contents, thus obtaining the variation of the binding energy according to the water content. Due to the fact that small uncertainties in the isotherms greatly affect these calculations, the net isosteric heats of sorption should be considered carefully (Chirife and Iglesias, 1992). Thus, estimations should be compared with experimental values obtained from calorimetric methods (Sanjuan et al., 1994). As an example, is shown in Fig. 6.2 the comparison of isosteric heats of sorption for chufa/tigernuts (*Cyperus esculentus*)

$$\left[\frac{d \ln a_w}{dT}\right]_W = \frac{q_{sn}}{RT^2} \qquad ...(2)$$

where q_{sn} is the net isosteric heat of sorption, T is the absolute temperature and R the universal gas constant.

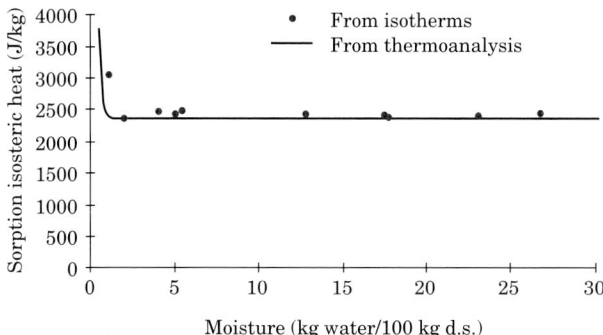

Fig. 6.2. Isosteric heat of desorption in chufa, from equilibrium isostherms and calorimetric data (Sanjuan et al., 1994)

3. INTERNAL AND EXTERNAL RESISTANCE

During drying simultaneously heat and mass transfer take place. For water being transported from the inside to the food surface and for evaporation enough energy should be supplied. As a consequence transport phenomena involve both external and internal resistance to heat and/or mass transfer.

It is frequently assumed that outside the food particle property transport occur by convection and molecular transport inside. Convective transfer is depicted in the most usual way in Table 6.1 (Sherwood et al., 1975; Treybal, 1980).

Table 6.1. External heat and mass transfer

	Heat	Mass
Flux	$Q = hA(T_\infty - T_s)$	$N_A = kA(C_s - C_\infty)$
External Resistance	$\dfrac{1}{hA}$	$\dfrac{1}{kA}$

Internal molecular transport needs more analysis, combining Fick's and Fourier's law and microscopic balance are obtained mass and heat diffusion equations, respectively. These equations are shown in Table 6.1 for infinite slab geometry. The resistance to heat and mass transfer is linked to diffusion coefficient.

Table 6.2. Internal heat and mass transfer

	Heat	Mass
Diffusion equation	$\dfrac{\partial T_p(x,t)}{\partial x} = \alpha \left(\dfrac{\partial^2 T_p(x,t)}{\partial x^2} \right)$	$\dfrac{\partial W_p(x,t)}{\partial x} = D_e \left(\dfrac{\partial^2 W_p(x,t)}{\partial x^2} \right)$
Internal Resistance	$\dfrac{L}{\alpha A}$	$\dfrac{L}{D_e A}$

The complexity degree of differential equations describing internal heat and mass transfer depends strongly of boundary conditions to be assumed. Actually, the external resistance, in the gas phase, is frequently neglected and only the internal one considered. Under this circumstances equilibrium conditions at the surface are assumed, the surface is at the ambient temperature and solid moisture corresponds to equilibrium. In fact temperature is lower at the beginning, wet bulb temperature, and increase during drying reaching the ambient temperature by the end of the process, if there was no temperature difference the heat transfer will not occur.

Because the foods are solid, internal resistance to heat transfer could be assumed to be lower than the external one. This means that temperature gradients inside the food particle are much less important than outside. By measuring the temperature variation and weight losses the external heat transfer coefficient can be estimated (Simal et al., 1993) by assuming internal heat transfer resistance negligible [Eq. (3)].

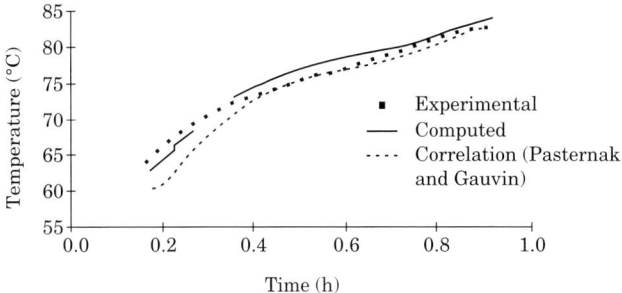

Fig. 6.3. Experimental, computed and theoretical temperature variation in the center of 10 mm potato cube during drying. Air temperature 90°C, air velocity 3 m/s (Simal et al., 1993)

Due to the fact that inside a solid heat is usually transported easily and not mass, many times only mass transfer is considered for kinetic determination purposes. In fact both heat and mass transfer should be considered and ways to lower internal/external resistances analyzed for improving the process.

External heat transfer takes place by convection or radiation. If convection is the main mechanism the boundary layer shows that if air velocity increase the rate of heat transfer also increase, there is an increase of the heat transfer coefficient. Of course the hotter the air the higher the heat transferred.

If radiation is involved this takes place independently of air velocity. Electromagnetic waves are absorbed by the food. In this case the nature of the radiation is of paramount importance, the energy is transferred directly to the solid. As a consequence if heat transfer is to be improved one should think about the source and the way of applying the radiation. Inside the food the internal resistance is not always negligible, when drying pores appears and this difficult heat transfer. The internal resistance is expected to vary during the drying process.

As long as mass transfer transport similar comments could be made, although external resistance is expected to be lower than the internal one. A gas diffuses easily while mass transport inside a solid is difficult. If external mass transfer resistance is important one of the most effective ways to affect the resistance is to act on the mass transfer coefficient through air velocity. Decreasing air relative moisture has less effect on mass transport, if the decrease is attained by increasing temperature, it could be harmful due to quality loss.

The internal resistance will only be affected through the solid temperature. As a consequence if heat is transferred by radiation this can be of great importance in cases where heat conduction is difficult. As the solid dries the food structure changes and this can affect the drying process. Whilst initially evaporation takes place at the surface or near the surface, when drying advances evaporation occurs mainly inside the solid and water vapor is transported to the surface.

These phenomena show that one should consider different drying periods. Initially if there is free or slightly bound water at the surface, evaporation takes place at a constant rate limited by heat transfer, this is known as the constant rate period, the temperature being the wet bulb temperature corresponding to air if velocity is high enough. It should be noticed that in food drying the constant rate drying period is seldom observed, only when external transport resistance is important then water can move at a high enough rate to maintain free water at the surface. This was observed for instance in the drying of peppers (Sanjuan et al., 2003) Fig. 6.4. The epidermis of the fruits acts an external resistance on the drying of the whole products (Sanjuan et al., 2003; Mulet et al., 2005).

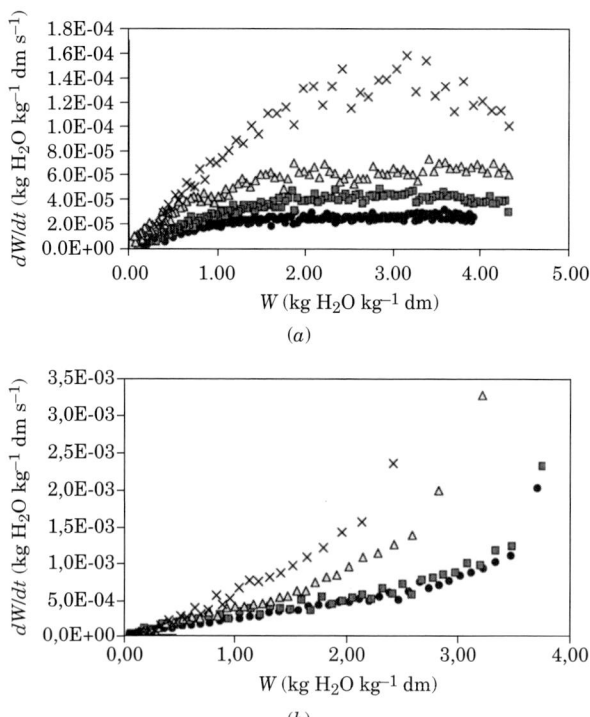

Fig. 6.4. Variation in dehydration rate of (a) whole and (b) shredded peppers with moisture content (Sanjuan et al., 2003)

When free water at the surface depletes water should travel from the inside to the surface and the drying rate decreases with time, this is known as the falling rate period [Fig. 6.4 (b)]. During the falling rate more than one period could be considered due to food structure changes, this will be illustrated by model analysis later.

From this comments one can imagine that establishing the prevailing resistance is very important for drier design and operation in order to improve the energy efficiency and care about product quality. It is also important for checking kinetics data from literature, in order to show the relative importance of the resistances some ways can be considered, this will be addressed later on. All the changes observed in transport resistances will suggest ways and means to establish drying strategies.

4. MODELING

Modeling constitutes an approach for analyzing drying and drier's operation. Process optimization needs modeling as a first step. Modeling in a drying process involves several aspects:

- Vapor/air equilibrium (Psychrometrics)
- Moisture equilibrium (isotherms)
- Drying kinetics
- Residence time and drier conditions
- Cost analysis.

Evidently all these aspects are interconnected, although very often the kinetics constitute the starting point. Models are built according to some assumptions, the most common by large is considering mass transport inside the food as being solely driven by moisture concentration differences, that is a Fickian mechanism. It is known that other mechanisms coexist like capillarity and also that the food is an structured material, nevertheless Fickian models if properly defined usually depict well the process. Models of this kind can easily be found in the literature, mostly for regular shaped geometries (Simal *et al.*, 1998). In fact the complexity advisable in a model depends on the use it is intended for.

Because the kind of water more or less bound to the material changes during drying as well as the food structure a model allowing to describe these changes could be necessary. Shrinkage occurs during food drying, also at microscopic level important changes take place, thus affecting the ratio internal/external resistance (Mayor and Sereno, 2004). Considering these effects one can imagine that for drying optimization (both energy and quality) several steps could be required.

Nevertheless very often simplified diffusional models are used and the effective diffusivity is identified as the value that gives the better fit to experimental data. Under these circumstances this parameter will include other effects. Based on these premise if the diffusion parameter varies this will constitute an indication of those effects, in that way if the effective diffusivity is plotted against air velocity it can be deduced when the internal resistance prevails (Blasco et al., 2006) as shown in Fig. 6.5. From this kind of plots one can establish a tentative limit for the air velocity during drying.

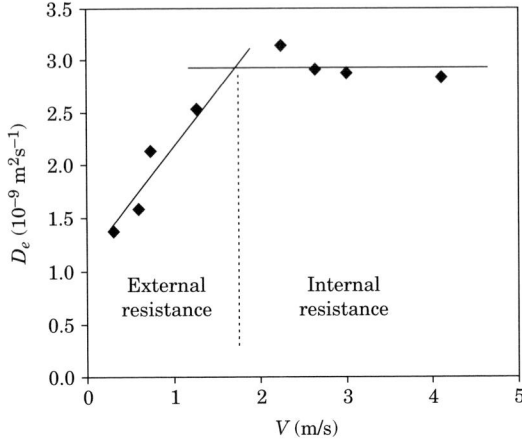

Fig. 6.5. Effect of air flow rate on effective moisture diffusivity identified in turmeric drying (Blasco et al., 2006)

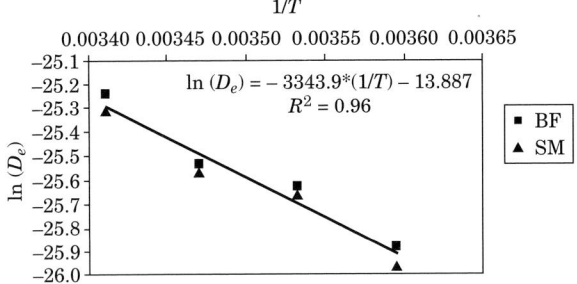

Fig. 6.6. Effect of temperature on effective moisture diffusivity identified in drying of frozen and salted *Biceps femoris* (BF) and *Semimembranosus* (SM) pork meat muscles (Clemente et al., 2007)

The effect of air temperature on the drying kinetics is usually described by using the Arrhenius law (Fig. 6.6) [Eq. (3)]. This effect is usually well established regardless of the complexity of the model (Mulet, 1994; Mulet et al., 2005; Blasco et al., 2006; Bon et al., 2007; Clemente et al., 2007).

The product structure changes will affect the effective diffusivity as well as the state of the diffusing water. The moisture content being an indicator of the drying stage it is expected to find an effective diffusivity dependence on moisture. This fact was already pointed out by Sherwood in 1931, who observed "the diffusion constant decrease with moisture concentration" (Sherwood et al., 1975). To consider the moisture effect Crank (1975) suggested Eq. (4), although further attempts have been made to describe better this effect [Eq. (5)] (Simal et al., 2006). This way of considering diffusion coefficient variation is helpful for further optimization calculations

$$D_e = D_0 \cdot \exp\left(\frac{-E_a}{RT}\right) \qquad \ldots(3)$$

$$D_e = D_0 \cdot \exp\left(a + \frac{b}{T} + cW\right) \qquad \ldots(4)$$

$$D_e = D_0 \cdot \exp\left(\frac{-\alpha}{RT}\left(\frac{W_p}{W_0}\right)^n\right) \qquad \ldots(5)$$

where D_e is the effective moisture diffusivity, D_0 the pre-exponential Arrhenius factor, E_a the activation energy, W_0 the initial moisture content, W_p the local moisture content and a, b, c, α and n fitting parameters.

To determine the prevalence of internal resistance a plot of $d(\ln \Psi)/dt$ with dimensionless moisture content (Ψ) could be considered (Mulet, 1994; Blasco et al., 2006) (Fig. 6.7). The slope of this curve depends on the predominant resistance, internal or external. In the case of internal resistance being predominant, a negative slope is found, while if external resistance is predominant, the slope is constant or slightly positive (Mulet, 1994). In Fig. 6.7, drying experiments conducted at high air flow rates showed a negative slope pattern all the time. In those carried out at low air flow rates, no variation was found for moisture contents (Ψ) higher than 0.25. Mass transfer is always controlled by internal resistance at the last stage of drying, and a negative trend is found in all the curves at low moisture contents. These results also showed that the air velocity threshold

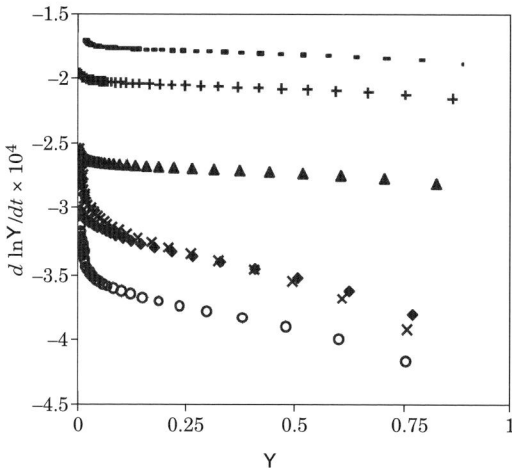

Fig. 6.7. Evaluation of internal or external control on drying rate. Drying experiments of turmeric at different air velocities (♦ 4, × 2.6, O 2.1, ▲ 0.7, + 0.5 and – 0.2 m/s) and 100°C (Blasco et al., 2006).

between 1–2 m/s, also seen in Fig. 6.5, limited each one of the zones controlled by external or internal resistance. They also suggest that without affecting the drying rate, the air velocity can be lowered at low moisture contents, thus saving energy.

5. OPTIMIZATION

The primary goal of drying is preservation and for that purpose the end point should be established according to the water activity concepts already mentioned. Simultaneously to attain the preservation goal should be considered the quality aspects (Banga and Singh, 1994; Femenia et al., 2009). There is a trade-off between quality and operating conditions, and cost (either environmental or monetary) involved (Perez-Francisco et al., 2008). In any case suitable, drying conditions (temperatures issue) could be established for product quality considering different factors (Banga and Singh, 1994).

As previously stated the food characteristics like moisture will influence the drying kinetics as well as the food temperature during drying. The moisture content also will affect the exhaust air characteristics. As a consequence recycling hot exhaust air will make sense in some instances and the recycling rate would also be different according to the product characteristics.

For operating optimization purpose a model of the drier, mass and energy balances, should be built (Garcia-Perez et al., 2009a) and then an objective must be defined (Banga and Singh, 1994). The definition of the objective function is a crucial step because this reflects the concerns to be considered. The other crucial step to be overcome is to solve the optimization problem. Optimization using response surface modeling (RSM) has been, and continues to be, the most common approach (Banga et al., 2003; Perez-Francisco et al., 2008). RSM methods are based on empirical models, thus, there exists most recently trends to consider more rigorous models. A review of optimization models was done by Banga et al. (2003).

Coming back to the objective function choice, the heart of the optimization problem, there are many ways of defining this function. In order to minimize costs the most common variables considered are recycle air ratio, air flow rate and drying air temperature (Mulet, 1994). If only costs are considered the energy cost makes vary largely the operating conditions (Aghbashlo et al., 2008). For food the energy cost is important, but less than for other industries like ceramics or pulp paper, maybe more important in foods are the quality of the product (primary goal) (Banga and Singh, 1994). Evidently the increased quality not always overcome the increased costs from a consumer point of view, but this is not necessarily true because increased quality frequently comes from sound processing on the light of better knowledge. Besides energy use that is well known and recognized as a factor to be improved the quality factors are less evident to be established. Due to the different behaviour of the product during drying, for quality improvement it seems that drying conditions need to be adapted along the drying process. This leads to different conditions to be considered and drier design involving different sections (Garcia-Perez et al., 2009a). Also the energy use improvement needs to be considered by allowing variable temperature in the sections (Hernandez-Diaz et al., 2008), a control problem could arise.

The pretreatments play a key role in keeping quality, addressing mainly keeping color or texture. An example for color keeping is the use of sulphur dioxide, traditionally employed for apricot drying (Simal et al., 1996; Miranda et al., 2009). The fact that this compound is harmful for asthmatic people brings the need of a careful use or even avoiding (Krokida et al., 1998). Blanching is also a traditional pre-treatment in order to maintain some quality properties of fresh vegetables after drying (Adedeji et al., 2009). For texture improvement the use of low temperature blanching could increase firmness after processing of many vegetables like cauliflower or broccoli (Garcia-Reverter et al., 1994; Sanjuan et al., 2000a, 2000b; Sanjuan et al., 2001). That fact is showed in Fig. 6.8, where better textural properties were obtained when blanching at low temperatures. Nowadays, the application of new technologies as pretreatments is becoming most popular, among others the use of power ultrasound (Fernandes and Rodrigues, 2008; Deng and Zhao, 2008), microwave (Dev et al., 2008) and pulse electric fields (Arevalo et al., 2004; Gachovska et al., 2008; Shynkaryk et al., 2008) seem to be promising according to literature.

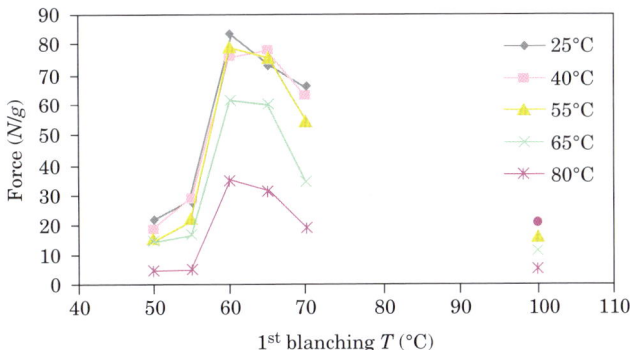

Fig. 6.8. Rehydration temperature effect on texture of rehydrated broccoli florets submitted to different blanching treatments (Sanjuan et al., 2001)

The storage conditions also need to be optimized for keeping quality (Dev Raj et al., 2006; Clemente et al., 2009). It is important for vegetables not to be subject to temperatures higher than 30°C to avoid heavy quality losses. As an example are shown in Fig. 6.9 some results for broccoli storage (Sanjuan et al., 2000). This was linked to a glass transition observed at around 34°C. Glass transitions can play a role on the quality keeping of dried vegetables.

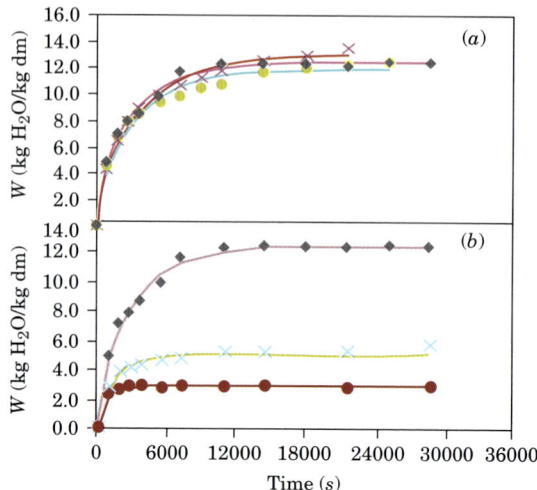

Fig. 6.9. Experimental and calculated rehydration curves for broccoli stems. (a) Storage temperature: 5°C. (b) Storage temperature: 40°C. ♦ 0 days. × 128 days. • 427 days (Sanjuan et al., 2000b)

Another point that should be kept in mind when addressing product quality optimization is rehydration conditions. It has been shown for broccoli that this process affects cell wall peptic substances (Femenia et al., 2000) and as a consequence a different texture is obtained. Rehydration carried out at low temperature, around 70–80°C, seems to be adequate for keeping texture.

As can be observed quality optimization is a difficult matter because quality should be preserved at every step and usually people care about the industrial process but not to others equally important like storage or rehydration. Considering the whole process is a must for product optimization, in order to address this optimization all the steps need to be modeled and their relationships with the previous ones and the effects on the following steps quantified. As far as environmental impact is concerned a global tool is needed to compare different alternatives and choose environmentally optimized processes. For that purpose the LCA technique seems promising (Sanjuan et al., 2005).

6. DRYING TECHNOLOGIES

Because air drying is a very old technology many developments took place in the past, mainly driven by trial and error methodologies or by necessity of feeding people in special situations.

Sun drying was the oldest one linked to the natural, atmospheric, drying of grains, fruits and vegetables. The oldest version of sun drying was at open sun, this method of drying is still practiced in many countries in different forms like shown in Fig. 6.10, where grapes were dried outdoor on trays to facilitate handling.

Fig. 6.10. Open sun drying of grapes in Gata (Alicante, Spain)

Sun drying is still widely used in many countries although is losing ground against more controlled drying processes. Nevertheless the "surname" sun-dried carries an environmentally friendly message and those products can constitute a market niche, although careful attention should be paid to product quality a main concern for consumers.

For quality control purpose open-sun drying should be avoided because dust and insect contamination of the food as well as danger of rain or dew at night. For that purpose different driers using sun energy have been development in the world (Das and Sarma, 2001; Rossello *et al.*, 1990). An exhaustive review of traditional and new sun drying technologies has been carried out by Murthy (2009). As example, Fig. 6.11 shows an experimental industrial size drier built (late 80) in Majorca (Spain) for drying fruits and vegetables. In many places, like in Majorca, drying activity has almost disappeared due to high labor cost. More industrialized activity for handling large quantities of raw materials remain in place.

Fig. 6.11. Sun drier in Mallorca (Spain)

There is a continued interest in the literature for the study of sun dried products both in driers or open sun. Open sun drying experiments could be analyzed by considering the radiation collected by the food thus allowing to establish kinetic that can be used in different parts of the world and different food shapes.

Although not many attention is paid to solar drying for fruit and vegetables in developed countries it can constitute a cost effective environmentally friendly way of drying, although many times it is more labour intensive than other alternatives. For the drying of fruits and vegetables two main types of driers have been frequently considered, cabinet or tunnel, as for hot air drying is concerned.

Vega-Mercado et al. (2001) made a classification (generations) of drying technologies that could be useful just for the sake of classification, evidently there are a lot of possible variations. Air forced driers (i.e., tray, truck tray, rotary flow conveyor, tunnel) involve hot air flowing over an extensive area of the product to remove water from the surface. They are mostly suitable for solid materials such as grains, sliced fruits and vegetables, or chunked products. The food industry uses this type of dryer in a variety of processes of which several commercial alternatives are available. In the case of purees or slurries, spray and drum dryers are more convenient in order to obtain dry powders and flakes.

Hot air dried products present often high shrinkage, case-hardening, colour damage, nutritional and textural losses and low rehydration capacity (Ratti, 2001; Lewicki, 2006). Vacuum freeze drying emerged like an alternative to hot air drying in order to obtain high quality products. Conventional freeze drying is carried out at vacuum conditions to accelerate sublimation providing excellent quality properties compared to hot air drying due to water mainly lacking in liquid state and low temperatures used (frozen samples) (Chen et al., 2007; Rahman and Mujumdar, 2008). At freezing conditions, degradation reactions slow considerably down and product shape is well maintained. Otherwise, vacuum freeze drying requires batch operation and high investment costs due to operate at low pressure (Nijhuis et al., 1998; Claussen et al., 2007), for that reason, it is mainly used in products with high economical interest. Atmospheric freeze drying is considered an adequate alternative to vacuum one due to it facilitates operation by working at atmospheric pressure. The operation consists mainly in circulating an air flow through the product at a lower temperature than thawing point, as a consequence, it allows processing in a continuous way. Atmospheric freeze drying provides dried products with almost similar quality properties than vacuum freeze drying (Lokra et al., 2008), it has been applied to several products, such as pepper, carrot (Alves-Filho et al., 2007), shrimps and cod fish (Garcia-Perez et al., 2005). Despite the high interest associated to atmospheric freeze drying, the full development of this technology is limited by a very low drying rate and therefore, a high energy use. Drying times are extended by working below freezing point at atmospheric pressure, thus it would be interesting to increase heat and mass transfer without increasing product temperature. Drying rate could be improved by adequately combining other energy sources than air during drying (Chou and Chua, 2001), such as, microwave (Funebo et al., 2002; Andres et al., 2007), infrared radiation (Chua and Chou, 2005; Wang and Sheng, 2006), radio frequency (Jumah, 2005) and power ultrasound (Gallego-Juarez, 1998; Gallego-Juarez et al., 1999; Mulet et al., 2003; Garcia-Perez et al., 2006a; Garcia-Perez et al., 2007; Carcel et al., 2007a; Gallego-Juarez et al., 2007; Garcia-Perez et al., 2009b). These new drying technologies may be applied in a continuous or intermittent way (Jumah, 2005), alone or combined (hybrid) (Chou and Chua, 2001). An increase in drying rate is mainly due to the heating effect of microwave, radio-frequency and infrared radiation. In comparison to those technologies, power ultrasound does not lead to the product being heated to any significant degree, thus avoiding quality loss due to heating during drying (Gallego-Juarez, 1998). As a consequence, the use of ultrasound either to dry heat sensitive materials (Gallego-Juarez et al., 1999; Mulet et al., 2003) or to be applied in drying processes carried out at low temperatures, like atmospheric freeze drying, has a great potential.

7. NEW DRYING TECHNOLOGIES ULTRASONIC ASSISTED DRYING

The application of power ultrasound during drying may reduce boundary layer thickness by pressure variations, oscillating velocities and microstreaming affecting the solid-gas interfaces. The aforementioned effects would involve an improvement of the water transfer rate from the solid surface to the air medium (Gallego-Juarez et al., 2007). On the other hand, internal water transfer may be also affected by alternating expansion and compression waves produced by ultrasound in the material (a phenomenon referred to the "sponge effect"). Finally, ultrasound energy may also affect the strongest attached moisture in the solid matrix (Muralidhara et al., 1985) due to cavitation.

Power ultrasound have been applied to affect mass transfer processes in solid-liquid treatments, like meat (Carcel et al., 2007b) and cheese brining (Sanchez et al., 2000), the osmotic dehydration of fruits (Carcel et al., 2007c) and several extraction processes (Riera et al., 2004). Nevertheless, applications on solid-gas systems, like air-forced drying, are much less frequent due to some technical difficulties. Among other things, the high impedance mismatch between the application systems and air, which makes the acoustic wave transmission difficult, and the high acoustic energy absorption of the air must be considered. These limitations may be overcome by adequately designing the ultrasonic application system. In this sense, Gallego-Juarez et al. (1999) developed a stepped plate ultrasonic transducer to apply power ultrasound during hot air drying. Different prototypes were developed for the frequency range 10–40 kHz and power between 100 W and 1 kW being used to dehydrate carrots, potatoes and mushrooms (Gallego-Juarez *et al.*, 2007). A very intense effect of power ultrasound was found using direct contact between vibrating elements and materials being dried, even, the effect increased applying a low static pressure. The effect of power ultrasound on drying was reduced when the application was carried out using an air borne technique. In order to develop more efficient air-borne ultrasonic devices, a cylindrical one (21.8 kHz) was designed constituting the drying chamber (Fig. 6.12) (Garcia-Perez *et al.*, 2006a). The design was based on the idea of using the drying chamber as the element to irradiate acoustic energy to the material being dried. In this way, no additional elements are needed to apply ultrasound during drying.

Fig. 6.12. Air forced drier assisted by power ultrasound. 1. Vibrating cylindrical drying chamber, 2. Mechanical amplifier, 3. Ultrasonic transducer

The system has been found to be very effective to improve drying rate of different products, providing significant drying time reductions. Figure 6.13 shows air forced drying kinetics of eggplant with and without power ultrasound application. A drying time reduction over 70% was observed in Fig. 6.13 by ultrasonic effect, which confirms a significant improvement of drying kinetic. The effect of power ultrasound is affected by process variables, previous works have addressed the influence of air velocity (Carcel *et al.*, 2007a; Garcia-Perez *et al.*, 2007) and temperature (Garcia-Perez *et al.*, 2006b), acoustic energy applied (Garcia-Perez *et al.*, 2009b) and product characteristics (Garcia-Perez *et al.*, 2007). According to the magnitude of these variables, the effect of power ultrasound on drying kinetics may be large or even disappear. As an example, the influence of the applied acoustic power on drying of lemon peel is shown on Fig. 6.14.

Food Dehydration under Forced Convection Conditions 171

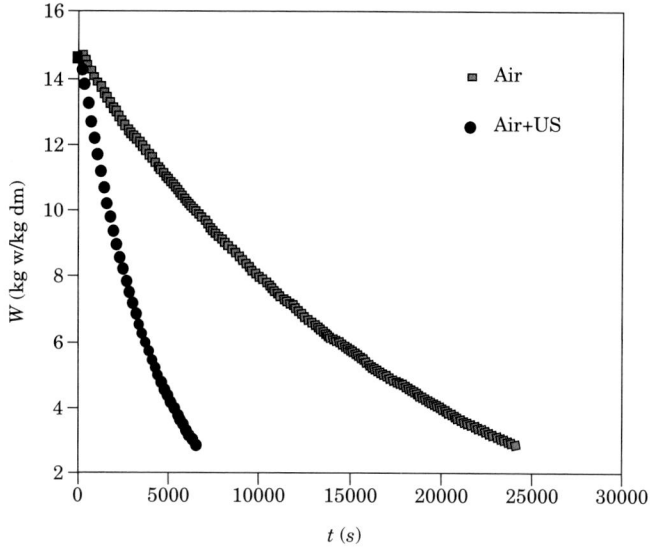

Fig. 6.13. Drying kinetics of eggplant cylinders (height 20 mm and diameter 20.4 mm) at 40°C and 1 m/s with (AIR+US, 37 kW/m^3) and without (AIR) power ultrasound application

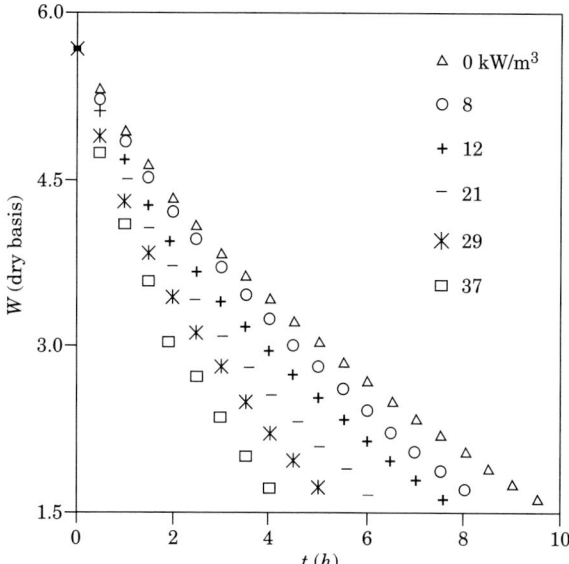

Fig. 6.14. Drying kinetics of lemon peel slabs (thickness 7 mm) at 40°C and 1 m/s applying different acoustic energy (Garcia-Perez et al., 2009b)

From drying kinetics modeling, it was observed that the effects associated to power ultrasound may reduce both external and internal resistance to mass transfer. The effect of the acoustic energy applied on the effective moisture diffusivity and mass transfer coefficient are depicted in Fig. 6.15 for the drying kinetics shown in Fig. 6.14.

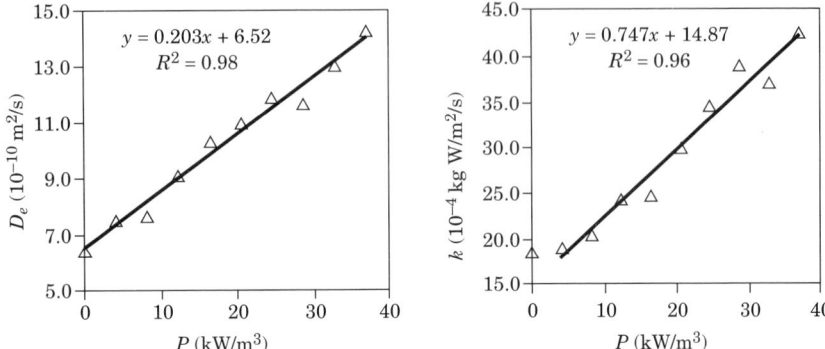

Fig. 6.15. Influence of applied acoustic energy on effective moisture diffusivity and mass transfer coefficient for lemon peel drying (Mulet et al., 2007; Garcia-Perez et al., 2009b)

8. FUTURE PROSPECTS AND NEEDS

There is a large amount of research, and available knowledge, carried out by academia, but the industry is not taking full advantages even in the more developed countries. A lack of interaction between researchers and industry is observed worldwide (Mujumdar, 2006). As a consequence industrial advancement is slow because there is a mismatch between research and industrial needs.

In less developed countries where industry is not very important there is a general feeling that drying is an easy operation and not too much input is needed and anybody can do it. A consequence of this attitude has been the failure of many drying projects. Drying of foods is a complex business and a mere translation from other fields is not often advisable.

Research is nowadays mainly focused on modeling and the understanding of basic processes linked not only to drying itself but also to storage (glass transition, reaction kinetics, microstructure, etc.).

Energy efficiency linked to environmentally friendly processes and products also appear as a growing trend mostly in some developed countries.

There is a need to check existing data on drying kinetics in order to establish their applicability regarding the prevailing resistance. Also there is a need to study products (isotherms and drying kinetics) not common in more developed countries where the main research in the field has been carried out. This will lead to discover new niches of activity. This is also true for countries where drying research is an active fields where the use of off-quality fresh market products is often disregarded.

One can say that there is a general concern towards improving the final consumer products quality and search for environmentally friendly processes. As a consequence, pretreatments, (including enrichment), optimization and control of operating conditions, combination of drying strategies according to the drying stage for a better migration of water, storage and consumption are addressed.

ACKNOWLEDGEMENTS

The authors acknowledge to the Ministerio de Educacion y Ciencia de España and the Polytechnic University of Valencia by the financial support of the Projects DPI2009-14549-C04-04 and PAID-06-08-3180, respectively.

REFERENCES

1. Achanta, S. and Okos, M.R. 1996. Predicting the quality of dehydrated foods and biopolymers, Research needs and opportunities, *Dry. Technol.* **14**: 1329–68.
2. Adedeji, A.A., Gachovska, T.K., Ngadi, M.O. and Raghavan, G.S.V. 2008. Effect of pretreatments on drying characteristics of okra, *Dry. Technol.* **26**: 1251–6.
3. Aghbashlo, M., Kianmehr, M.H. and Arabhosseini, A. 2008. Energy and exergy analyses of thin-layer drying of potato slices in a semi-industrial continuous band dryer, *Dry. Technol.* **26**: 1501–8.
4. Alves, O., Eikevik, T., Mulet, A., Garau, C. and Rossello, C. 2007. Kinetics and mass transfer during atmospheric freeze drying of red pepper, *Dry. Technol.* **25**: 1155–61.
5. Anderson, R.B. 1946. Modification of the Brunauer, Emmett and Teller equation, *J. Am. Chem. Soc.* **68**: 686–91.
6. Andres, A., Fito, P., Heredia, A. and Rosa, E.M. 2007. Combined drying technologies for development of high-quality shelf-stable mango products, *Dry. Technol.* **25**: 1857–66.
7. Arevalo, P., Ngadi, M.O., Bazhal, M.I. and Raghavan, G.S.V. 2004. Impact of pulsed electric fields on the dehydration and physical properties of apple and potato slices, *Dry. Technol.* **22**: 1233–46.
8. Banga, J.R. and Singh, R.P. 1994. Optimization of air drying of foods, *J. Food Eng.* **23**: 189–211.
9. Banga, J.R., Balsa-Canto, E., Moles, C.G. and Alonso, A.A. 2003. Improving food processing using modern optimization methods, *Trends Food Sci. Tech.* **14**: 131–44.
10. Blasco, M., Garcia-Perez, J.V., Bon, J., Carreres, J.E. and Mulet, A. 2006. Effect of blanching and air flow rate on turmeric drying, *Food Sci. Technol. Int.* **12**: 315–23.
11. Bon, J., Rossello, C., Femenia, A., Eim, V. and Simal, S. 2007. Mathematical modeling of drying kinetics for apricots: Influence of the external resistance to mass transfer, *Dry. Technol.* **25**: 1829–35.
12. Carcel, J.A., Garcia-Perez, J.V., Riera, E. and Mulet, A. 2007a. Influence of high intensity ultrasound on drying kinetics of persimmon, *Dry. Technol.* **25**: 185–93.
13. Carcel, J., Benedito, J., Bon, J. and Mulet, A. 2007b. High intensity ultrasound effects on meat brining, *Meat Sci.* **76**: 611–9.
14. Carcel, J., Benedito, J., Rossello, C. and Mulet, A. 2007c. Influence of ultrasound intensity on mass transfer in apple immersed in a sucrose solution, *J. Food Eng.* **78**: 472–9.
15. Chen, J.P., Tai, C.Y. and Chen, B.H. 2007. Effects of different drying treatments on the stability of carotenoids in Taiwanese mango (*Mangifera indica* L.), *Food Chem.* **100**: 1005–10.

16. Chirife, J. and Iglesias, H.A. 1992. Estimation of the precision of isosteric heats of sorption determined from the temperature dependence of food isotherms, *Lebensmittel Wissenschaft und Technologie* **25**: 83–84.
17. Chou, S.K. and Chua, K.J. 2001. New hybrid drying technologies for heat sensitive foodstuffs, *Trends Food Sci. Tech.* **12**: 359–69.
18. Chua, K.J. and Chou, S.K. 2005. A comparative study between intermittent microwave and infrared drying of bioproducts, *Int. J. Food Sci. Tech.* **40**: 23–29.
19. Claussen, I.C., Andresen, T., Eikevik, T.M. and Strommen, I. 2007. Atmospheric freeze drying-Modeling and simulation of a tunnel dryer, *Dry. Technol.* **25**: 1959–65.
20. Clemente, G., Bon, J., Garcia-Perez, J.V. and Mulet, A. 2007. Natural convection drying at low temperatures of previously frozen salted meat, *Dry. Technol.* **25**: 1885–91.
21. Clemente, G., Bon, J., Benedito, J. and Mulet, A. 2009. Desorption isotherms and isosteric heat of desorption of previously frozen raw pork meat, *Meat Sci.* **82**: 413–8.
22. Cohen, J.S. and Yang, T.C.S. 1995. Progress in food dehydration, *Trends Food Sci. Tech.* **6**: 20–25.
23. Crank, J. 1975. The Mathematics of diffusion. Clarendon Press.
24. Das, P. and Sarma, S.K. 2001. Drying of ginger using solar cabinet dryer, *J. Food Sci. Tech.-India* **38**: 619–21.
25. De Boer, J.M. 1953. The dynamic character of adsorption. Clarendon Press.
26. Deng, Y. and Zhao, Y. 2008. Effect of pulsed vacuum and ultrasound osmopretreatments on glass transition temperature, texture, microstructure and calcium penetration of dried apples (Fuji), *LWT - Food Sci. Tech.* **41**: 1575–85.
27. Dev, S.R.S., Padmini, T., Adedeji, A., Gariepy, Y. and Raghavan, G.S.V. 2008. A comparative study on the effect of chemical, microwave, and pulsed electric pretreatments on convective drying and quality of raisins, *Dry. Technol.* **26**: 1238–43.
28. Femenia, A., Bestard, M.J., Sanjuan, N., Rossello, C. and Mulet, A. 2000. Effect of rehydration temperature on the cell wall components of broccoli (*Brassica oleracea* L. var. italica) plant tissues, *J. Food Eng.* **46**: 157–63.
29. Femenia, A., Sastre–Serrano, G., Simal, S., Eim, V.M., Garau, C. and Rossello, C. 2009. Effects of air-drying temperature on the cell walls of kiwifruit processed at different stages of ripening, *LWT-Food Sci. Tech.* **42**: 106–12.
30. Fernandes, F.A.N. and Rodrigues, S. 2008. Dehydration of sapota (*Achras sapota* L.) using ultrasound as pre-treatment, *Dry. Technol.* **26**: 1232–7.
31. Funebo, T., Ahrne, L., Prothon, F., Kidman, S., Langton, M. and Skjoldebrand, C. 2002. Microwave and convective dehydration of ethanol treated and frozen apple-physical properties and drying kinetics, *Int. J. Food Sci. Tech.* **37**: 603–14.
32. Gachovska, T.K., Adedeji, A.A., Ngadi, M. and Raghavan, G.V.S. 2008. Drying characteristics of pulsed electric field-treated carrot, *Dry. Technol.* **26**: 1244–50.
33. Gallego-Juarez, J.A. 1998. Some applications of air-borne power ultrasound to food processing. In: *Ultrasound in Food Processing.* Eds. Povey, M.J.W. and Mason, T.J., Chapman & Hall.
34. Gallego-Juarez, J.A., Rodriguez-Corral, G., Galvez-Moraleda, J.C. and Yang, T.S. 1999. A new high intensity ultrasonic technology for food dehydration, *Dry. Technol.* **17**: 597–608.
35. Gallego-Juarez, J.A., Riera, E., De la Fuente, S., Rodriguez-Corral, G., Acosta-Aparicio, V.M. and Blanco, A. 2007. Application of high-power ultrasound for dehydration of vegetables: processes and devices, *Dry. Technol.* **25**: 1893–1901.

36. Garcia-Perez, J.V., Alves-Filho, O., Eikevik, T.M., Strommen, I. and Mulet, A. 2005. Effect of drying air temperature on heat pump fluidized bed drying of cod fish. Proceedings AFSIA/EFCE Drying Conference, Paris, France.
37. Garcia-Perez, J.V., Carcel, J.A., De la Fuente, S. and Riera, E. 2006a. Ultrasonic drying of foodstuff in a fluidized bed. Parametric study, *Ultrasonics* **44**: e539–e43.
38. Garcia-Perez, J.V., Rossello, C., Carcel, J.A., De la Fuente, S. and Mulet, A. 2006b. Effect of air temperature on convective drying assisted by high power ultrasound, *Defect Diff. Forum* **258–260**: 563–74.
39. Garcia-Perez, J.V., Carcel, J.A., Benedito, J. and Mulet, A. 2007. Power ultrasound mass transfer enhancement in food drying, *Food Bioprod. Process.* **85**: 247–54.
40. Garcia-Perez, J.V., Carcel, J.A., Clemente, G. and Mulet, A. 2008. Water sorption isotherms for lemon peel at different temperatures and isosteric heats, *LWT-Food Sci. Tech.* **41**: 18–25.
41. Garcia-Perez, J.V., Carcel, J.A., Garcia-Alvarado, M.A. and Mulet, A. 2009a. Simulation of grape stalk deep bed drying, *J. Food Eng.* **90**: 308–14.
42. Garcia-Perez, J.V., Carcel, J.A., Riera, E. and Mulet, A. 2009b. Influence of the applied acoustic energy on the drying of carrots and lemon peel, *Dry. Technol.* **27**: 281–7.
43. Garcia-Reverter, J., Bourne, M.C. and Mulet, A. 1994. Low temperature blanching affects firmness and rehydration of dried cauliflower florets, *J. Food Sci.* **59**: 1181–3.
44. Guggenheim, E.A. 1966. Application of statistical mechanics. Clarendon Press.
45. Iglesias, H.A. and Chirife, J. 1982. Handbook of food isotherms. Academic Press.
46. Hernandez-Diaz, W.N., Ruiz-Lopez, I.I., Salgado-Cervantes, M.A., Rodriguez-Jimenes, G.C. and Garcia-Alvarado, M.A. 2008. Modeling heat and mass transfer during drying of green coffee beans using prolate spheroidal geometry, *J. Food Eng.* **86**: 1–9.
47. Jumah, R. 2005. Modeling and simulation of continuous and intermittent radio frequency-assisted fluidized bed drying of grains, *Food Bioprod. Process.* **83**: 203–10.
48. Lewicki, P.P. 2006. Design of hot air drying for better foods, *Trends Food Sci. Technol.* **17**: 153–63.
49. Lewicki, P.P. and Michaluk, E. 2004. Drying of tomato pretreated with calcium, *Dry. Technol.* **22**: 1813–27.
50. Lokra, S., Helland, M.H., Claussen, I.C., Straetkvern, K.O. and Egelandsdal, B. 2008. Chemical characterization and functional properties of a potato protein concentrate prepared by large-scale expanded bed adsorption chromatography, *LWT-Food Sci. Tech.* **41**: 1089–99.
51. Maroulis, Z.B., Saravacos, G.D., Panagiotou, N.M. and Krokida, M.K. 2001. Moisture diffusivity data compilation for foodstuffs: effect of material moisture content and temperature, *Int. J. Food Prop.* **4**: 225–37.
52. Mayor, L. and Sereno, A.M. 2004. Modeling shrinkage during convective drying of food materials: a review, *J. Food Eng.* **61**: 373–86.
53. Miranda, G., Berna, A., Salazar, D. and Mulet, A. 2009. Sulphur dioxide evolution during dried apricot storage, *LWT-Food Sci. Tech.* **42**: 531–3.
54. Mujumdar, A.S. 2006. Handbook of industrial drying. CRC Press.
55. Mulet, A. 1994. Drying modeling and water diffusivity in carrots and potatoes, *J. Food Eng.* **22**: 329–48.
56. Mulet, A., Garcia-Pascual, P., Sanjuan, N. and Garcia-Reverter, J. 2002. Equilibrium isotherms and isosteric heats of morel (*Morchella esculenta*), *J. Food Eng.* **53**: 75–81.

57. Mulet, A., Carcel, J.A., Sanjuan, N. and Bon, J. 2003. New food drying technologies- Use of ultrasound, *Food Sci. Tech. Int.* **9**: 215–21.
58. Mulet, A., Blasco, M., Garcia-Reverter, J. and Garcia-Perez, J.V. 2005. Drying kinetics of Curcuma longa rhizomes, *J. Food Sci.* **7**: E318–E23.
59. Mulet, A., Carcel, J.A., De la Fuente, S., Riera, E. and Garcia-Perez, J.V. 2007. Convective drying of lemon peel assisted by power ultrasound: influence of ultrasonic power applied. *Proceedings of the International Congress on Ultrasonics (ICU)*, Vienna, Austria.
60. Muralidhara, H.S., Ensminger, D. and Putnam, A. 1985. Acoustic dewatering and drying (low and high frequency): State of the art review, *Dry. Technol.* **3**: 529–66.
61. Murthy, M.V.R. 2009. A review of new technologies, models and experimental investigations of solar driers, *Renew. Sust. Energ. Rev.* **13**: 835–44.
62. Nijhuis, H.H., Torringa, H.M., Muresan, S., Yuksel, D., Leguijt, C. and Kloek, W. 1998. Approaches to improving the quality of dried fruit and vegetables, *Trends Food Sci. Tech.* **9**: 13–20.
63. Perez-Francisco, J.M., Cerecero-Enriquez, R., Andrade-Gonzalez, I., Ragazzo-Sanchez, J.A. and Luna-Solano, G. 2008. Optimization of vegetal pear drying using response surface methodology, *Dry. Technol.* **26**: 1401–5.
64. Rahman, S.M.A. and Mujumdar, A.S. 2008. A novel atmospheric freeze-drying system using a vibro-fluidized bed with adsorbent, *Dry. Technol.* **26**: 393–403.
65. Raj, D., Subanna, V.C., Ahlawat, O.P., Gupta, P. and Huddar, A.G. 2006. Effect of pre-treatments on the quality characteristics of dehydrated onion rings during storage, *J. Food Sci. Tech.* **43**: 571–4.
66. Ratti, C. 2001. Hot air and freeze-drying of high-value foods: a review, *J. Food Eng.* **49**: 311–9.
67. Riera, E., Golas, Y., Blanco, A., Gallego, J.A., Blasco, M. and Mulet, A. 2004. Mass transfer enhancement in supercritical fluids extraction by means of power ultrasound, *Ultrason. Sonochem.* **11**: 241–4.
68. Rossello, C., Berna, A. and Mulet, A. 1990. Solar drying of fruits in a Mediterranean climate, *Dry. Technol.* **8**: 305–21.
69. Sanchez, E.S., Simal, S., Femenia, A. and Rossello, C. 2000. Effect of acoustic brining on the transport of sodium chloride and water in Mahon cheese, *Eur. Food Res. Tech.* **212**: 39–43.
70. Sanjuan, N., Clemente, G., Bon, J. and Mulet, A. 2000a. The influence of blanching pretreatments on the quality of dehydrated broccoli stems, *Food Sci.Tech. Int.* **6**: 227–34.
71. Sanjuan, N., Benedito, J., Bon, J. and Mulet, A. 2000b. Changes in the quality of dehydrated broccoli stems during storage, *J. Sci. Food Agr.* **80**: 1589–94.
72. Sanjuan, N., Clemente, G., Bon, J. and Mulet, A. 2001. The effect of blanching on the quality of dehydrated broccoli florets, *Eur. Food Res. Tech.* **213**: 474–9.
73. Sanjuan, N., Lozano, M., Garcia-Pascual, P. and Mulet, A. 2003. Dehydration kinetics of red pepper (*Capsicum annuum* L var Jaranda), *J. Sci. Food Agr.* **83**: 697–701.
74. Sanjuan, N., Ubeda, L., Clemente, G., Mulet, A. and Girona, F. 2005. LCA of integrated orange production in the Comunidad Valenciana (Spain), *Int. J. Agr. Resour. Gov. Ecol.* **4**: 163–77.
75. Sanjuan, R., Garcia-Reverter, J., Bon, J. and Mulet, A. 1994. Moisture retention in chufa (*Cyperus esculentus* L.) equilibrium isotherms and isosteric heats, *Revista Española de Ciencia y Tecnologia de Alimentos* **34**: 653–62.
76. Sherwood, T.K., Pigford, R.L. and Wilke, C.K. 1975. Mass transfer. McGraw-Hill, Inc.

77. Shynkaryk, M.V., Lebovka, N.I. and Vorobiev, E. 2008. Pulsed electric fields and temperature effects on drying and rehydration of red beetroots, *Dry. Technol.* **26**: 695–704.
78. Simal, S., Berna, A., Mulet, A. and Rossello, C. 1993. A method for the calculation of heat transfer coefficient in potato drying, *J. Sci. Food Agr.* **63**: 365–7.
79. Simal, S., Rossello, C., Sanchez, E. and Cañellas, J. 1996. Quality of raisins treated and stored under different conditions, *J. Agr. Food Chem.* **44**: 3297–3302.
80. Simal, S., Rossello, C. and Mulet, A. 1998. Modeling of air drying in regular shaped bodies, *Trends Chem. Eng.* **4**: 171–80.
81. Simal, S., Garau, M.C., Femenia, A. and Rossello, C. 2006. A diffusional model with a moisture-dependent diffusion coefficient, *Dry. Technol.* **24**: 1365–72.
82. Simal, S., Femenia, A., Castell-Palou, A. and Rossello, C. 2007. Water desorption thermodynamic properties of pineapple, *J. Food Eng.* **80**: 1293–1301.
83. Spiess, W.E.L. and Wolf, W. 1983. The result of the COST 90 project on water activity. *In*: Jowitt, R., Escher, F., Hallstrom, B., Meffert, H.F.Th., Spiess, W.E.L. and Vos, G. *Eds*., Physical Properties of Foods, Applied Science Publishers.
84. Treybal, R.B. 1980. Mass transfer operations. McGraw-Hill, Inc.
85. Vega-Mercado, H., Gongora-Nieto, M.M. and Barbosa-Canovas, G.V. 2001. Advances in dehydration of foods, *J. Food Eng.* **49**: 271–89.
86. Wang, J. and Sheng, K. 2006. Far-infrared and microwave drying of peach, *LWT-Food Sci. Tech.* **39**: 247–55.
87. Zhang, X.W., Liu, X., Gu, D.X., Zhou, W., Wang, R.L. and Liu, P. 1996. Desorption isotherms of some vegetables, *J. Sci. Food Agr.* **40**: 303–6.

7

Dispersion in Porous Media

J.M.P.Q. Delgado[*]

ABSTRACT

The phenomenon of dispersion, transverse or radial and longitudinal or axial, in porous media is reviewed for a great deal of information from the literature. Dispersion plays an important part, for example, in contaminant transport in ground water flows, in miscible displacement of oil and gas and in reactant and product transport in packed bed reactors. There are several variables that must be considered, in the analysis of dispersion in porous media, like the viscosity and density of the fluid, particle size distribution, particle shape, effect of fluid velocity and effect of temperature or Schmidt number, etc. Empirical correlations are presented for the prediction of the dispersion coefficients over the entire range of practical values of Schmidt number, Sc, and Peclet number, Pe_m.

NOTATION

C	concentration of solute
d	average diameter of inert particles
D	diameter of packed bed
D_L	longitudinal dispersion coefficient
D_m	molecular diffusion coefficient
D'_m	apparent molecular diffusion coefficient (= D_m/τ)
D_T	transverse dispersion coefficient
L	length
p	variable of Eq. (15)
Pe_f	Peclet number defined by Eq. (25)
$Pe_L(\infty)$	asymptotic value of Pe_L when Re → ∞

---

[*] LFC-Laboratório de Física das Construcções, Departamento de Engenharia Civil, Faculdade de Engenharia da Universidade do Porto, Rua Dr. Roberto Frias, s/n, 4200-465 Porto-PORTUGAL.
 Corresponding author : E-mail : jdelgado@fe.up.pt

$\text{Pe}_T(\infty)$ asymptotic value of Pe_T when $\text{Re} \to \infty$
r radial co-ordinate
R column radius
t time
T absolute temperature
U superficial fluid velocity
u average interstitial fluid velocity
z cartesian coordinates

GREEK LETTERS

ε bed voidage
μ dynamic viscosity
θ dimensionless time
ρ density
τ tortuosity
ξ_c variable defined by Eqs. [17 (d)–(e)]

DIMENSIONLESS GROUPS

Pe_a Peclet number based on longitudinal dispersion coefficient ($= uL/D_L$)
Pe_m Peclet number of inert particle ($= ud/D_m$)
Pe'_m Effective Peclet number of inert particle ($= ud/D'_m$)
Pe_L Peclet number based on longitudinal dispersion coefficient ($= ud/D_L$)
Pe_T Peclet number based on transversal dispersion coefficient ($= ud/D_T$)
Re Reynolds number ($= \rho U d/\mu$)
Sc Schmidt number ($= \mu/\rho D_m$)

1. INTRODUCTION

The problem of solute dispersion during underground water movement has attracted interest from the early days of this century (Slichter, 1905), but it was only since the 1950's that the general topic of hydrodynamic dispersion, or miscible displacement, became the subject of more systematic study. This topic has interested hydrologists, geophysicists, petroleum and chemical engineers, among others, and for some time now it is treated at length in books on flow through porous media (*e.g.*, Scheidegger, 1974; Bear, 1972). Some books on chemical reaction engineering (*ex*: Wen and Fan, 1975; Froment and Bischoff, 1990) treat the topic of dispersion (longitudinal and lateral) in detail and it is generally observed that data for liquids and gases do not overlap, even in the "appropriate" dimensionless representation.

Since the early experiments of Slichter (1905) and particularly since the analysis of dispersion during solute transport in capillary tubes, developed by Taylor (1953) and Aris (1956, 1959), a lot of work has been done on the description of the principles of solute transport in porous media of inert particles and in packed bed reactors (see Bear, 1972; Dullien, 1979).

Gray (1975) and Bear (1972) derived the proper form of the transport equation for the average concentration of solute in a porous medium, by using the method of volume or spatial averaging.

Brenner (1980) developed a general theory for determining the transport properties in spatially periodic porous media in the presence of convection, and showed that dispersion models are valid asymptotically in time for the case of dispersion in spatially periodic porous media, while Carbonell and Whitaker (1983) demonstrated that this should be the case for any porous medium. These authors presented a volume-average approach for calculating the dispersion coefficient and carried out specific calculations for a two-dimensional spatially periodic porous medium. Eidsath et al. (1983) have computed axial and lateral dispersion coefficients in packed beds based on these spatially periodic models, and have compared the results to available experimental data. The axial dispersion coefficient calculated by Eidsath et al. shows a Peclet number dependence that is too strong, while their transverse dispersion. However, in soils or underground reservoirs, large scale nonuniformities lead to values of dispersion coefficients that differ much from those measured in packed beds, and for these cases spatially periodic models cannot be expected to provide excellent results without modifications.

There have been other attempts at correlating and predicting dispersion coefficients based on a probabilistic approach (ex: Saffman, 1960) where the network of pores in the porous medium is regarded as an array of cylindrical capillaries with parameters governed by probability distribution functions.

Dispersion in porous media has been studied by a significant number of investigators; using various experimental techniques. However, measurements of axial and radial dispersion are normally carried out separately, and it is generally recognised that "experiments on lateral dispersion are much more difficult to perform than those on longitudinal dispersion" (Scheidegger, 1974). When a fluid is flowing through a bed of inert particles, one observes the dispersion of the fluid in consequence of the combined effects of molecular diffusion and convection in the spaces between particles. Generally, the dispersion coefficient in longitudinal direction is superior to the dispersion coefficient in radial direction by a factor of 5, for values of Reynolds number larger than 10. For low values of the Reynolds number (say, Re < 1), the two dispersion coefficients are approximately the same and equal to molecular diffusion coefficient.

The detailed structure of a porous medium is greatly irregular and just some statistical properties are known. An exact solution to characterize flowing fluid through one of these structures is basically impossible. However, by the method of volume or spatial averaging it is possible to obtain the transport equation for the average concentration of solute in a porous medium (Bear, 1972).

At a "macroscopic" level, the quantitative treatment of dispersion is currently based on Fick's law, with the appropriate dispersion coefficients; cross stream dispersion is related to the transverse dispersion coefficient, D_T, whereas streamwise dispersion is related to the longitudinal dispersion coefficient, D_L. If a small control volume is considered, a mass balance on the solute, without chemical reaction, leads to

$$D_L \frac{\partial^2 C}{\partial z^2} + \frac{1}{r}\frac{\partial}{\partial r}\left(D_T r \frac{\partial C}{\partial r}\right) - u \frac{\partial C}{\partial z} = \frac{\partial C}{\partial t} \qquad ...(1)$$

2. LONGITUDINAL OR AXIAL DISPERSION

Over the past five decades, longitudinal dispersion in porous media has been measured and correlated extensively for gaseous and liquid systems. Many publications are available for a variety of applications, including: packed bed reactors (Levenspiel and Smith, 1957; Edwards and Richardson, 1968; Gunn, 1969; Tsotsas and Schlunder, 1988) and soil column systems (Pfannkuch, 1963; Perkins and Johnston, 1963; Bear, 1972).

One of the first results published about longitudinal dispersion in packed beds of inertial particles was in the 1950's by Danckwerts (1953), who published his celebrated paper on residence time distribution in continuous contacting vessels, including chemical reactors, and thus provided methods for measuring axial dispersion rates. The author studied dispersion along the direction of flow for a step input in solute concentration in a bed of Raschig rings, crossed by water with a value of Re $\cong$ 25 and obtained a $Pe_L = 0.52$.

Kramers and Alberda (1953) followed Danckwerts's study with a theoretical and experimental investigation by the response to a sinusoidal input signal. These authors proposed that packed beds could be represented as consecutive regions of well-mixing rather than a sequence of stirred tanks (mixing-cell model) and suggested a $Pe_L \cong 1$, for Re $\to \infty$. McHenry and Wilhelm (1957) assumed the axial distance between the mixing-cells in a packing to be equal to particle diameter and showed that Pe_L must be about 2 for high Reynolds number. The difference in the two results may be explained on the basis of experimental results of Kramers and Alberda (1953) while are obtained with $L/D \approx 4.6$, a value lesser than $L/D > 20$ (Gunn, 1968). Bruinzeel et al. (1962) show that transverse dispersion can be neglected for a small ratio of column diameter to length and large fluid velocity.

Brenner (1962) presented the solution of a mathematical model of dispersion for a bed with finite length, L, and the most relevant conclusion of his work was that for $\text{Pe}_a\ (= uL/D_L) \geq 10$, the equations obtained by Danckwerts (1953) for an input step in solute concentration and Levenspiel and Smith (1957) for a pulse in solute concentration, in an infinite bed, are corrected.

Hiby (1962) proposed a better empirical correlation to cover the range of Reynolds numbers to 100. He reported experimental results with the aid of photographs to compare the two dispersion mechanisms presented: diffusional model in turbulent flow and the mixing-cell model.

Sinclair and Potter (1965) used a frequency response technique applied to the flow of air through beds of glass ballotini in a Reynolds number range between 0.1 and 20. A further investigation in the intermediate Reynolds number region has been carried out by Evans and Kenney (1966) who used a pulse response technique in beds of glass spheres and Raschig rings.

Experiments reported by Gunn and Pryce (1969) showed that axial dispersion coefficients given by the theoretical equation for the diffusional model and the theoretical equation for the mixing-cell model are very similar. The authors also showed that neither the mixing-cell model nor the axially dispersed plug flow model could describe axial dispersion phenomena.

Typically, the boundary conditions adopted have corresponded to the semi-infinite bed, *i.e.*, L is sufficiently large ($L/D > 20$). Dispersion of the given tracer was measured at two points in the outlet and the distortion of a tracer forced by a pulse input (*ex*: Carberry and Bretton, 1958), frequency response (*ex*: Ebach and White, 1958) and step input (*ex*: Hiby, 1962). Figure 7.1 illustrates some experimental data points for longitudinal dispersion in liquid and gas systems.

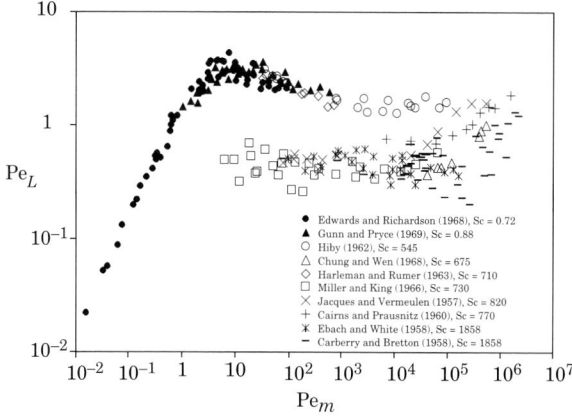

Fig. 7.1. Experimental data points for longitudinal dispersion in liquids and gas systems

2.1. Parameters Influencing Longitudinal Dispersion–Porous Medium

Perkins and Johnston (1963) in their article review showed some of the variables that influence longitudinal and radial dispersion. However, before attempting in the parameters influencing dispersion, it is important to consider the effect of the packing of the bed on dispersion coefficients. Gunn and Pryce (1969) and Roemer et al. (1962) showed that when particles in packed beds are not well packed the dispersion coefficient is increased. The results of Gunn and Pryce (1969) showed that different re-packing of the bed gave deviations of 15% in Peclet values. These experiments confirm that fluid mechanical characteristics are not only defined by the values of the porosity and tortuosity, but depend of the quality of packing in the bed.

A rigorous measurement of the porosity in a packed bed is fundamental to minimize the errors in the experimental measurements, because the porosity between the inert particles of the bed helps the diffusion of a tracer and gradually increases dispersion.

A more coherent interpretation of the experimental data may be obtained through the use of dimensional analysis. As a starting point it is reasonable to accept the functional dependence

$$D_L = \phi\,(L, D, u, d, \rho, \mu, D_m) \qquad \ldots(2)$$

for randomly packed beds of mono sized particles with diameter d, where ρ and μ are the density and viscosity of the liquid, respectively, and D_m is the coefficient of molecular diffusion of the solute. Making use of Buckingham's p theorem, Eq. (2) may be rearranged to give

$$\frac{D_L}{D_m} \text{ or } Pe_L = \Phi\left(\frac{L}{D}, \frac{D}{d}, \frac{ud}{D_m}, \frac{\mu}{\rho D_m}\right) \qquad \ldots(3)$$

and it is useful to define $Pe_m = ud/D_m$ and $Sc = \mu/\rho D_m$. This result suggests that experimental data be plotted as (D_L/D_m) vs. Pe_m.

2.1.1. *Column length*

One first aspect to be considered, as a check on the experimental method (infinite medium), is the influence of the length of the bed (L) on the measured value of axial dispersion. In reality, if an experimental method is valid, values of the dispersion coefficient measured with different column lengths, under similar conditions, should be equal, within the reproducibility limits.

The dependence of the axial dispersion coefficient on the position in packed beds was first examined by Taylor (1953). The author showed that, in laminar flow, dispersion approximation would be valid if the following equation is satisfied,

$$\theta = D_m t/R^2 \gg 0.14 \qquad \ldots(4)$$

where R is the tube radius. Carbonell and Whitaker (1983) concluded that the axial dispersion coefficient becomes constant if the following expression is satisfied

$$\theta = \left(\frac{1-\varepsilon}{\varepsilon}\right)^2 \frac{D_m t}{d^2} \gg 1 \qquad ...(5)$$

Han et al. (1985) showed that values of, for uniform size packed beds, measured at different positions in the bed are function of bed location unless the approximate criterion

$$\frac{L}{d}\frac{1}{\text{Pe}_m}\left(\frac{1-\varepsilon}{\varepsilon}\right)^2 \geq 0.3 \text{ or } \theta = \frac{D_m t}{d^2} \geq 0.15 \qquad ...(6)$$

is satisfied. The authors showed that for $\text{Pe}_m < 700$, longitudinal dispersion coefficients were nearly identical for all values of $x = L$, and for $\text{Pe}_m > 700$ observed an increase in the value of dispersion coefficients with increasing distance down the column.

2.1.2. *Ratio of column diameter to particle diameter*

It is well known (*e.g.*, Vortmeyer and Schuster, 1983) that the voidage of a packed bed (and therefore the fluid velocity) is higher near a containing flat wall. The effects of radial variations of porosity and velocity on axial and radial transport of mass in packed beds were analytically quantified by several authors as Vortmeyer and Winter (1982) and Delmas and Froment (1988).

Schwartz and Smith (1953) were the first to present experimental data showing zones of high porosity extending two or three particle diameters from the containing flat wall. The results indicated that unless $D/d > 30$ important velocity variations exist across the packed bed. Other studies showed that packed bed velocity profiles significantly differ from flows with large diameter particles in small diameter tubes (*ex*: Cairns and Prausnitz, 1959).

Hiby (1962) showed that the effect of D/d is not significant in the measured of longitudinal dispersion coefficient when the ratio is greater than 12.

Stephenson and Stewart (1986) showed that the area of high fluid velocities limits to the area of high porosities, and this area does not extend more than a particle diameter of the wall and the assumption of a flat velocity profile is reasonable. This work confirms the earlier experiments reported by Schuster and Vortmeyer (1980) and Vortmeyer and Schuster (1983).

A similar effect was observed in measuring pressure drops across packings, so an empirical rule can be considered that the variations, in radial position, of the fluid velocity, porosity and dispersion coefficient can be negligible, if $D/d > 15$ (Gunn, 1968).

2.1.3. Ratio of column length to particle diameter

Strang and Geankoplis (1958) and Liles and Geankoplis (1960) make much of the effect of but the evidence from fluid mechanical studies (Gunn and Malik, 1966) was that the effect is confined to a dozen layers of particles and is not very important.

Experimental results of Guedes de Carvalho and Delgado (2003) with two different spherical particles diameter and the same length of the packed bed showed that axialal dispersion coefficient does not increased with particle diameter, as long as the condition $D/d > 15$ is satisfied (see Vortmeyer and Schuster, 1983; Ahn et al., 1986, for wall effects).

2.1.4. Particle size distribution

Another aspect of dispersion in packed beds that needs to receive attention is the effect of porous medium structure. In a packed bed of different particle sizes, the small particles accumulate in the interstices between large particles, and porosity tends to decrease.

Raimondi et al. (1959) and Niemann (1969) studied the effect of particle size distribution on longitudinal dispersion and concluded that D_L increases with a wide particle size distribution. Eidsath et al. (1983) showed a strong effect of particle size distribution on dispersion; if the ratio of particle diameters went from a value of 2 to 5 the axial dispersion increased by a factor of 1.5.

Han et al. (1985) showed that for a size distribution with a ratio of maximum to minimum particle diameter equal to 7.3, D_L values are 2 to 3 times larger than the uniform size particles.

Wronski and Molga (1987) studied the effect of particle size non-uniformities on axial dispersion coefficients during laminar liquid flow through packed beds (with a ratio of maximum to minimum particle diameter equal to 2.13) and proposed a generalized function to determine the increase of the axial dispersion coefficients in non-uniform beds relative to those obtained in uniform beds. Guedes de Carvalho and Delgado (2003) obtained the same conclusion in their experiments, with ballotini and a ratio of maximum to minimum particle diameter equal to 3.5 in comparison with glass ballotini that have the same size.

2.1.5. Particle shape

Bernard and Wilhelm (1950); Ebach and White (1958); Carberry and Bretton (1958); Strang and Geankopolis (1958); Hiby (1962) and Guedes de Carvalho and Delgado (2003) using beds of spheres, cubes, Raschig rings, sand, saddles and other granular material have concluded that generally longitudinal dispersion coefficient tend to be greater with packs of non-spherical particles than with packs of spherical particles, with the same size.

Particle shape is a significant parameter, with higher values of D_L (i.e., lower Pe_L values) being observed in packed beds of sand and Raschig rings comparatively with the results obtained with spherical beds. Therefore, increased particle sphericity correlates with decreased dispersion, with a sphericity defined as the surface area of a particle divided by the surface area of a sphere of volume equal to the particle.

2.2. Parameters Influencing Longitudinal Dispersion–Fluid Properties

2.2.1. *Viscosity and density of the fluid*

Some investigators, as Hennico *et al.* (1963), used glycerol and obtained a significant effect of viscosity, at large Reynolds number, on values of D_L. In vertical miscible displacements, if a less viscous fluid displaced another fluid viscous fingers will be formed (Perkins and Johnston, 1963). However, if a more viscous fluid displaced a different fluid the dispersion mechanisms are unaffected, but the situation will tend to reduce convective dispersion. This leads to increased dispersion relative to the more viscous fluid displacing a less viscous one.

The importance of density gradients has investigated by Benneker *et al.* (1996) and their experiments showed that the D_L value is considerably affected by fluids with different densities due the action of gravity forces. Fluid density creates similar effects to fluid viscosity. In a displacement with a denser fluid above the less-dense fluid, gravity forces cause redistribution of the fluids. However, if a denser fluid is on the bottom, usually, a stable displacement occurs.

2.2.2. *Fluid velocity*

The first two groups of Eq. (3) have importance only when D/d is less than 15 and L/D is so small that the characteristics of dispersion are affected by changing velocity distributions. So, for packed beds we will usually have $D_L/D_m = \Phi (Pe_m, Sc)$.

In order to understand the influence of fluid velocity on the dispersion coefficient, it is important to consider the limiting case where $u \to 0$. If D_L was defined based on the area open to diffusion, in the limit $u \to 0$, solute dispersion is determined by molecular diffusion, with $D_L = D'_m = D_m/\tau$ (τ is equal to $\sqrt{2}$ as suggested by Sherwood *et al.*, 1975).

As the velocity of the fluid is increased, the contribution of convective dispersion becomes dominant over that of molecular diffusion (see Wilhelm, 1962) and $D_L = ud/Pe_L (\infty)$, where u is the interstitial fluid velocity and $Pe_L (\infty) \cong 2$ for gas or liquid flow through beds of (approximately) isometric particles, with diameter d (Cairns and Prausnitz, 1960; Jacques and Vermeulen, 1958). Assuming that the diffusive and convective components of dispersion are additive, the same authors suggest that which may be written in dimensionless form (Gunn, 1968) as

$$\frac{D_L}{D_m} = \frac{1}{\tau} + \frac{1}{2}\frac{ud}{D_m} \quad \text{or} \quad \frac{1}{Pe_L} = \frac{1}{\tau}\frac{\varepsilon}{Re\,Sc} + \frac{1}{2} \quad ...(7)$$

This equation is expected to give the correct asymptotic behaviour, in gas and liquid flow through packed beds, at high and low values of Pe_m.

2.2.3. *Fluid temperature (or Schmidt number)*

In recent years, data on longitudinal dispersion have been made available for values of Sc between the two extremes of near ideal gas (Sc $\cong$ 1) and cold water (Sc > 550). Such data were obtained with either supercritical carbon dioxide (1.5 < Sc < 20) or heated water (55 < Sc < 550). The experimental results show a consistent increase in Pe_L with a decrease in Sc and it may be seen that the dependence is slight for the higher values of Sc (say for Sc of order 750 and above). At the lower end of the range of Pe_m investigated there seems to be a tendency for Pe_L to become independent of Sc, even if the values of D_L are still significantly above D_m. In the intermediate range, $100 < Pe_m < 5000$, values of Pe_L are very nearly constant, for each value of Sc. The convergence of the different series of points at about $Pe_m \cong 20$ seems to suggest that Pe_L is insensitive to Sc below this value of Pe_m.

A good additional test of the consistency of the data of Guedes de Carvalho and Delgado (2003) is supplied by the plot in Fig. 7.2, where it may be seen that all the series of points converge at high Re, as would be expected for turbulent flow. The agreement with the data of Jacques and Vermeulen (1958) and Miller and King (1966), for cold water, is worth stressing.

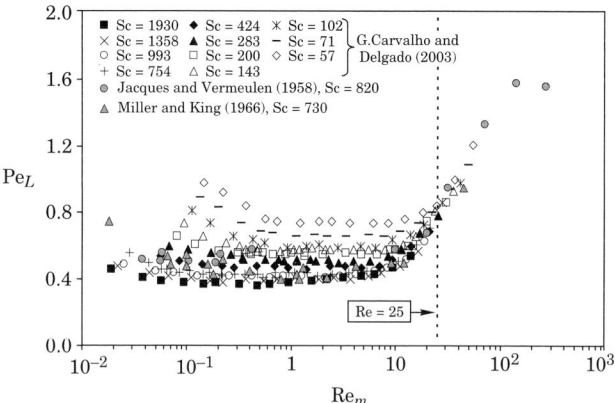

Fig. 7.2. Dependence of Pe_L on Re for different values of Sc

Recently, some workers have measured axial dispersion for the flow of supercritical carbon dioxide through fixed beds and this provides important new data in the range 1.5 < Sc < 20. However, the various authors fail to recognize the direct dependence of Pe_L on Sc. Catchpole *et al.* (1996)

represent their data and those of Tan and Liou (1989) in a single plot of Pe_L vs. Re. The majority of points are in the range $1 < Re < 30$ and the data of both groups, together, define a horizontal cloud with mid line at about $Pe_L \cong 0.8$, spreading over the approximate range $0.3 < Pe_L < 1.1$. The data of Yu et al. (1999) are for $0.01 < Re < 2$ and $2 < Sc < 9$. It is worth referring here that the modelling work of Coelho et al. (1997) gives theoretical support to experimental findings for low Re, both for spherical and non-spherical particles. No influence of Sc on Pe_L is detected, but unfortunately the results are not very consistent, particularly in the range $Pe_m < 20$, where the scatter is high and the values of are much too low.

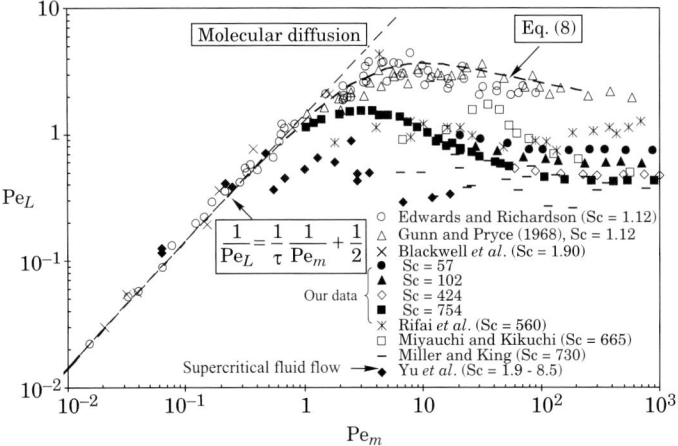

Fig. 7.3. Dependence of on for Stokes flow regime

Figure 7.3 shows that for low values of Pe_m (Stokes flow regime) there seems to be a tendency for Pe_L to become independent of Sc. The values of Pe_L reported by Miller and King (1966), for $6 < Pe_m < 100$, are much too low; this may be because the particles used in most experiments are too small and this is known to yield enhanced dispersion coefficients, possibly due to particle agglomeration (see Gunn, 1987). The data reported by Miyauchi and Kikuchi (1975) and plotted in Fig. 7.3, for $6 < Pe_m < 300$, are higher than our experimental data.

There are considerable experimental difficulties in the measurement of axial dispersion in the liquid phase at small Reynolds number, because the usual method of obtaining low Reynolds number is to reduce particle size and this is known to yield enhanced dispersion coefficients.

2.3. Correlations

After an exhaustive compilation and a critical analysis of the dispersion data, for beds of mono-sized particles of constant voidage, the results obtained suggest that our experimental data, more than five hundred values

(Guedes de Carvalho and Delgado, 2005; Guedes de Carvalho and Delgado, 2003; Delgado and Guedes de Carvalho, 2001), are consistent and accurate with the vast amount of data available in literature. And, since they span very wide ranges of the parameters involved, they were taken as the reference data to help identify simple mathematical expressions for the prediction of dispersion coefficients.

The coefficient of axial dispersion for gas flow (Sc ≅ 1) is predicted with good accuracy by Eq. (7), except in the approximate range $0.5 < Pe_m < 100$, where experimental values may be more than twice those given by the equation (see Fig. 7.4). Several equations have been proposed to represent the data in this intermediate range and the equation of Hiby (1962),

$$\frac{D_L}{D'_m} = 1 + \frac{0.65 Pe'_m}{1 + 7\sqrt{\tau/Pe'_m}} \quad \text{(valid for Re < 100)} \quad ...(8)$$

is shown to fit the data points reasonably well.

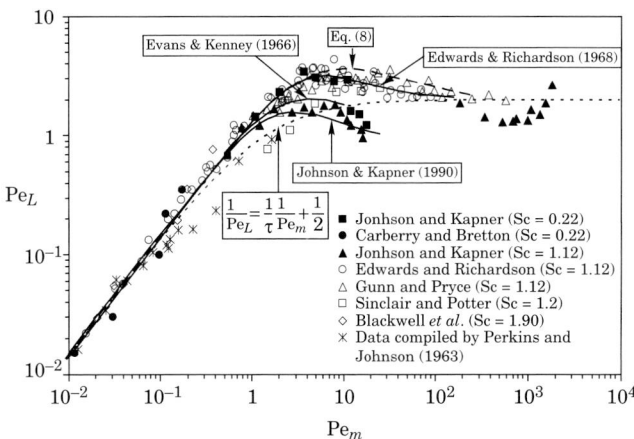

Fig. 7.4. Longitudinal dispersion in gaseous systems

For liquid flow, a large number of data are available, that were obtained with different solutes in water at near ambient temperature, corresponding to values of Sc in the range $500 < Sc < 2000$. Most of the data reported in the literature, for this range of Sc, form a "thick cloud" running parallel to the line defined by Eq. (7), though somewhat below it (at approximately, $0.3 < Pe_L < 2$).

For the case of liquid flow in a porous media, we used the division in five dispersion regimes to obtain the expressions presented below.

1. Diffusion regime (valid for $Pe_m < 0.1$):

$$\frac{D_L}{D'_m} = 1 \quad ...(9)$$

2. Predominant diffusional regime (valid for $0.1 < \text{Pe}_m < 4$):

$$\frac{D_L}{D'_m} = \frac{\text{Pe}'_m}{0.8/\text{Pe}'_m + 0.4} \qquad \ldots(10)$$

with a an average relative deviation lesser than 14%.

3. Predominant mechanical dispersion (valid for $4 < \text{Pe}_m$ and $\text{Re} < 10$):

$$\frac{D_L}{D'_m} = \frac{\text{Pe}'_m}{\sqrt{18\text{Pe}'^{-1.2}_m + 2.35\text{Sc}^{-0.38}}} \qquad \ldots(11)$$

with a deviation lesser than 11%, over the entire range of Pe'_m and Sc.

4. Pure mechanical dispersion (valid for $10 < \text{Re}$ and $\text{Pe}_m < 10^6$):

$$\frac{D_L}{D'_m} = \frac{\text{Pe}'_m}{25\text{Sc}^{1.14}/\text{Pe}'_m + 0.5} \qquad \ldots(12)$$

with a an average relative deviation lesser than 16%, over the entire range of Pe'_m and Sc.

5. Dispersion out of Darcy domain (valid for $\text{Pe}_m > 10^6$):

$$\frac{D_L}{D'_m} = \frac{\text{Pe}'_m}{2} \qquad \ldots(13)$$

The correlations proposed are shown (see Fig. 7.5) to be significantly more accurate than previous correlations (see Fig. 7.6) and they cover the entire spectrum of values of Pe_m and Sc expected to be useful. It is important to bear in mind that Eqs. (9)–(13) are recommended only for random packings of approximately "isometric" particles.

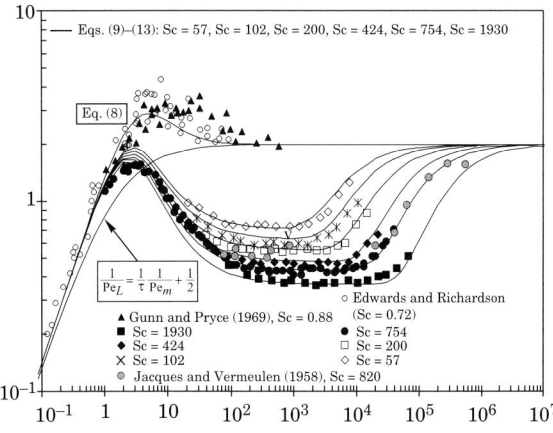

Fig. 7.5. Comparison between experimental data and Eqs. (9) and (13)

It is worth considering here the predicting accuracy of alternative correlations. Gunn (1969) admitted the existence of two regions in the packing, one of fast flowing and the other of nearly stagnant fluid, to deduce the following expression for the longitudinal dispersion coefficient in terms of probability theory

$$\frac{1}{\text{Pe}_L} = \frac{\varepsilon \text{Pe}_m}{4\alpha_1^2(1-\varepsilon)}(1-p)^2 + \left[\frac{\varepsilon \text{Pe}_m}{4\alpha_1^2(1-\varepsilon)}\right]^2 p(1-p)^3$$

$$\left\{\exp\left[-\frac{4(1-\varepsilon)\alpha_1^2}{p(1-p)\varepsilon \text{Pe}_m}\right] - 1\right\} + \frac{1}{\tau \text{Pe}_m} \quad ...(14)$$

where α_1 is the first zero of equation $J_0(U) = 0$ and p is defined, for a packing of spherical particles, by

$$p = 0.17 + 0.33 \times \exp\left(-\frac{24}{\text{Re}}\right) \text{ for spheres, } \tau = \sqrt{2} \quad ...[15(a)]$$

$$p = 0.17 + 0.29 \times \exp\left(-\frac{24}{\text{Re}}\right) \text{ for solid cylinders, } \tau = 1.93$$
$$...[15(b)]$$

$$p = 0.17 + 0.20 \times \exp\left(-\frac{24}{\text{Re}}\right) \text{ for hollow cylinders, } \tau = 1.8$$
$$...[15(c)]$$

Tsotsas and Schlunder (1988) deduced an alternative correlation for the prediction of Pe_L. The authors defining two zones in a simple flow model consisting of a fast stream (central zone in the model capillary) and a stagnant fluid, but the mathematical expressions associated with it are a little cumbersome,

$$\frac{1}{\text{Pe}_L} = \frac{1}{\tau}\left[\frac{1}{\text{Pe}_{z,1}} + \frac{1}{\text{Pe}'_m}(1-\xi_c^2)\right] + \frac{1}{32}\left(\frac{D_c}{d}\right)^2$$

$$[\text{Pe}_{r,1}\,\xi_c^2\,f_1(\xi_c) + \text{Pe}'_m\,f_2(\xi_c)] \quad ...(16)$$

where the longitudinal and radial Peclet number of the fast stream is

$$\frac{1}{\text{Pe}_{z,1}} = \frac{1}{\text{Pe}'_1} + \frac{1}{1.14(1+10/\text{Pe}'_1)} \quad ...[17(a)]$$

$$\frac{1}{\text{Pe}_{r,1}} = \frac{1}{\text{Pe}'_1} + \frac{1}{8} \quad ...[17(b)]$$

$$\text{Pe}'_1 = \frac{u_1 d}{D'_m} \quad ...[17(c)]$$

and $u_1 = u/\xi_c^2$ is the interstitial velocity of the fast stream, with ξ_c (the dimensionless position of the velocity jump, i.e., the ratio between the radius of the zone of high velocities and the radius of packed bed) equal to

$$\text{Re} \leq 0.1 \rightarrow \xi_c = 0.2 + 0.21 \exp(2.81y) \qquad \ldots[17\,(d)]$$
$$\text{Re} \geq 0.1 \rightarrow \xi_c = 1 - 0.59 \exp[-f(y)] \qquad \ldots[17\,(e)]$$
with
$$y = \log(\text{Re}) + 1 \qquad \ldots[17\,(f)]$$
$$f(y) = y(1 - 0.274y + 0.086y^2) \qquad \ldots[17\,(g)]$$

Finally, the distributions functions $f_1(\xi_c)$ and $f_2(\xi_c)$ are defined by:
$$f_1(\xi_c) = (1 - \xi_c^2)^2 \qquad \ldots[17\,(h)]$$
$$f_2(\xi_c) = 4\xi_c^2 - 3 - 4\ln(\xi_c) - \xi_c^4 \qquad \ldots[17\,(i)]$$

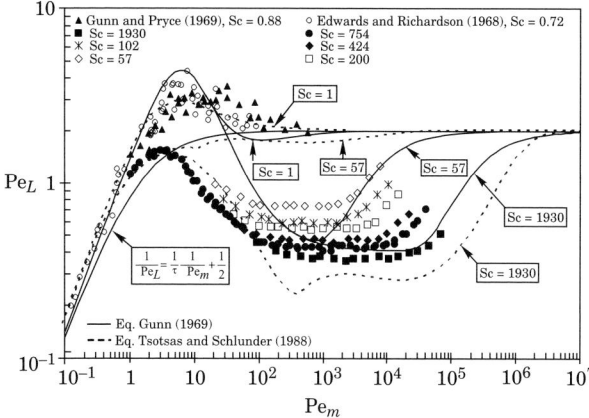

Fig. 7.6. Comparison between experimental data and correlations presented in literature

In Fig. 7.6, the lines corresponding to the correlations of Gunn (1969) and of Tsotsas and Schlunder (1988) are represented, for the higher and lower values of Sc in our experiments (Sc = 57 and Sc = 1930), as well as for gas flow (Sc = 1). It may be seen that the correlation of Gunn (1969) is not sensitive to changes in Sc, for $\text{Pe}_m < 10^3$, and the correlation of Tsotsas and Schlunder (1988) is much too sensitive to variations in Sc; however, this correlation describes dispersion in gas flow with good accuracy.

3. TRANSVERSE OR RADIAL DISPERSION

Generally, radial dispersion coefficients are measured in non-reactive conditions, because the rate of mass transfer, observed experimentally, is directly related to the coefficient of radial dispersion in the bed. The most popular technique for the measurement of transverse dispersion consists in feeding a continuous stream of tracer from a "point" source somewhere in the bed and measuring the radial variation of tracer concentration at one or more downstream locations. The first study of mass transfer by radial dispersion in gaseous systems was carried out by Towle and Sherwood (1939). The results presented were very important for packed bed dispersion because they showed that dispersion was not influenced by the tracer molecular weight.

Bernard and Wilhelm (1950) reported the first measurements, in liquid systems, of experimental values of transverse dispersion coefficients in packed beds of inerts by a Fickian model. The authors took into account the wall effect condition and their experiments suggested that for high values of Reynolds number the value of is constant and between 11 and 13.

Baron (1952) proposed a new model of radial dispersion in which a particle of tracer executes a simple random-walk displacement of ± ½ particle diameter to give a transversal Peclet number between 5 and 13, when Re ≥ ∞. The basis for this prediction is the random-walk theory, in which a statistical approach is employed. This method does not take into account effects of radial variations in velocity and void space. Latinen (1951) has extended the random-walk concept to three dimensions and predicted a value of 11.3, for $Pe_T(\infty)$.

Plautz and Johnstone (1955) used the equation derived by Wilson (1904), for heat transfer, and suggested a Pe_T between 11 and 13, for Re ≥ ∞. Fahien and Smith (1955) assumed that for Reynolds numbers in the range between 40 and 100, the Peclet number is independent of fluid velocity and equal to 8. The authors were the first to consider that the tracer pipe can be of significant diameter compared to the diameter of the bed.

Dorweiler and Fahien (1959) used the equation derived by Fahien and Smith (1955) to study mass transfer in laminar and transient flows. The results showed that for Re < 200, the Peclet number based on the transverse dispersion coefficient is a linear function of the fluid velocity and for Re > 200, at room temperature, the Peclet number is constant as also shown by Bernard and Wilhelm (1950), Plautz and Johnstone (1955) and Fahien and Smith (1955). The authors have demonstrated a difference in the Peclet number with radial position. The transversal Peclet number is constant from the axis to 0.8 times the radius and then rises near the wall.

Hiby and Schummer (1960), and later Roemer et al. (1962), presented the solution of the mass balance equation [Eq. (1)], considering the tracer pipe to be of significant diameter compared to the diameter of the packed bed.

Saffman (1960) considered the packed bed as a network of capillary tubes randomly orientated with respect to the main flow. At high Pe_m and at very long time, Saffman found that the dispersion never becomes truly mechanical, with zero velocity of the fluid at the capillary walls, the time required for a tracer particle to leave a capillary would become infinite as its distance from the walls goes to zero. The author proposed that $D_T = (3/16)\,ud$ when Re → ∞, but this prevision of transverse dispersion coefficient is higher than observed experimentally.

Hiby (1962) and Blackwell (1962) presented an experimental technique in which they divided the sampling region into two annular regions and calculated the transversal dispersion coefficient from the averaged concentrations of each of the two samples.

The experimental data points of Wilhelm (1962) suggested that $Pe_T (\infty) = 12$, for beds of closely sized particles, and this value is accepted for the majority of the investigators (ex: Hiby, 1962; Wilhelm, 1962; Gunn, 1968; Coelho and Guedes de Carvalho, 1988).

Roemer et al. (1962) studied radial mass transfer in packed beds at low flow rates, Re < 100. The authors considered the tracer pipe to be of significant diameter compared to the diameter of the bed ("finite source" model) and longitudinal and transverse dispersion are equal. In this work the authors compared the values of Pe_T obtained with two methods ("instantaneous finite source" and "point source") and concluded that the values of Pe_T obtained with the "point source" method were 10% less that the values obtained with the "instantaneous finite source" method. The authors estimated that the neglecting the longitudinal dispersion in calculations of D_T, for low values of Reynolds numbers, can cause errors of 10%.

Coelho and Guedes de Carvalho (1988) developed a new experimental technique, based on the measurement of the rate of dissolution of planar or cylindrical surfaces, buried in the bed of inert particles and aligned with the flow direction. This alternative technique is simple to use, allows the determination of the coefficient of radial dispersion in packed beds over a wide range of flow rates, and it is easily adaptable to work over a range of temperatures above ambient, as shown Guedes de Carvalho and Delgado (2000) and Delgado and Guedes de Carvalho (2001).

In recent years, nuclear magnetic resonance has been used to determine both diffusion and dispersion coefficients (e.g., Gibbs et al., 1992; Baumeister et al., 1995), with significant advantages, but this technique were limited to low fluid velocities.

It is important to remember that, at high Reynolds numbers, the main mechanism of transverse dispersion is the fluid deflection caused by deviations in the flow path caused by the particles in the bed (axial dispersion is caused by differences in fluid velocity in the flow), i.e., dispersion is caused by hydrodynamic mechanisms (macroscopic scale) and not by molecular diffusion (Brownian motion).

The result is a poor mixture at the "microscopic scale". In fact, there are detected different values of solute concentration over a distance of the order of a particle diameter or less, what explains the convenience of use of an efficient averaging procedure (Gunn, 1968). This is probably one of the reasons that explain the difference observed in some experimental results of dispersion (see Fig. 7.7). Gunn and Pryce (1969) showed that the standard deviation without repacking in the measurement of Pe_T was 5%, while when the bed was repacked each time of measurement, the standard deviation found was 15%.

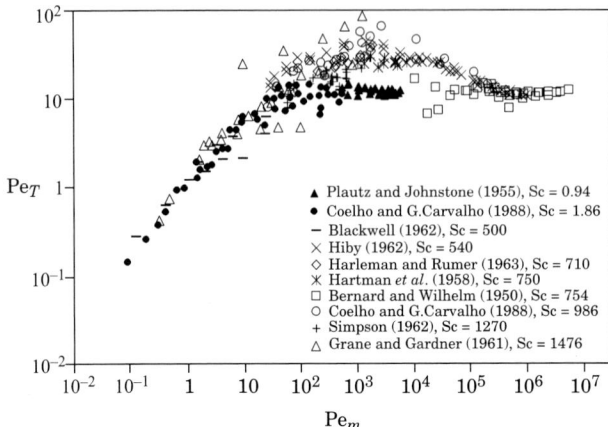

Fig. 7.7. Experimental data points for transverse dispersion in liquids and gas systems

3.1. Parameters Influencing Transverse Dispersion–Porous Medium

3.1.1. *Length of the packed column*

Han *et al.* (1985) showed that values of the radial dispersion coefficient, for uniform size packed beds, measured at different positions in the bed are not function of bed location, *i.e.*, they observed no time dependent behaviour for radial dispersion, because transverse dispersion is caused by mechanical mechanism alone.

An important aspect to be considered, as a check on the experimental method of Coelho and Guedes de Carvalho (1988), is the influence of the length of the test cylinder on the measured value of D_T. In reality, the two variables are independent, provided that the criterion given by the following equation is satisfied.

$$\frac{L}{d} \geq 0.62 \left(\frac{ud}{D_m} \right) \qquad \ldots(18)$$

3.1.2. *Ratio of column diameter to particle diameter*

Several investigators, as Fahien and Smith (1955) and Latinen (1951), have studied the wall effect on transverse dispersion coefficient. The experiments suggested that in a packing structure characterized by significant variations of void fraction in radial direction, up to distance of about two particle diameters from the wall, a non-uniform radial velocity profile is induced, with a maximum just near the wall. As result, wall effects occur due large voidage fluctuations near the wall. Their, also showed, that the increase in radial dispersion in the laminar region would be the same order of magnitude as in the turbulent region.

3.1.3. Particle size distribution

Eidsath *et al.* (1983) studied the effect of particle size distribution on dispersion. As the ratio of particle diameter went from a value of 2 to 5, the radial dispersion decreased by a factor of 3, but perhaps the results were a cause of the simple geometry employed in these computations (packed bed of cylinders). Steady-state measurements of radial dispersion reported by Han *et al.* (1985), with same void fraction and the same mean particle diameter, but different particle size range (ratio of maximum to minimum particle diameter equal to 2.2 and 7.3), showed that there was no evidence to indicate a change in radial dispersion with particle size distribution.

3.1.4. Particle shape

The effect of particle shape on the radial dispersion coefficient has been given attention by several investigators both for gaseous and liquid systems. England and Gunn (1970) measured the dispersion of argon in beds of solid cylinders and beds of hollow cylinders and have concluded that D_T tend to be greater with packs of hollow cylinders than with packs of solid cylinders, and these results were greater than obtained with packs of spherical particles.

The same conclusion, in liquid systems, has been obtained by Hiby (1962), who used packed beds of glass spheres and Rachig rings, and Bernard and Wilhem (1950), who used packed beds of cubes, cylinders and glass spheres. Radial dispersion coefficient tends to be greater in packed beds of non-spherical particles.

However, Blackwell (1962), Guedes de Carvalho and Delgado (2000) and others reported results with packed beds of sand and showed that D_T obtained with "glass ballotini" are very close to those for sand (not pebble or gravel) and the conclusion seems to be that particle shape has only a small influence on radial dispersion, for random packings of "isometric" particles.

3.2. Parameters Influencing Transverse Dispersion–Fluid Properties

3.2.1. Viscosity and density of the fluid

An effect of fluid densities and viscous forces on transverse dispersion has been studied by Grane and Garner (1961) and Pozzi and Blackwell (1963) and the authors concluded that when a fluid is displaced from a packed bed by a less viscous fluid, the viscous forces create an unstable pressure distribution and the less viscous fluid will penetrate the medium in the form of fingers, unless the density has an opposing effect.

3.2.2. Fluid velocity

For very low fluid velocities, u, dispersion is the direct result of molecular diffusion, with $D_T = D'_m$. As the velocity of the fluid is increased, the contribution of convective dispersion becomes dominant over that of molecular diffusion and D_T becomes less sensitive to temperature. According to several authors (see Hiby, 1962; Wilhelm, 1962; Gunn, 1968; Coelho and Guedes de Carvalho, 1988) $D_T \to ud/\text{Pe}_T (\infty)$, for high enough values of u, where d is particle size and $\text{Pe}_T (\infty) \cong 12$ for beds of closely sized particles. Assuming that the diffusive and convective components of dispersion are additive, the same authors suggest that $D_T = D'_m + ud/K$, which may be written in dimensionless form as

$$\frac{D_T}{D_m} = \frac{1}{\tau} + \frac{1}{12}\frac{ud}{D_m} \quad \text{or} \quad \frac{1}{\text{Pe}_T} = \frac{1}{\tau}\frac{\varepsilon}{\text{ReSc}} + \frac{1}{12} \quad \ldots(19)$$

This equation has been shown to give a fairly accurate description of transverse dispersion in gas flow through packed beds, but it is not appropriate for the description of dispersion in liquids, over an intermediate range of values of ud/D_m.

Delgado and Guedes de carvalho (2001) show that the value of the transverse dispersion coefficient is seen to increase with fluid velocity and comparison between the two plots shows that D_T also increases with particle size (d).

3.2.3. *Fluid temperature (or Schmidt number)*

The dependence of D_T on liquid properties and velocity is best given in plots of Pe_T vs. Pe_m, for different values of Sc. Not surprisingly, Fig. 7.8 shows that the variation of Pe_T with Pe_m gets closer to that for gas flow as the value of Sc is decreased. For the lowest Sc tested (Sc = 54; $T = 373$ K), Pe_T does not differ by more than 30% from the value given by Eq. (19), with $\text{Pe}_T (\infty) = 12$, over the entire range of Pe_m. But for the higher values of Sc, the experimental values of Pe_T may be up to four times the values given by Eq. (19).

Delgado and Guedes de Carvalho (2001) had studied the dependence of D_T/D_m on Sc, up to $\text{Pe}_m \cong 1350$, and they reported a smooth increase in D_T/D_m with Pe_m, for all values of Sc. But the data show that there is a sudden change in the trend of variation of Pe_T with Pe_m, somewhere above $\text{Pe}_m \cong 1350$, a maximum being reached in the approximate range $1400 < \text{Pe}_m < 1800$ (depending on Sc). The fact that the change in trend corresponds to a much enhanced increase in D_T (i.e., a decrease in Pe_T), in response to a small increase in u (i.e., in Pe_m), strongly suggests a connection with the transition from laminar to turbulent flow in the interstices of the packing. The plot of Pe_T vs. Re, shown in Fig. 7.8, seems to support this view, since the maxima in Pe_T are reached for $0.3 < \text{Re} < 10$ (depending on Sc) and this is the approximate range of values of Re for the transition from laminar to turbulent flow. The range $1 < \text{Re} < 10$ is often indicated for that transition (see for example Bear, 1972), but Scheidegger (1974) as giving Re = 0.1 for the lower limit of that transition.

The plot in Fig. 7.8 also suggests that "purely mechanical" fluid dispersion will be observed above about Re = 100; this value is estimated as the convergence of the data points for liquids with the line representing Eq. (19).

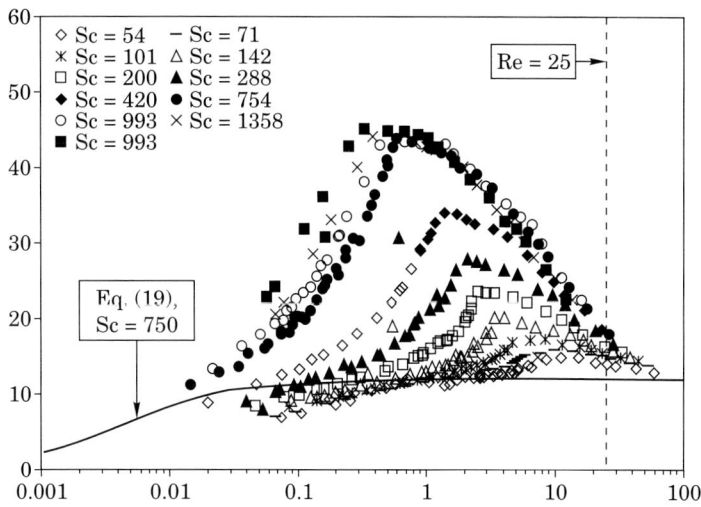

Fig. 7.8. Dependence of Pe_T on Re for different values of Sc

3.3. Correlations

For gas flow, by simply adding the contributions of molecular diffusion and turbulent dispersion, we suggested Eq. (19) (in dimensionless form)

$$\frac{D_T}{D'_m} = 1 + \frac{Pe'_m}{12} \qquad ...(19)$$

with a an average relative deviation lesser than 12%. Equation (19) give the correct asymptotic behaviour (both for very high and very low Pe_m) for both gases and liquids, as reported by several workers (see Gunn, 1969; Wilhelm, 1962). In the intermediate range of Pe_m they are still a reasonable approximation for gases, the wider deviation being observed in the intervals $3 < Pe_m < 300$, as shown in Fig. 7.9.

For the case of liquid flow in a porous media, we used the division in four dispersion regimes to obtain the expressions presented below:

1. Diffusion regime (valid for $Pe_m < 1$):

$$\frac{D_T}{D'_m} = 1 \qquad ...(20)$$

2. Predominant mechanical dispersion (valid for $1 < Pe_m < 1600$):

$$\frac{D_T}{D'_m} = 1 + \frac{1}{2.7 \times 10^{-5} Sc + 12/Pe'_m} \quad \text{for Sc} < 550 \qquad ...[21(a)]$$

$$\frac{D_T}{D'_m} = 1 + \frac{1}{0.017 + 14/\text{Pe}'_m} \quad \text{for Sc} \geq 550 \qquad ...[21\,(b)]$$

with a deviation lesser than 8% and 5%, respectively, over the entire range of Pe'_m and Sc.

3. Pure mechanical dispersion (valid for $1600 < \text{Pe}_m < 10^6$):

$$\frac{D_T}{D'_m} = \frac{\text{Pe}'_m}{(0.058\text{Sc} + 14) - (0.058\,\text{Sc} + 2)\exp\left(-\dfrac{500\text{Sc}^{0.5}}{\text{Pe}'_m}\right)} \quad \text{for Sc} < 550 \quad ...[22\,(a)]$$

$$\frac{D_T}{D'_m} = \frac{\text{Pe}'_m}{45.9 - 33.9 \times \exp\left(-\dfrac{21\,\text{Sc}}{\text{Pe}'_m}\right)} \quad \text{for Sc} \geq 550 \quad ...[22\,(b)]$$

and the experimental data do not deviate by more than 6% and 4% from the values given by Eqs. [22 (a)] and [22 (b)], respectively.

4. Dispersion out of Darcy domain (valid for $\text{Pe}_m > 10^6$):

$$\frac{D_T}{D'_m} = \frac{\text{Pe}'_m}{12} \qquad ...(23)$$

The experimental data are shown in Fig. 7.9, alongside the solid lines corresponding to Eqs. (20)–(23), for the values of Sc indicated in the figure. The agreement is seen to be generally very good, even when the values of Pe_T are represented on a linear scale. For Sc > 550 (see experiments with alues of Sc of 754 and 1930), the above equations representing the data must take into account that Pe_T is only dependent on Pe_m, in the ascending part of the curve Pe_T vs. Pe_m and that Pe_T only depends on Re, in the descending part of the same curve.

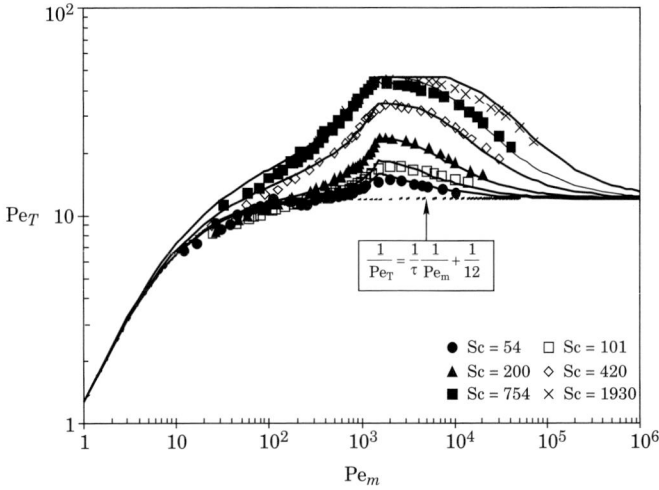

Fig. 7.9. Comparison between experimental data and Eqs. (20) and (23)

In this context it is interesting to consider the predicting accuracy of some alternative empirical correlations that have been proposed to represent the experimental data in liquid flow, as the equation of Gunn (1969):

$$\frac{1}{Pe_T} = \frac{1}{Pe_f} + \frac{1}{\tau}\frac{\varepsilon}{Re\,Sc} \qquad ...(24)$$

where the fluid-mechanical Peclet number, Pe_f, is defined by,

$Pe_f = 40 - 29e^{-7/Re}$ for spheres, $\tau = \sqrt{2}$ ...[25 (a)]
$Pe_f = 11 - 4e^{-7/Re}$ for solid cylinders, $\tau = 1.93$ ...[25 (b)]
$Pe_f = 9 - 3.3e^{-7/Re}$ for hollow cylinders, $\tau = 1.8$ ...[25 (c)]

And the empirical equation proposed by Wen and Fan (1975),

$$Pe_T = \frac{17.5}{Re^{0.75}} + 11.4 \text{ (for high values of } Pe_m) \qquad ...(26)$$

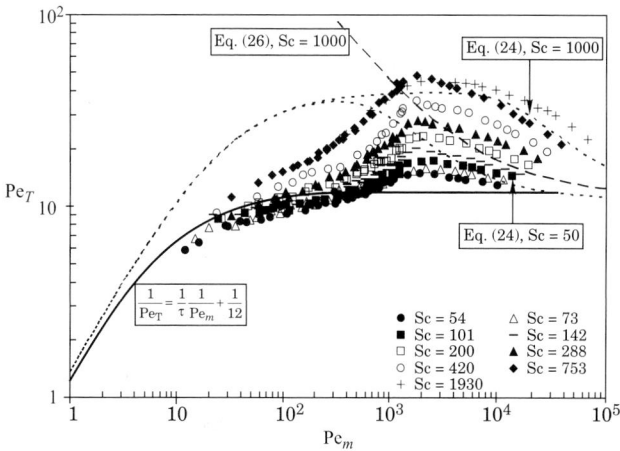

Fig. 7.10. Comparison between experimental data and correlations of other investigators

In Fig. 7.10, the solid line corresponding to Eq. (19), with $Pe_T(\infty) = 12$, is again represented and so are two dashed lines corresponding to a correlation proposed by Gunn (1969) for the extreme values of Schmidt number obtained in Guedes de Carvalho and Delgado (2005) experiments. Comparison with the experimental points shows that the correlation does not account for the influence of Sc on Pe_T, for $Pe_m < 600$, and it may be seen to be very inadequate at low values of Sc, for $600 < Pe_m < 10^5$. The empirical correlation proposed by Wen and Fan (1975), for high values of Pe_m and Sc, is also very inadequate, because it is only based in experimental data of Bernard and Wilhelm (1950) and on the Hartman et al. (1958).

4. CONCLUSIONS

The present work increases our knowledge about dispersion in packed beds by providing a critical analysis on the effect of fluid properties and porous medium on the values of axial and radial dispersion coefficients.

A large number of experimental data on dispersion available in literature for packed beds were examined to pave the way for the formulation of new correlations for the prediction of Pe_T and Pe_L. Empirical correlations are presented for the prediction of the dispersion coefficients (D_T and D_L) over the entire range of practical values of Schmidt number and Peclet number. The simple mathematical expressions represent the data available, in literature, with good accuracy and they are shown to be a significant improvement over previous correlations. The axial dispersion coefficient can be calculated by Eqs. (9) and (13) and the radial dispersion coefficient by Eqs. (20) and (23).

REFERENCES

1. Ahn, B.J., Zoulalian, A. and Smith, J.M. 1986. Axial Dispersion in Packed Beds with Large Wall Effect, *AIChE J.* **32(1)**: 170–4.
2. Aris, R. 1959. On the Dispersion of a Solute by Diffusion, Convection and Exchange between Phases, *Proc. Royal Soc. A* **252(1271)**: 538–50.
3. Aris, R. 1956. On the Dispersion of a Solute in a Fluid flowing Through a Tube, *Proc. Royal Soc. A* **235(1200)**: 67–77.
4. Baron, T. 1952. Generalized Graphical Method for the design of Fixed Bed Catalytic Reactors, *Chem. Eng. Progr.* **48(3)**: 118–24.
5. Baumeister, E., Klose, U., Albert, K. and Bayer, E. 1995. Determination of the Apparent Transverse and Axial Dispersion Coefficients in a Chromatographic Column by Pulsed Field Gradient Nuclear Magnetic Resonance, *J. Chromatogr. A* **694(2)**: 321–31.
6. Bear, J. 1972. Dynamics of Fluids in Porous Media. Dover Publications, New York.
7. Benneker, A.H., Kronberg, A.E., Post, J.W., Van der Ham, A.G.J. and Westerterp, K.R. 1996. Axial Dispersion in Gases Flowing Though a Packed Bed at Elevated Pressures, *Chem. Eng. Sci.* **51(10)**: 2099–2108.
8. Bernard, R.A. and Wilhelm, R.H. 1950. Turbulent Diffusion in Fixed Beds of Packed Solids, *Chem. Eng. Progr.* **46(5)**: 233–44.
9. Blackwell, R.J. 1962. Laboratory Studies of Microscopic Dispersion Phenomena, *Soc. Petrol. Engrs. J.* **225(1)**: 1–8.
10. Blackwell, R.J., Rayne, J.R. and Terry, W.M. 1959. Factors Influencing the Efficiency of Miscible Displacement, *Petroleum Transactions AIME* **216**: 1–8.
11. Brenner, H. 1962. The Diffusion Model of Longitudinal Mixing in Beds of Finite Length. Numerical Values, *Chem. Eng. Sci.* **17(4)**: 229–43.
12. Brenner, H. 1980. Dispersion Resulting From Flow Through Spatially Periodic Porous-Media, *Philos. T. Roy. Soc. A* **297**: 81–133.
13. Bruinzeel, C., Reman, G.H. and van der Laan, E.T. 1962. Eddy Diffusion. in Particulate Fluidized Beds: Model Experiments for the Design of a Large-Scale. 3th *Congress of the European Federation of Chemical Engineering*, Olympia, London.
14. Cairns, E.J. and Prausnitz, J.M. 1960. Longitudinal Mixing in Packed Beds, *Chem. Eng. Sci.* **12(1)**: 20–34.
15. Carberry, J.J. and Bretton, R.H. 1958. Axial Dispersion of Mass in Flow Through Fixed Beds, *AIChE J.* **4(3)**: 367–75.
16. Carbonell, R.G. and Whitaker, S. 1983. Dispersion in Pulsed Systems. Part. II–Theoretical Developments for Passive Dispersion in Porous Media, *Chem. Eng. Sci.* **38(11)**: 1795–1801.
17. Catchpole, O.J., Berning, R. and King, M.B. 1996. Measurement and Correlations of Packed Bed Axial Dispersion Coefficients in Supercritical Carbon Dioxide, *Ind. Engng. Chem. Res.* **35(3)**: 824–8.

18. Chung, S.F. and Wen, C.Y. 1968. Longitudinal Dispersion of Liquid Flowing Through Fixed and Fluidized Beds, *AIChE J.* **14(6)**: 857–66.
19. Coelho, D., Thovert, J.F. and Adler, P.M. 1997. Geometrical and Transport Properties of Random Packings of Spheres and Aspherical Particles, *Phys. Rev. E* **55(2)**: 1959–78.
20. Coelho, M.A.N. and Guedes de Carvalho, J.R.F. 1988. Transverse Dispersion in Granular Beds: Part I–Mass Transfer from a Wall and the Dispersion Coefficient in Packed Beds, *Chem. Eng. Res. Des.* **66(2)**: 165–77.
21. Danckwerts, P.V. 1953. Continuous Flow Systems, *Chem. Eng. Sci.* **2(1)**: 1–13.
22. Delgado, J.M.P.Q. and Guedes de Carvalho, J.R.F. 2001. Measurement of the Coefficient of Transverse Dispersion in Packed Beds over a Range of values of Schmidt Number (50–1000), *Transp. Porous Med.* **44(1)**: 165–80.
23. Delmas, H. and Froment, G.F. 1988. Simulation Model Accounting for Structural Radial Nonuniformities in Fixed Bed Reactors, *Chem. Eng. Sci.* **43(8)**: 2281–7.
24. Dorweiler, V.P. and Fahien, R.W. 1959. Mass Transfer at Low Flow Rates in a Packed Column, *AIChE J.* **5(2)**: 139–44.
25. Dullien, F.A.L. 1979. Porous Media: Fluid Transport and Pore Structure. Academic Press, San Diego.
26. Ebach, E.A. and White, R.R. 1958. Mixing of Fluids Flowing Through Beds of Oacked Solids, *AIChE J.* **4(2)**: 161–9.
27. Edwards, M.F. and Richardson, J.F. 1968. Gas Dispersion in Packed Beds, *Chem. Eng. Sci.* **23(2)**: 109–23.
28. Eidsath, A., Carbonell, R.G., Whitaker, S. and Herrmann, L.R. 1983. Dispersion in Pulsed Systems-III Comparison Between Theory and Experiments for Packed Beds, *Chem. Eng. Sci.* **38(11)**: 1803–16.
29. England, R. and Gunn, D.J. 1970. Dispersion, Pressure Drop, and Chemical Reaction in Packed Beds of Cylindrical Particles, *Trans. Inst. Chem. Eng.* **48**: T265–T275.
30. Evans, E.V. and Kenney, C.N. 1966. Gaseous Dispersion in Packed Beds at Low Reynolds Numbers, *Trans. Inst. Chem. Eng.* **44**: T189–T197.
31. Fahien, R.W. and Smith, J.M. 1955. Mass transfer in Packed Beds, *AIChE J.* **1(1)**: 28–37.
32. Froment, G.F. and Bischoff, K.B. 1990. Chemical Reactor Analysis and Design. 2nd edition, John Wiley & Sons.
33. Gibbs, S.J., Lightfoot, E.N. and Root, T.W. 1992. Protein Diffusion in Porous Gel Filtration Chromatography Media Studied by Pulsed Field Gradient NMR Spectroscopy, *J. Phys. Chem.* **96(18)**: 7458–62.
34. Grane, F.E. and Gardner, G.H.F. 1961. Measurements of Transverse Dispersion in Granular Media, *J. Chem. Eng. Data* **6(2)**: 283–7.
35. Gray, W.G. 1975. A Derivation of the Equations for Multi-Phase Transport, *Chem. Eng. Sci.* **30(2)**: 229–33.
36. Guedes de Carvalho, J.R.F. and Delgado, J.M.P.Q. 2005. Overall map and correlation of dispersion data for flow through granular packed beds, *Chem. Eng. Sci.* **60(2)**: 365–75.
37. Guedes de Carvalho, J.R.F. and Delgado, J.M.P.Q. 2003. The Effect of Fluid Properties on Dispersion in Flow Through Packed, *AIChE J.* **49(8)**: 1980–5.
38. Guedes de Carvalho, J.R.F. and Delgado, J.M.P.Q. 2000. Lateral Dispersion in Liquid Flow Through Packed Beds at Pem < 1400, *AIChE J.* **46(5)**: 1089–95.
39. Gunn, D.J. 1987. Axial and Radial Dispersion in Fixed Beds, *Chem. Eng. Sci.* **42(2)**: 363–73.
40. Gunn, D.J. 1969. Theory of Axial and Radial Dispersion in Packed Beds, *Trans. Inst. Chem. Eng.* **47**: T351–T359.
41. Gunn, D.J. 1968. Mixing in Packed and Fluidised Beds, *Chem. Eng. J.* pp. CE153–CE172.

42. Gunn, D.J. and Malik, A.A. 1966. Flow Through Expanded Beds of Solids, *Trans. Inst. Chem. Eng.* **44**: T371–T379.
43. Gunn, D.J. and Pryce, C. 1969. Dispersion in Packed Beds, *Trans. Inst. Chem. Eng.* **47**: T341–T350.
44. Han, N.W., Bhakta, J. and Carbonell, R.G. 1985. Longitudinal and Lateral Dispersion in Packed Beds: Effect of Column Length and Particle Size Distribution, *AIChE J.* **31(2)**: 277–88.
45. Harleman, D.R.F. and Rumer, R. 1963. Longitudinal and Lateral Dispersion in an Isotropic Porous Medium, *J. Fluid Mech.* **16(3)**: 385–94.
46. Hartman, M.E., Wevers, C.J.H. and Kramers, H. 1958. Lateral Diffusion with Liquid Flow Through a Packed Bed of Ion-Exchange Particles, *Chem. Eng. Sci.* **9(1)**: 80–82.
47. Hennico, A., Jacques, G. and Vermeulen, T. 1963. Longitudinal Dispersion in Single-Phase Liquid Flow through Ordered and Random Packings. Lawrence Rad Lab Rept UCRL 10696.
48. Hiby, J.W. 1962. Longitudinal and transverse mixing during single-phase flow through granular beds. Interact between Fluid & Particles (London Instn Chem Engrs). pp. 312–25.
49. Hiby, J.W. and Schummer, P. 1960. Zur Messung der Transversalen Effektiven Diffusion in Durchstromten Fullkorpersaulen, *Chem. Eng. Sci.* **13(2)**: 69–74.
50. Jacques, G.L. and Vermeulen, T. 1958. Longitudinal Dispersion in Solvent-Extraction Columns: Peclet Numbers for Random and Ordered Packings. Univ California Rad Lab Rep No 8029, US Atomic Energy Commission, Washington, DC.
51. Johnson, G.W. and Kapner, R.S. 1990. The Dependence of Axial-Dispersion on Non-Uniform Flows in Beds of Uniform Packing, *Chem. Eng. Sci.* **45(11)**: 3329–39.
52. Kramers, H. and Alberda, G. 1953. Frequency Response Analysis of Continuous Flow Systems, *Chem. Eng. Sci.* **2(4)**: 173–81.
53. Latinen, G.A. 1951. Mechanism of Fluid-Phase Mixing in Fixed and fluidised Beds of Uniformly Sized Spherical Particles. PhD Dissertation, Princeton University.
54. Levenspiel, O. and Smith, W.K. 1957. Notes on the Diffusion–Type Model for the Longitudinal Mixing of fluids in Flow, *Chem. Eng. Sci.* **6(5)**: 227–33.
55. Liles, A.W. and Geankopolis, C.J. 1960. Axial Diffusion of Liquids in Packed Beds and End Effects, *AIChE J.* **6(4)**: 591–5.
56. McHenry, J.R. and Wilhelm, R.H. 1957. Axial Mixing of Binary Gas Mixtures flowing in a Random Bed of Spheres, *AIChE J.* **3(1)**: 83–91.
57. Miller, S.T. and King, C.J. 1966. Axial Dispersion in Liquid Flow Through Packed Beds, *AIChE J.* **12(4)**: 767–73.
58. Miyauchi, T. and Kikuchi, T. 1975. Axial dispersion in Packed Beds, *Chem. Eng. Sci.* **30(3)**: 343–8.
59. Niemann, E.H. 1969. Dispersion During Flow Nonuniform Heterogeneous Porous Media. MS Thesis, Chem Eng Dept, Purdue University, Lafayette.
60. Perkins, T.K. and Johnston, O.C. 1963. A Review of Diffusion and Dispersion in Porous Media, *Soc. Petrol. Engrs. J.* **3(3)**: 70–84.
61. Pfannkuch, H.O. 1963. Contribution a L'Etude des Déplacements de Fluids Miscibles dans un Milieu Poreux, *Rev. Inst. Fr. Pétrole* **18(2)**: 215–9.
62. Plautz, D.A. and Johnstone, H.F. 1955. Heat and Mass Transfer in Packed Beds, *AIChE J.* **1(2)**: 193–9.
63. Pozzi, A.L. and Blackwell, R.J. 1963. Design of Laboratory Models for Study of Miscible Displacement, *Soc. Petrol. Engrs. J.* **3(1)**: 28–40.

64. Raimondi, P., Gardner, G.H.F. and Petrick, C.B. 1959. Effect of Pore Structure and Molecular Diffusion on the Mixing of Miscible Liquids Flowing in Porous Media. *AIChE-SPE Joint Symposium*, San Francisco.
65. Rifai, M.N.E., Kaufman, W.J. and Todd, D.K. 1956. Dispersion Phenomena in Laminar Flow Through Porous Media, Sanit Engrg Rept 3, *Inst. Eng. Res. Series* **90(1)**: 1–157.
66. Roemer, G., Dranoff, J.S. and Smith, J.M. 1962. Diffusion in Packed Beds at Low Flow Rates, *I&EC Fundaments* **1(4)**: 284–7.
67. Saffman, P.C. 1960. Dispersion in Flow Through a Network of Capillaries, *J. Fluid Mech.* **7(2)**: 194–207.
68. Scheidegger, A.E. 1974. The Physics of Flow Through Porous Media. 3rd edition, University of Toronto Press.
69. Schuster, J. and Vortmeyer, D. 1980. Ein einfaches Verfahren zur näherungsweisen Bestimmung der Porosität in Schttungen als Funktion des Wasdabstandes, *Chem. Eng. Tech.* **52(8)**: 848–55.
70. Schwartz, C.E. and Smith, J.M. 1953. Flow Distribution in Packed Beds, *Ind. Eng. Chem.* **45(6)**: 1209–18.
71. Sherwood, T.K., Pigford, R.L. and Wilke, C.R. 1975. Mass Transfer. International Student Edition, McGraw-Hill Kogakusha.
72. Simpson, E.S. 1962. Transverse Dispersion in Liquid Flow Through Porous Media, *US Geological Survey Professional Paper* **411–C**: 1–30.
73. Sinclair, R.J. and Potter, O.E. 1965. The Dispersion of Gas in Flow Through a Bed of Packed Solids, *Trans. Inst. Chem. Eng.* **43**: T3–T9.
74. Slichter, C.S. 1905. Field Measurement of the Rate of Movement of Underground Waters. US Geological Survey, *Water Supply Paper* pp. 140.
75. Stephenson, J.L. and Stewart, W.E. 1986. Optical Measurements of Porosity and Fluid Motion in Packed Beds, *Chem. Eng. Sci.* **41(8)**: 2161–70.
76. Strang, D.A. and Geankopolis, C.J. 1958. Longitudinal Diffusivity of Liquids in Packed Beds, *Ind. Eng. Chem.* **50(9)**: 1305–8.
77. Tan, C.S. and Liou, D.C. 1989. Axial Dispersion of Supercritical Carbon Dioxide in Packed Beds, *Ind. Engng. Chem. Res.* **28(8)**: 1246–50.
78. Taylor, G. 1953. Dispersion of Soluble Matter in Solvent Flowing Slowly Through a Tube, *Proc. Royal Soc. A* **219(1137)**: 186–203.
79. Towle, W.L. and Sherwood, T.K. 1939. Studies in Eddy Diffusion, *Ind. Eng. Chem.* **31(4)**: 457–67.
80. Tsotsas, E. and Schlunder, E.U. 1988. On Axial Dispersion in Packed Beds with Fluid Flow, *Chem. Eng. Process* **24(1)**: 15–31.
81. Vortmeyer, D. and Schuster, J. 1983. Evaluation of Steady Flow Profiles in Rectangular and Circular Packed Beds by a Variational Method, *Chem. Eng. Sci.* **38(10)**: 1691–9.
82. Vortmeyer, D. and Winter, R.P. 1982. Impact of Porosity and Velocity Distribution of the Theoretical Prediction of Fixed-Bed Chemical Reactor Performance: Comparison with Experimental Data, ACS Symposium Series. pp. 49–61.
83. Wen, C.Y. and Fan, L.T. 1975. Models for Systems and Chemical Reactors. Marcel Dekker.
84. Wilhelm, R.H. 1962. Progress Towards the A Priori Design of Chemical Reactors. *Pure Appl. Chem.* **5(3–4)**: 403–21.
85. Wilson, H.A. 1904. On Convection of Heat, *Proc. Cambridge Philos. Soc.* **12(5)**: 406–23.
86. Wronski, S. and Molga, E. 1987. Axial Dispersion in Packed Beds: The Effect of Particle Size Non-Uniformities, *Chem. Eng. Process* **22(3)**: 123–35.
87. Yu, D., Jackson, K. and Harmon, T.C. 1999. Dispersion and Diffusion in Porous Media Under Supercritical Conditions, *Chem. Eng. Sci.* **54(3)**: 357–67.

8

Nanostructured Materials

A. Shokuhfar and M. Mohebali[*]

ABSTRACT

This chapter reviews synthesis methods, properties and applications of nanostructured materials. Mechanical alloying, sol-gel and CVD as some of the dominant and promising synthesis methods along with sonochemistry as a relatively novel one are presented through a concise description of the subject and illustrated by the results obtained from some of the research projects performed in our Advanced Materials and Nanotechnology Research Laboratory in K.N. Toosi University of Technology. Phase transients, diffusion and mechanical properties of nanostructured materials including strength, hardness, ductility and superplasticity are discussed with the main focus laid on the structure-properties relationship. Although it was not possible to summarize all the applications of nanostructured materials, some inspiring examples of biomedical, catalytic and gas sensing applications are selectively described, in order to give the reader an insight into the vast known and potential applications of nanostructured materials.

1. INTRODUCTION

During the last decades, not only nanotechnology has evolved from an imaginary concept into a promising and fast-growing field of science, but has found its way into our everyday life. You have undoubtedly heard of sunscreens or refrigerators featuring some tiny particles which though can't be seen, but are supposed to improve your state of health or the quality of the product. That's true; at the heart of nanotechnology, lie so vast and versatile, nanomaterials. These are no new class of materials, but the same ones used for so many years by humans, manipulated at the finest levels by means of nanotechnology.

Today, we can engineer the finest building blocks of materials, *i.e.*, to move single atoms and molecules and place them in the desired locations, to achieve an improved quality in one or more characteristic properties of

---

[*] Department of Mechanical Engineering, K.N. Toosi University of Technology, Tehran-IRAN.
[*] *Corresponding author* : E-mail : mohebali@sina.kntu.ac.ir

that material and create the so called "Nanostructured materials"[1]. Nanostructured materials include atomic clusters, layered (lamellar) films, filamentary structures, and bulk nanostructured materials. More precisely, nanostructured materials are materials with at least one dimension in range of 1–100 nanometers. They consist of atoms of elements or clusters of mixed elements, all packed together to form a nanoparticle or at a larger scale, a bulk nanostructured material with nanosized crystallites.

The type of atoms and their arrangement, *i.e.*, composition and structure, are the key elements which determine the properties of a material; however, when the dimensions become smaller than a certain size, then the size effects also influence the material properties; here, the size of nanoparticle or the crystallite size is equal or smaller than a characteristic length concerning a certain property. The other reason for extraordinary properties of nanostructured materials is the high surface to volume ratio; that is, the presence of a large fraction of atoms at surfaces or interfaces, which indicates that the inter-atomic forces and chemical bonds play an important role in the behavior of material. In polycrystalline nanostructures, for example, more than half of the atoms could be located in grain boundaries.

Since now, a fair number of books and reviews[2–8] have been published on the subject and have covered one or more aspects of nanomaterials; yet, this field is developing so fast that writing even a book chapter, some new concepts may have emerged. This chapter aims at reviewing synthesis methods, properties and applications of nanostructured materials, while providing the reader with an up-to-date reference and presenting the findings of research teams in our laboratory up to present. Since a full description of all these aspects was not possible in a book chapter, the emphasis has been made on the most important synthesis methods which indeed would affect the final structure of nanomaterials and their properties. In the light of the methods described, the diffusion and mechanical behavior of nanostructured materials as well as some examples of their applications in different fields will be discussed.

2. SYNTHESIS METHDOS

Material properties depend on a number of factors which are indeed controlled through the synthesis method and its variables. Today, a considerable number of synthesis and preparation methods of nanostructured materials are present; some like self-assembly and chemical methods are still under development; some like HPT, have remained an academic approach and some like mechanical alloying, ECAP and Sol-Gel have entered industrial scale productions. Nevertheless, to a great extent, it's the material which determines the synthesis method; for instance, mechanical alloying is the number one choice for producing nanostructured solid solutions of immiscible alloy systems, while Sol-Gel and chemical routes are capable of producing nanoparticles of desired shape and size.

Methods of preparation of nanomaterials could be categorized in two different approaches, namely, Top-Down and Bottom-Up. Top-down approach has first been employed by scientists in Bronze Age to make things out of wood or stone. Proposed for the first time by Feynman as a means to make systems in nano scale, the top-down approach takes advantage of bigger systems which would build smaller systems; these new smaller systems, then, build even much smaller systems until nano-scale systems are produced. This paradigm dictates that you begin with a larger material and slowly process it by removing matter and leaving behind nanoscale features. In this method, the dimensional tolerance is a direct result of the tools' precision utilized. The biggest problem with top-down approach is the imperfection of the surface structure, which reduces the surface-to-volume ratio, and has a diverse effect on a broad range of properties. Regardless of the surface imperfections and other defects that top-down approaches may introduce, they will continue to play an important role in the synthesis and fabrication of nanostructured materials[6]. Typical examples of this approach include severe plastic deformation methods, lithography and mechanical attrition.

In the Bottom-up approach, materials are built-up from the bottom: atom-by-atom, molecule-by-molecule, or cluster-by-cluster. The control of arrangement of atoms from the nano scale to the macro scale is the exciting feature of this approach and the lacking feature of top-down approach, because in nano scale, there's little chance for top-down methods to modify the nanostructure. An important method in this category is self-assembly which employs chemical and biological processes to produce desired structures.

Molecular self-assembly is a strategy for nanofabrication that involves designing molecules and supramolecular entities so that shape-complementarity causes them to aggregate into desired structures. Self-assembly has a number of advantages as a strategy: First, it carries out many of the most difficult steps in nanofabrication, those involving atomic-level modifications of structure, using the very highly developed techniques of synthetic chemistry. Second, it draws from the enormous wealth of examples in biology for inspiration: self-assembly is one of the most important strategies used in biology for the development of complex, functional structures. Third, it can incorporate biological structures directly as components in the final systems. Fourth, because it requires that the target structures be the thermodynamically most stable ones open to the system, it tends to produce structures that are relatively defect-free and self-healing[9].

2.1. Mechanical Alloying

Mechanical alloying is a method that makes use of a high energy ball mill to grind and/or alloy blended elemental powder mixtures in atomic scale. It can be used to process a wide range of materials from simply pure elemental powders to Intermetallics, ceramics, polymers and also to produce alloys and composites. This technique was first developed by J. Benjamin and his colleagues[10] around 1966 in an attempt to produce oxide dispersion strengthened nickel-base superalloy, for gas turbine applications. In the 70's, the term "mechanical alloying" was introduced and a number of new alloys and phase mixtures were developed. In the 80's, mechanical alloying was considered a potential nonequilibrium processing technique and subsequently, amorphous structures, supersaturated solid solutions and nanocrystalline phases were obtained and inducement of chemical reactions at lower temperatures was introduced[11]. In recent years, nanostructures of brittle ceramics, polymer blends and metal-ceramic nanocomposites have been investigated, which demonstrates the high flexibility of this process.

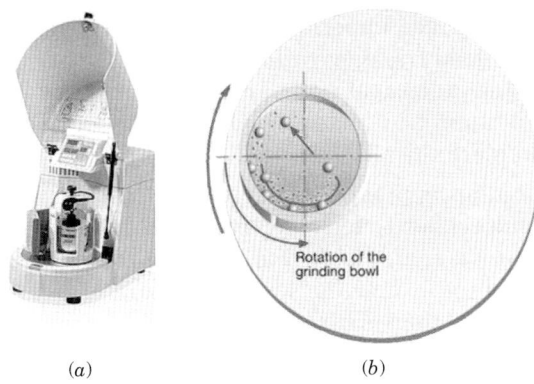

(a) (b)

Fig. 8.1. (a) A High Energy Ball mill machine (b) A schematic representation of the movements of grinding medium in the container[12]

The process starts by mixing powders of different elements with the desired proportion and particle size of 1–200 μm; then the mixture along with the grinding medium is loaded into a sealed container which, depending upon the type of milling equipment employed, the container is moved or agitated. Collision of the charge and the balls, delivers a high energy to the powder, and the material undergoes a severe plastic deformation. The powder deforms plastically and the rest of energy is converted into heat and stored in the metal, raising its internal energy. Plastic deformation at high strain rates within particles refines the grain size to nanoscale after prolonged milling. The grain refinement mechanism by the ball milling process has been proposed by Fecht[2] and include three basic stages:

- Initially, the deformation is localized in shear bands consisting of an array of dislocations with high density.
- At a certain strain level, these dislocations annihilate and recombine to small angle grain boundaries separating the individual grains. The subgrains formed via this route are already in the nanometer size range (about 20–30 nm).

- The orientations of the single-crystalline grains with respect to their neighboring grains become completely random, and the high-angle grain boundaries replace low-angle grain boundaries.

Basically, there are two types of this process. The first one is mechanical milling in which a high energy ball mill is used to crush and refine powders. The second type is reaction milling which involves *in situ* solid state chemical reactions between different powders during mixing and milling[1].

Like any other method, mechanical alloying has its own advantages and drawbacks. Advantages include simplicity, relatively inexpensive equipment, versatility in production of a wide range of materials and possibility of producing large quantities, that can be scaled up to several tons. A serious problem, concerning mechanical alloying is the surface contamination from the milling media (balls and vial) and/or atmosphere. Though it's not possible to completely overcome this problem, evacuating the container or using inert gas atmospheres, may help. It should be noted that the degree of contamination depends on a number of factors like mechanical properties of the powder and its chemical affinity for milling media. Other drawbacks of mechanical alloying include rough structures of produced powder, non-homogeneity in particle size and inhomogeneous chemical composition. In order to improve the roughness of the final structure, Cryomilling, in which the milling operation is carried out at cryogenic (very low) temperatures and/or milling of materials is done in cryogenic media such as liquid nitrogen, can be employed to modify the deformation behavior of materials, *i.e.,* increasing the brittleness. This could also help in reducing the degree of contamination. By adjusting the energy received by the powder through a number of variables like Ball-Powder ratio (BPR), and the milling time, homogeneity of particle size and chemical composition should be improved.

As already indicated, this method can be used to produce amorphous or nanocrystalline structures, whether in pure elements or equilibrium and nonequilibrium alloys. Nanocrystalline structure of a large number of elements has been investigated, and crystallite size of the final nanostructure has been determined by standard X-Ray analysis methods. The minimum grain size achieved is however, dependent upon a number of variables as well as the properties of the element, alloy, or compound being milled. The minimum grain size obtainable by milling, d_{min}, has been attributed to a balance between the defect/dislocation structure introduced by the plastic deformation of milling and its recovery by thermal processes. It has been found that the minimum grain size induced by milling scales inversely with the melting temperature of a group of face-centered cubic (fcc) structure metals studied[13]. These data are plotted in Fig. 8.2 along with data for other metallic elements and carbon (graphite)[14]. For these data, only the lower-melting-point metals show a clear inverse dependence of minimum grain size on melting temperature. The minimum grain size for elements with higher melting temperatures ($>T_m$ for Ni), exhibit essentially constant values with melting temperature for given crystal structure classes. For these elements it appears that d_{min} is in the order: fcc < bcc < hcp[8].

Equilibrium and nonequilibrium solid solutions with nanocrystalline structures in Ti[15–20], Al[21–25], Cu[18, 26–28] and Fe[29–34] systems along with many others have been successfully produced via mechanical alloying and the solid solution formation mechanisms, structure-property relationship and many other facts have been investigated.

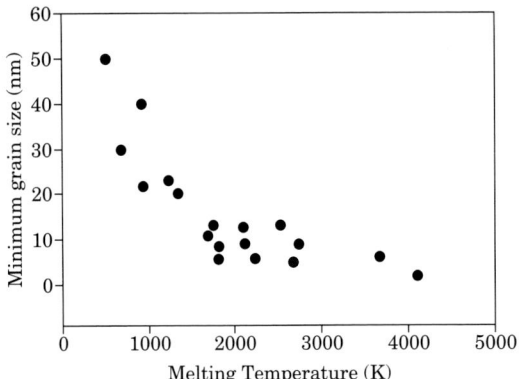

Fig. 8.2. Minimum grain size *vs.* melting temperature[8]

Besides these ordinary solid solutions, in which a principal component occupies at least 50 at% in the composition, alloys composed of more principal elements with similar molar fractions have been investigated by mechanical alloying. High-entropy alloys (HEAs), with at least five principal elements, in which high mixing entropies in the liquid or regular solid solution state could enhance the formation of solid solution phases, have been investigated by mechanical alloying to enable investigation of kinetic behavior of elements and the phase evolution. Yeh and coworkers[35–36] investigated some multi-component systems and reported that the alloying rate correlates best with the melting point of the elements among metallurgical factors. The mechanism for this correlation is explained through the effect of melting point on solid-state diffusion and mechanical disintegration which are critical for the final alloying.

A great number of nanocrystalline intermetallics and nanocomposites are also produced by mechanical alloying. By controlling milling environment atmosphere, contamination could be controlled, and more purposefully, chemical reactions could be induced to produce metal-ceramic nanocomposites.

The metal powders (Ti, Fe, V, Zr, W, Hf, Mo, etc) transform to a nanocrystalline nitride by high-energy ball milling under nitrogen gas flow.

This solid-state interdiffusion reaction during reactive ball milling is triggered by fragmentation of the starting powder, thus creating new surfaces. These freshly created surfaces react with the flowing nitrogen gas to form a nitride surface layer over the unreacted core particle. With further milling, this reaction continues and a homogeneous nitride phase is formed and the unreacted core of metal disappears resulting in a nanostructured (often metastable) metal-nitride with a typical grain size of 5 nm[3].

Mechanical alloying is a complex process and hence involves optimization of a number of variables to achieve the desired product phase and/or microstructure[37–38]. Some of the important parameters that have an effect on the final constitution of the powder are:

- type of mill,
- milling container,
- milling speed,
- milling time,
- type, size, and size distribution of the grinding medium,
- ball-to-powder weight ratio,
- extent of filling the vial,
- milling atmosphere,
- process control agent, and
- temperature of milling.

All these process variables are not completely independent. For example, the optimum milling time depends on the type of mill, size of the grinding medium, temperature of milling, ball-to-powder ratio, etc.[11].

Reviewing a number of investigations by our research teams, some examples of different applications, underlying mechanisms in formation of nanocrystalline structures and the effects of a number of process variables are studied in the following sections.

2.1.1. Nanocrystalline Ni_3Al-based alloy

Nazarian et al.[39], have studied the production of nanocrystalline Ni_3Al-based alloy using mechanical alloying. In this research, a new alloy with nominal compositions of Ni–8.14Al–7.83Cr–1.45Mo–0.01B (wt%) was produced by mechanical alloying and hot pressing processes. Mechanical alloying was carried under three different milling conditions, series I-III in Table 8.1.

Based on XRD data, Williamson-Hall method was employed to calculate the mean crystallite size and mean lattice strain. The evolution of the transformations occurring during milling process was performed by X-ray analysis as shown in Fig. 8.3. According to Figs. 8.3 (a) and 8.3 (c), the peaks related to raw materials (Al, Ni, Cr and Mo) can be seen in starting materials. After 5 h of milling, the peaks related to Al, Cr and Mo vanished but only Ni peaks remained visible. After 10 h of milling, the Ni peaks shifted to lower angles that can be described by the diffusion of the elements Al, Cr and Mo into Ni structure and the formation of a Ni-based solid solution phase. A Ni_3Al intermetallic compound was formed after 15 h of milling. Lack of superlattice diffraction peaks suggests that the crystalline Ni_3Al phase has a disordered structure. There is no major difference among the peaks of the materials produced after milling of raw materials from 15 to 40 h except broadening of the peaks and decreasing of their intensity.

Table 8.1. Mechanical alloying conditions[39]

Series	Rotation speed (rpm)	Milling media
I	300	Hexane
II	150	Hexane
III	300	–

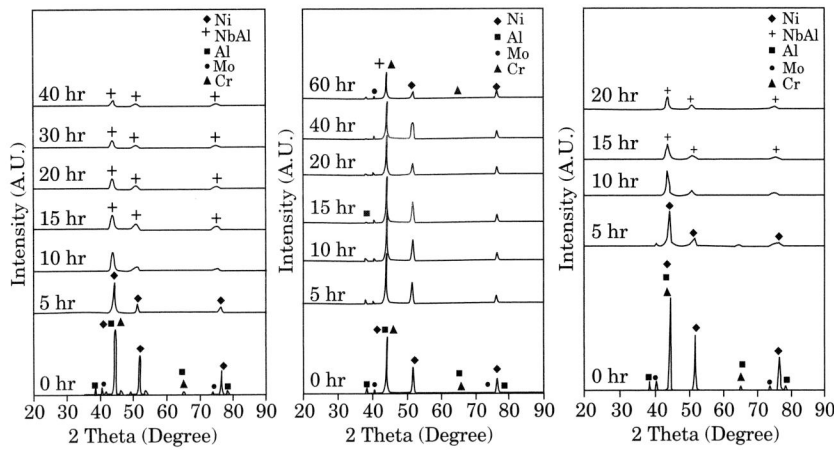

Fig. 8.3. X-ray patterns of (a) series I, (b) series II and (c) series III after different milling times[39]

When the milling speed was decreased to 150 rpm in samples of series II [Fig. 8.3 (b)], the peaks related to Al were still observed even after milling for 15 h and those related to Cr, Mo and Ni were observed even after 60 h of milling. However, the decreasing peak intensities and broadening of the peaks were not considerable in the samples of series II.

Variations in the crystallite size versus milling time in the samples of series I, II and III are shown in Fig. 8.4. According to this figure, crystallite size initially decreased sharply with prolonged milling time but it took a decreasing trend for longer milling times. Finally, with enough longer times, milling was observed to have no considerable effect on crystallite size. It has been established that particle strain increases while particle size decreases at the onset of the milling process when recovery and recrystallization of powder particles occur simultaneously. With prolonged milling times, it is possible to create an equilibrium between the above-mentioned processes so that particle size becomes nearly fixed. On the other hand, decreasing of particles size leads to increased energy of the material. Therefore, decreasing particle size is by itself a limiting factor for further decreasing of particle size.

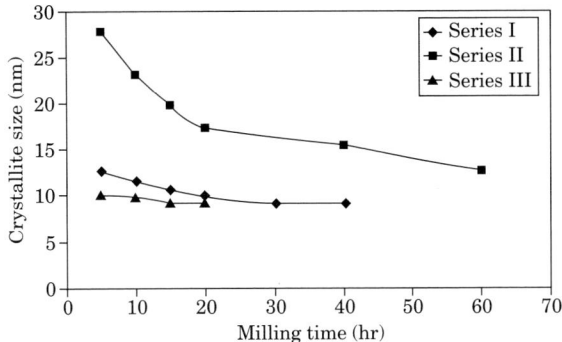

Fig. 8.4. Variation of crystallite size versus milling time under different conditions[39]

It should be noted that a greater peak broadening due to milling was observed in the samples of series III than in those of series I. This is due to the fact that impacts experienced by and among powder particles, the milling vial, and the balls in dry milling are more intensive than those under wet conditions. Therefore, the particles in samples of series III had a smaller size than those in series I.

Lattice Strain for milled powders of series I, II and III, versus milling time is shown in Fig. 8.5. The variations in lattice strain in the samples of series III were more intensive than those in series I. But in the samples of series II, variations in lattice strain versus milling time were not considerable due to low energy of impacts.

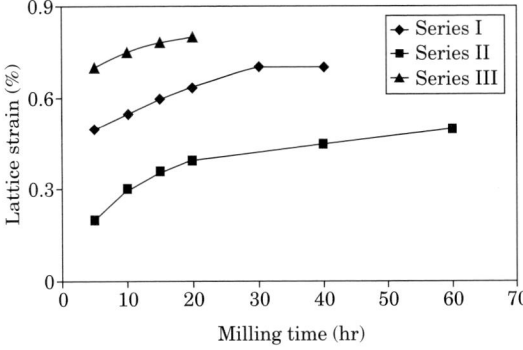

Fig. 8.5. Variation of lattice strain versus milling time under different conditions[39]

Variations in the lattice parameter of the milled powders versus milling time are presented in Fig. 8.6. For the samples of series I and III, lattice parameter of milled powders increased sharply at the onset of milling but its rate considerably decreased over longer milling times. Ball milling had no considerable effect on the lattice parameter in the samples of series II. These events can be described with regard to XRD patterns (Fig. 8.3). Shifting of Ni peaks to lower angles after milling of raw materials for 10 h in the samples of series III is evidence showing the solution of the alloying elements in the Ni lattice. Therefore, the lattice parameter of nickel increases. The atomic radius of Al (0.143 nm) and Mo (0.136 nm) are larger than that of Ni (0.125 nm), whereas the atomic radii of Cr (0.125 nm) is almost equal to that of Ni. Therefore, diffusion of Al and Mo atoms into the Ni lattice leads to an increase in the Ni lattice parameter and the transfer of X-ray peaks of Ni to lower angles. According to Fig. 8.6, in the range of 5 to 10 h of milling in the samples of series I and III, the lattice parameter increased sharply, this is due to the diffusion of the alloying elements into the Ni lattice. This diffusion also led to the formation of a Ni-based solid solution phase after 10 h of milling.

Lattice parameter in the sample of series III after 15 h of milling was less than that in the samples of series I, which is evidence of the lower dissolution of the alloying elements in the Ni lattice in the sample of series III. This may be due to the excessive adhesion of powders to the milling vial and balls during the process. Therefore, a non-homogenate Ni_3Al-based material can be obtained after 15 h of dry milling. According to Fig. 8.6, this non-homogenate structure transformed to a homogenate one after 20 h of dry milling and the lattice parameter, therefore, reached its relatively fixed quantity.

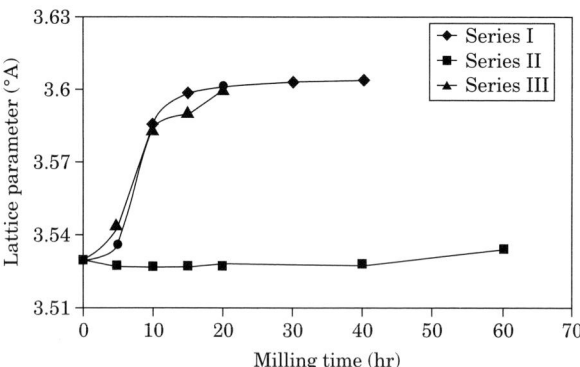

Fig. 8.6. Variation of lattice strain versus milling time under different conditions[39]

EDX analysis was used to gain a better understanding of the XRD results obtained for the samples in series I and III, as shown in Table 8.2. The samples of Series I had a homogenate structure after 15 h of milling with a good agreement with the nominal compositions of alloy (Ni–8.14

Al–7.83Cr–1.45Mo–0.01B (wt%)), where after 20 h of milling Fe contamination appeared in EDX analysis due to excessive wearing of the steel vial and balls. Fe contamination increased over prolonged milling time. In the samples of series III milled for 15 and 20 h, Fe contamination was not detected due to the excessive adhesion of powders to the vial and balls. But the chemical composition of 15-h-milled powder under this condition did not have a good homogeneity and showed a considerable difference with the nominal compositions of the considered alloy. A good homogeneity was, however, achieved after 20 h of milling. Therefore, better results can be obtained via wet milling of the starting materials.

Table 8.2. EDX analysis of milled powders of series I and III after different milling times[39]

Series	Milling time (h)	Element (wt. %)				
		Ni	Al	Cr	Mo	Fe
I	15	82.26	8.09	7.98	1.67	–
	20	82.82	7.82	6.70	1.24	1.42
	30	82.29	7.61	7.16	1.41	1.53
	40	82.08	7.56	6.96	1.58	1.82
III	15	80.61	9.85	7.49	2.05	–
	20	82.30	7.59	7.98	2.13	–

B was not detected

2.1.2. Nanostructured Al-4.5%Cu Alloy

A number of nanostructured solid-solutions and immiscible alloy systems have been studied by mechanical alloying. Analysis of X-ray diffraction pattern is a common practice in investigation of solid-solution formation during mechanical alloying generally from changes in the lattice parameter values calculated from shifts in peak positions or even the absence of second phase peaks. Suryanarayana[11] discusses the difficulties associated with the determination of the increase in the solid solubility limit promoted by mechanical alloying, by the absence of peaks of the second phase in the XRD pattern. Several factors influence the detectability of a second phase by XRD, such as the broadening effect due to the crystalline refinement and the lattice strain, the very small size of the second phase and the defects introduced by cold deformation. Mechanical alloying also introduces impurities into the milled powder, from the materials of the container and the balls, and nitrogen, oxygen or carbon from the atmosphere or the PCA. This contamination can dilate or contract the lattice parameter, depending on the kind of impurity atom introduced, which can obscure the determination of the solid solution. Therefore, it is desirable to use other techniques along with XRD to investigate the process. SEM and TEM images are usually used to better investigate the structure of milled powders. Fogagnolo et al.[24] employed differential scanning calorimetry

(DSC) to investigate solid-solution formation during mechanical alloying of Al-4.5 wt% Cu.

Mostaed et al.[25, 40] employed XRD, SEM and HRTEM to investigate the formation of supersaturated Al-4.5 wt% Cu solid solution. Analysis of X-ray diffraction patterns of powders milled for 5 h with BPR (Ball to powder weight ratio) of 20:1, confirmed solid-solution formation, first by decrease in the intensity of Al and Cu peaks and then by appearance of Al_2Cu peaks after aging process at 250°C for 4 h.

Figure 8.7 shows the microstructure of Al-4.5 wt% Cu powder milled for 5 h at the BPR of 20:1. It can be clearly seen that the powders' morphology of the sample is flatted. It means that welding has dominated fracture during MA process. Presence of a flatted not rounded morphology and distribution of Al and Cu in the particles as a result of elemental diffusion indicate the fact that the mechanically alloyed powders' morphology is related to the intermediate (second) stage. Therefore, obtaining the uniform shape, size and microstructure of the powder require prolonging the milling time or increasing the impact force.

Fig. 8.7. SEM micrograph of Al-4.5% Cu after 5 h milling at the BPR of 20:1

According to Figs. 8.8 (b) and 8.8 (c), the mutual diffusion of Al and Cu occurred in this sample during MA process. But it must be mentioned that obtaining the uniform distribution of Al and Cu in the particles requires prolonging the milling time or increasing the impact force.

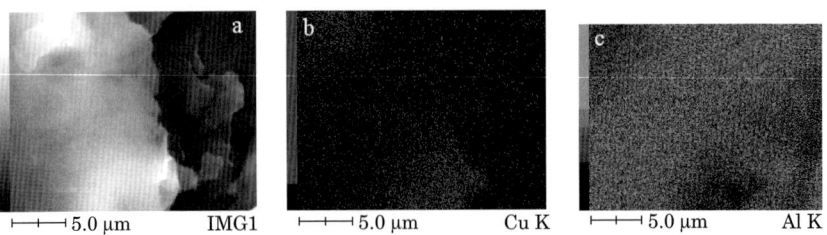

Fig. 8.8. (a) SEM micrograph of Al-4.5% Cu after 5 h milling at the BPR of 20:1; (b) and (c) elemental mapping images of (a)[78]

Figure 8.9 and 8.10 show HRTEM images of Al-4.5 wt% alloy after 5 hr of milling with BPR of 20:1. The fringes in Figure 8.9 are the rotational moiré fringes, because of the calculated rotational angle by equation $D = d/\theta$ (~ 9°7'), which is approximately equal to the angle between the Al (111) planes and fringes are schematically shown in Fig. 8.11 (b) (~ 9°). In fact, these fringes appeared as a result of the rotation of two Al nanocrystallites with respect to each other at the interface.

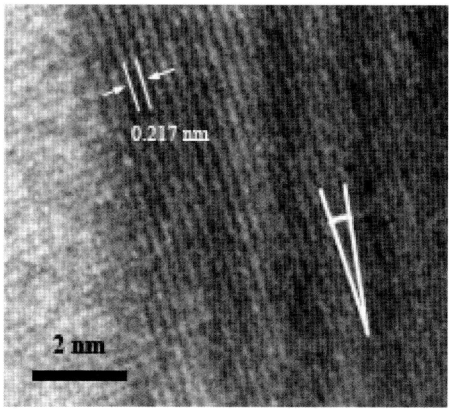

Fig. 8.9. HRTEM image of Al-4.5 wt.% Cu after 5 h milling at the BPR of 20:1 (a) circles indicate some of the edge dislocations (b) Showing the angle between Al (111) planes and fringes

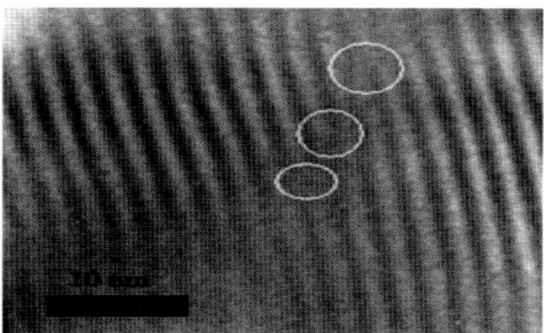

Fig. 8.10. HRTEM image of Al-4.5 wt.% Cu after 5 h milling at the BPR of 20:1 (a) circles indicate some of the edge dislocations (b) Showing the angle between Al (111) planes and fringes

If an extra half plane defined as an edge dislocation forms in the crystal, it will induce another mismatching. Therefore, at the region in which the edge dislocation presents the corresponding fringe's position is shifted immediately. Hence, the circles in Fig. 8.10, indicate some of the edge dislocations located in one of the upper or lower crystallite. On the other hand, presence of the Cu atoms in some of the Al unit cells and also other point defects slightly decrease the interplanar spacing in those unit cells. Hence, the interplanar spacing between two Al (111) planes would not be uniform along ⟨110⟩ directions. As a result of that, the corresponding fringes would not be straight (Fig. 8.9). The presence of a large number of edge dislocations and point defects in Fig. 8.10 signifies that mechanical alloying process is accompanied by plastic deformation and high levels of defects which are indeed the high diffusivity paths.

Mostaed et al.[40] also studied Al-4.5% Cu/SiC nanocomposite and found out that the addition of SiC particles enhanced solid solution formation and decreased crystallite size. They attributed this effect to the introduction of higher dislocation and vacancy densities which eventually lead to higher diffusion rates of copper atoms in the aluminum matrix. Figure 8.11 shows a bright field image of Al-4.5%Cu/SiC sample after 5 h milling at BPR of 20:1 and the corresponding electron diffraction pattern. The grain size is in nanometer range and SiC is also present in the Al-4.5% Cu matrix.

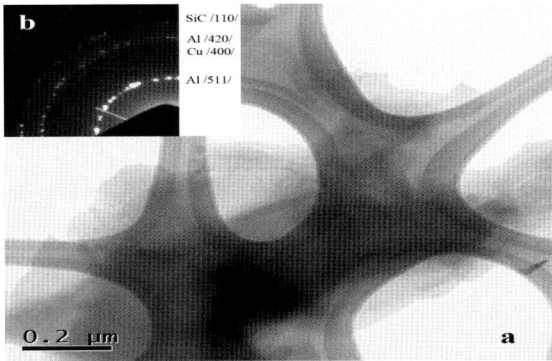

Fig. 8.11. TEM image of Al-4.5 wt.% Cu after 5 h milling at the BPR of 20:1. (a) Bright field image and (b) the corresponding electron diffraction pattern (EDP)

2.1.3. *Nanocrystalline Hydroxyapatite by mechanochemical treatment*

In a comprehensive set of experiments conducted by Nasiri, Shokuhfar et al.[41–43], a mechanochemical treatment was developed to obtain nanocrystalline Hydroxyapatite. The effect of milling media, chemical reaction and milling time on crystallite size, morphological properties and crystallinity degree was investigated. Hydroxyapatite is a biocompatible material with a great potential for bone substitution due to its tight bonding with the bone tissue and because it exhibits osteoconductive behavior and has no adverse side effects on the human organism.

Mechanochemical treatment has recently received particular attention as an alternative route to preparing materials characterized by better formability and biocompatibility with natural bones, wherein both nanoparticles (particles with average diameters less than 100 nm) and nanocrystalline micro particles (average crystallite sizes of less than 10 nm) are discussed. In this process, milling media are often selected on the basis of their high hardness, such as, WC or SiC, or their chemical inertness, e.g., hardened stainless steel. Unfortunately, these two attributes result in high melting points and/or chemically stable contaminants that are difficult to remove. Polyamide has already been proved to possess good biocompatibility with various human cells and tissues and widely used in biomaterials application. In this experiment, a mechanochemical process has been used successfully to produce nanocrystalline powders of HA using an experimental procedure:

$$4CaCO_3 + 6CaHPO_4 \rightarrow Ca_{10}(PO_4)_6(OH)_2 + 4H_2O + 4CO_2$$

For the first time, milling was performed in sealed polyamide6 vials and zirconia balls under air, with 600 rpm as rotation speed. The powder to ball mass ratio used in all the experiments was about 1/20. To avoid excessive heat, the milling was performed in 45 min milling steps with 15 min pauses. The mechanochemical process was performed for 20, 40, 60 and 80 h for the above reaction.

Using the Williamson-Hall method to analyze XRD data, crystallite size and lattice strain at different milling times are represented in Fig. 8.12 and Table 8.3. According to these data, by increasing the milling time, the crystallite size decreased and the lattice strain increased; but the rate of both of these variations (increase in the lattice strain and decrease in the crystallite size) decrease by increasing the milling time. The reason is twofold:

- The first reason is that by increasing the milling time the specific surface energy will rise and particles agglomeration can happen.

- The secondary reason is that by increasing the milling time, vials temperature increases, leading to the second recovery during mill treatment.

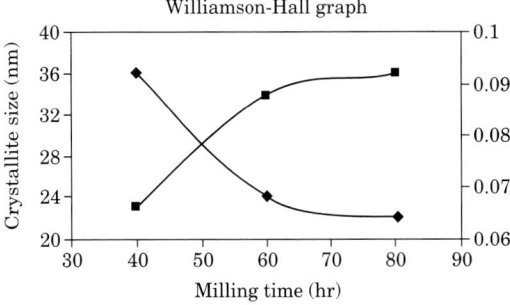

Fig. 8.12. The graph of obtained crystallite size and the lattice strain of HA phase[41]

Table 8.3. Crystallite size and lattice strain calculated for the samples at different milling times[41]

Milling time (hrs)	Crystallite size (nm)	Lattice strain (%)
40	36	0.066
60	24	0.088
80	22	0.092

To investigate the effect of reaction on the final structure, two different experimental processes were used, as follows:

$$6CaHPO_4 + 4Ca(OH)_2 \rightarrow Ca_{10}(PO_4)_6(OH)_2 + 6H_2O \quad (R1)$$
$$4CaCO_3 + 6CaHPO_4 \rightarrow Ca_{10}(PO_4)_6(OH)_2 + 4H_2O + 4CO_2 \quad (R2)$$

Figure 8.13 illustrates the schematic diagram of nanocrystalline HA synthesis via mechanochemical process in polyamide6 vials. The mechanically alloyed powder particles rarely have a perfect spherical shape. In the early stages of milling (and also at the late stages in some cases), powder may have a flaky shape. Figure 8.14 shows the scanning electron micrographs of the R1 sample. Here, with increasing milling time, the reduction of particle size occurs while the shape of product is flake like.

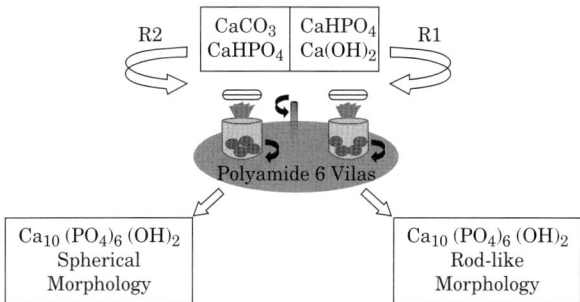

Fig. 8.13. Schematic diagram of nanocrystalline HAp synthesis via mechanochemical process in polyamide 6 vials[42]

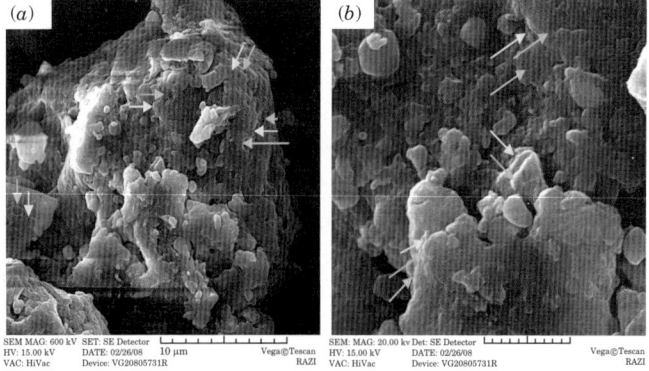

Fig. 8.14. SEM micrographs of the R1 reaction with flaky shape after 60 h milling time (a) x 10 and (b) x 20[42]

Figure 8.15 presents the scanning electron micrographs of the R2 sample. Here with increasing milling time, the reduction of particle size occurs similar to R1 sample but the shape of the production will be irregular. In this case, dominant phenomenons are the accretion and agglomeration.

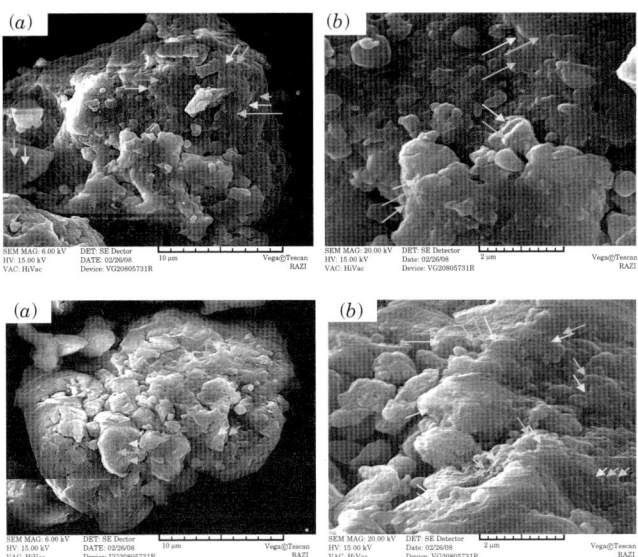

Fig. 8.15. SEM micrographs of the R1 reaction with flaky shape after 60 h milling time (a) x 10 and (b) x 20[42]

In Fig. 8.7 TEM micrographs of nanocrystalline HAp is shown after 60 h milling time in polymeric milling media for R1 reaction. The nanocrystalline HAp in a nanometer range can be easily distinguished. According to the TEM micrographs, with increasing milling time to 60 h, the reduction of crystallite size occurs for R1 reaction and the shape of product looks like a rod. In Fig. 8.7 (b), it can be seen that the morphology of nanocrystalline HAp after 80 h milling time, similar to 60 h milling, is also rod like. The TEM micrographs show that by increasing the mechanical activation time to 80 h in polymeric milling media, the crystallite size will further decrease. Moreover, with increasing milling time from 60 h to 80 h for R1 reaction, agglomeration is occurs. In Fig. 8.8, it can be seen that the morphology of nanocrystalline HAp for R2 reaction after 60 and 80 h milling time, is nearly spherical granules with a smooth geometry. Furthermore, with the same time of R2 reaction, the deagglomeration happens. Since spherical geometry is preferable for avoiding inflammation and achieving osteointegration, compared to irregular shape, obtained product is preferred for medical applications. The TEM observations reveals that, the particles are in the nanometer range, 17 ± 8 nm and 13 ± 7 nm for R1 reaction after 60 and 80 h milling time respectively, and 16 ± 9 nm and 15 ± 8 nm for R2 reaction after 60 and 80 h milling time. Therefore, we conclude that this method gives rise to HAp crystallite with their average size below 20 nm and 23 nm for R1 and R2 reactions after 80 h milling time, respectively. The values are close to the crystallite sizes which are estimated from the line broadening of the given X-ray diffraction peak.

In a last experiment, the effect of milling media on crystallinity degree and morphological properties was investigated. This time, millings were performed in sealed Polyamide6 and tempered Chrome Steel vials using zirconia balls.

By controlling the temperature during mechanical activation (45 min milling steps with 15 min pauses) and milling time, powders with three different crystallinity degrees were produced. Figures 8.16 (a) to 8.16 (c) and Figs. 8.17 (a) to 8.17 (c) present the XRD patterns for the powders of reactions R1 and R2, respectively, milled for different milling durations, compared with the diffraction peaks obtained from the literature. For both reactions, due to the lack of sufficient time for mechanical activation, no trace of HA was detected after 20 h of milling.

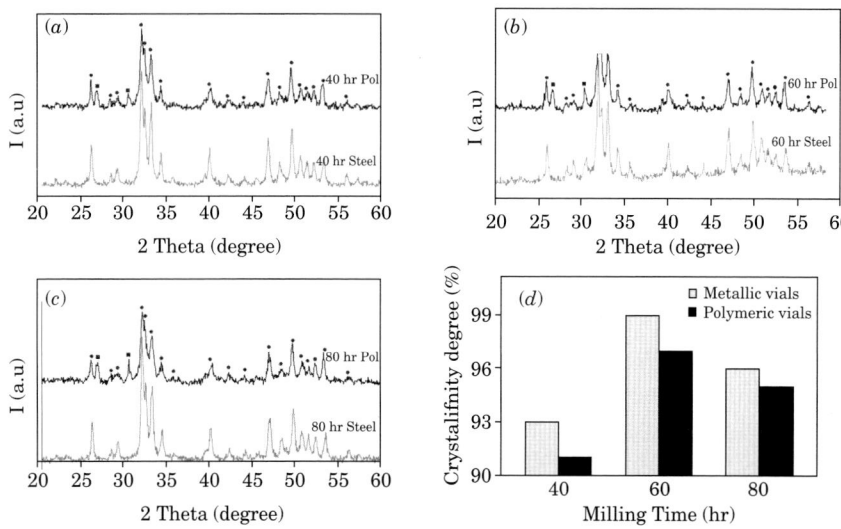

Fig. 8.16. XRD patterns of R1 samples, milled for 40 (a), 60 (b) and 80 h (c) in polymeric and metallic vials, CaHPO$_4$ (■), HA (•) and Bar chart show the crystallinity of R1 reaction samples for various milling times and media (d)[43]

Figure 8.16 (a) shows the XRD patterns of reaction R1 which had been milled for 40 h in polymeric and metallic vials. In this reaction, CaHPO$_4$ and Ca(OH)$_2$ were used as the initial compounds. The patterns show that the production of reaction R1 for polymeric and metallic media is HA. In these patterns the extra peaks represented by (n) were also observed in the XRD pattern of powders synthesized by using polymeric vials. It shows that CaHPO$_4$ is still presented in our sample as it was in the initial commercial production, whereas, these extra peaks in metallic vials are not present. Figures 8.16 (b) and (c) show the XRD patterns for the powder synthesized via the R1 reaction and milled for 60 and 80 h. It can be seen that the extra peaks are still present.

The crystallinity degree, relates to the fraction of crystalline phase which has been presented in the examined volume, was evaluated by the following formula:

$$X_c = 1 - (V\ 112/300/I\ 300) \qquad \ldots(1)$$

where $I\ 300$ is the intensity of (300) reflection and $V\ 112/300$ is the intensity of the hollow between (112) and (300) reflections, which completely disappears in non-crystalline samples. The obtained data shows that by increasing the milling time to 60 h for reaction R1, at first the crystallinity degree increased and reached to a maximum 60 h of milling, in both of vials, afterwards, by further increasing the milling time to 80 h, the crystallinity degree was decreased which can be seen in Fig. 8.16 (d). It can be seen that the increase in hydroxyapatite crystallinity compared to the increase in milling time is not linear; moreover, the crystallinity degree averages indicate that the average values of reaction R1 are 96% and 94% for metallic and polymeric vials, respectively. In fact, the crystallinity degree for reaction R1 in the metallic vials is higher than polymeric vials. Table 8.4 presents the comparison between crystallinity degree for the R1 reaction for different milling times in the polymeric and metallic vials.

Figures 8.17 (a) to 8.17 (c), illustrate the XRD patterns associated to reaction R2, milled for 40, 60 and 80 h. In this reaction, $CaHPO_4$ and $CaCO_3$ were used as the initial compounds. Based on XRD patterns, the extra peaks are not presented after 40, 60 and 80 h of milling in both of vials and the only detected phase was HA. The Comparison between the R1 and R2 reactions shows that CaHPO4 is a compound that must be avoided if the purpose is to produce HA without any extra phase present in the milling process. As can be seen in Fig. 8.17 (d) the fraction of crystalline phase (X_c) in the hydroxyapatite powders from reaction R2 indicates that by increasing the milling time from 40 to 80 h, the crystallinity degree decreases and reaches a minimum at 80 h of milling, for the R2 reaction in both vials. It can be observed that the decrease in hydroxyapatite crystallinity changes linearly by increasing the milling time. Furthermore, the calculated values indicate that the crystallinity degree averages for reaction R2 are 78% and 91% for metallic and polymeric vials, respectively. In fact, unlike reaction R1, the crystallinity degree averages for reaction R2 in the polymeric vials are higher than the metallic vials. The determined amounts of the degree of crystallinity for reaction R2 samples are given in Table 8.5. Comparisons between the obtained data from the two reactions show that the crystallinity degree averages for the obtained materials in the polymeric vials are higher than the metallic vials. Therefore, using polymeric vials can led to the production of nanocrystalline HA with higher crystallinity degrees.

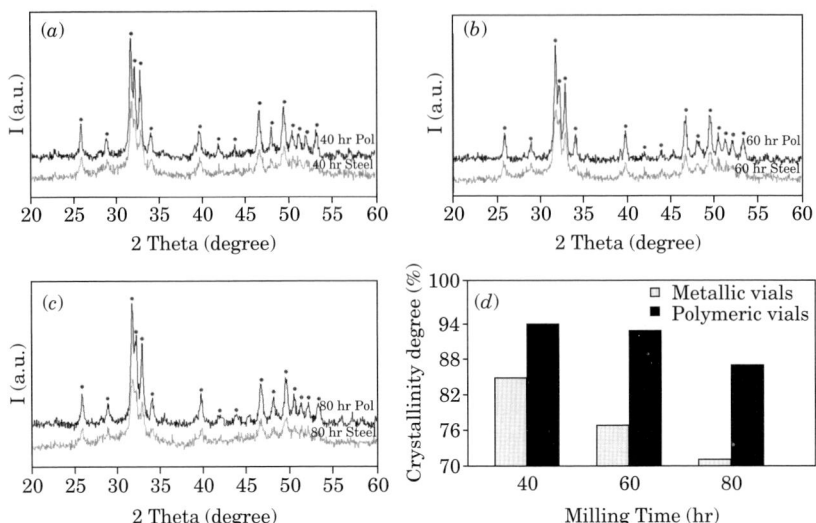

Fig. 8.17. XRD patterns of R2 samples milled for 40 (a), 60 (b) and 80 h (c) in polymeric and metallic vials, HA (•) and bar chart show the crystallinity of R2 reaction samples for various milling times and media (d)[43]

Table 8.4. Comparison between crystallinity degree for R1 reaction in polymeric and metallic vials[43]

Milling media	Milling time (hr)	Crystallinity, X_c (%)
Metallic vials	40	93
	60	99
	80	96
Polymeric vials	40	91
	60	97
	80	95

Table 8.5. Comparison between the crystallinity degree for reaction R1 in polymeric and metallic media[43]

Milling media	Milling time (hr)	Crystallinity, X_c (%)
Metallic vials	40	85
	60	77
	80	71
Polymeric vials	40	94
	60	93
	80	87

The size and shape of the powder particles may be determined accurately through transmission electron microscopy (TEM). Generally, very rarely, the mechanically alloyed powder particles have a perfect spherical shape. In the early stages of milling (and also in the final stages, in some cases) the powder may have a flaky shape. Typical TEM micrographs of HA have been shown in Fig. 8.3, after 60 h of milling time in polymeric and metallic milling media for reaction R1. As shown in Fig. 8.18, HA in the nanometer range can be easily distinguished. According to the TEM micrographs, milling for 60 h caused the reduction of crystallite size and also the shape of the product became rod like and irregular in polymeric and metallic vials, respectively. The broadening of the diffraction pattern of the nanocrystalline powder shown in Fig. 8.16 and Fig. 8.17 is in agreement with the TEM observation.

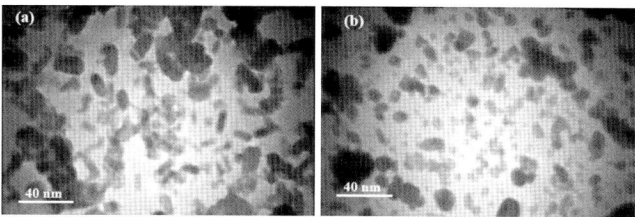

Fig. 8.18. Typical TEM micrographs of HA in the nanometer range, after a milling time of 60 h in polymeric (a) and metallic (b) milling media for reaction R1

In Fig. 8.19, it can be observed that the morphology of HA for reaction R2 after a milling time of 60 h was nearly spherical granules with a smooth geometry and chained hanks in polymeric and metallic vials, respectively. A higher magnification reveals that particles are in the nanometer range, 17 ± 8 (polymeric vial) and 6 ± 2 nm (metallic vial) for reaction R1, and 16 ± 9 (polymeric vial) and 12 ± 2 nm (metallic vial) for reaction R2, as shown in Fig. 8.20. Based on XRD patterns and TEM observations, the formation mechanism of nanocrystalline HA through the mechanochemical process in polymeric and metallic milling media is confirmed. Since spherical granules with a smooth geometry, compared to irregular shapes, cause less inflammatory response and facilitate faster bone growth, therefore obtained product in the polymeric vials is preferred for medical applications.

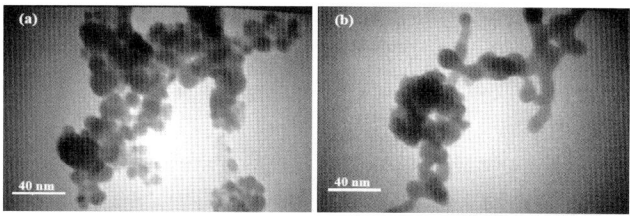

Fig. 8.19. TEM micrographs of HA, after a milling time of 60 h in polymeric (a) and metallic (b) milling media for reaction R2[43]

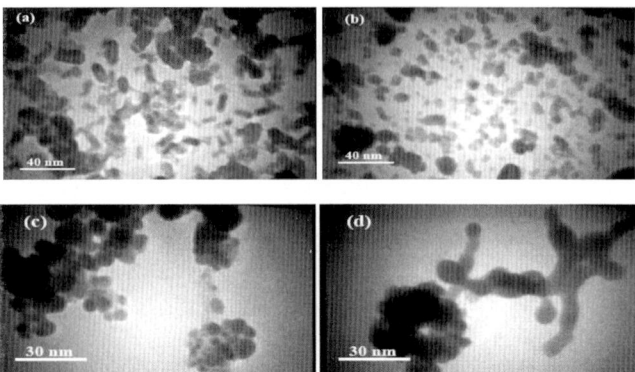

Fig. 8.20. The morphology of HA with higher magnification, reaction R1 milled for 60 h in polymeric (*a*) and metallic vials (*b*); reaction R2 milled for 60 h in polymeric (*c*) and metallic vials (*d*)[43]

2.1.4. Al/SiC nanocomposite by mechanical alloying

A nanocomposite material which consists of two or more components is usually defined by the dimension of reinforcement in the nanometer range. Al/SiC nanocomposites show a significant improvement in their tensile strength and elastic modulus as well as wearing resistance, compared to conventional metals and alloys. Although Al/SiC nanocomposites have been produced by mechanical alloying with as mixed Al and SiC particles in micrometer size, but the formation mechanism of the nanometric reinforcement in the presence of ductile phase such as Aluminum has not been cleared in the literature. Dashtbayazi *et al.*[44] employed a high-energy ball mill to prepare an Al/SiC nanocomposite by two different approaches. The first was high energy ball milling of SiC particles without Al matrix. From experimental observations, the size of SiC particles stayed at a relatively stable size after about 12 h of milling, and the experimental results showed that the production of SiC nanoparticles with the size under 100 nm was very difficult. The second approach involved milling of both SiC and Al particles simultaneously by high energy ball mill. In the latter, the SiC nanoparticles with the sizes under 100 nm were observed. Figure 8.21 shows the size variations of SiC in both routes applied to produce nanocomposite. The limiting size of SiC particles in the milling of SiC with and without Al matrix were under 100 and 500 nm, respectively. By milling, the Al particles deformed and fractured, while the SiC particles mainly fractured. Therefore, nanometric SiC particles can't be produced in the absence of ductile phase.

EDAX analysis of nanocomposite after 6 h of milling showed that the SiC particles incorporated into the Al matrix. Figure 8.22 (*a*) shows the morphology of the powder after 6 h of milling. As can been in this figure. Al particles deformed and SiC particles fractured and incorporated into the

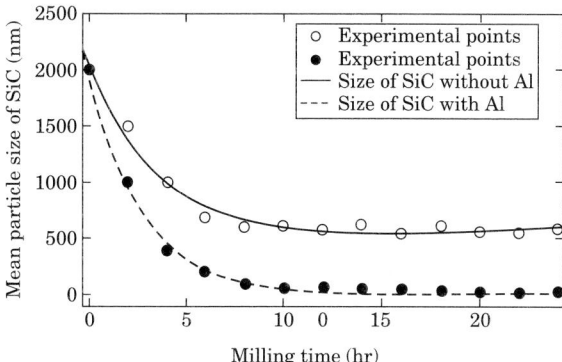

Fig. 8.21. The variations of mean particle size of SiC as a function of milling time in two routes were applied to produce Al/SiC nanocomposites[44]

Al matrix. By further milling, SiC particles completely penetrated the Al matrix and then fractured inside the Al particles as a result of ball to ball collisions. It seems that the ineffectiveness of ball milling process to produce SiC nanoparticles in range of below 100 nm is mainly due to the fact that the SiC particles run away the colliding balls without fracturing, while for SiC particles milled in the presence of ductile phase, this phase acts as a holder and do not allow the SiC particles to run away the colliding balls. Figure 8.22 (b) shows the SiC particles embedded in the Aluminum flake after 6 h of milling.

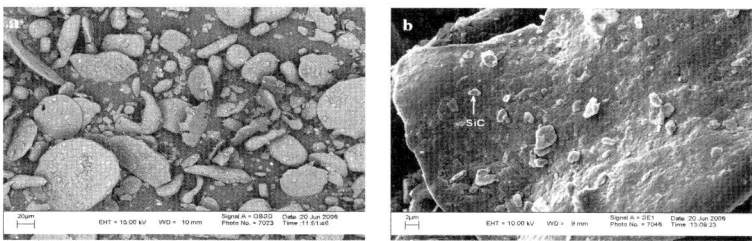

Fig. 8.22. SEM micrographs of Al/SiC mechanically-milled after 6 h milling time (a) Low magnification (b) High magnification[44]

The SiC particle size will keep decreasing until such a point that the fracture strength of the small particles will be equal to or greater than the stress caused by the collision. Often this balance point corresponds to the SiC particle size in the range of a few nanometers to 100 nm. Figure 8.23 shows TEM micrographs of nanocomposite sample after 24 h milling time. The electron diffraction pattern in Fig. 8.23 (a) shows that SiC particles in nanometer size in the Al matrix. Although electron diffraction pattern shows Al nanocrystallites in Fig. 8.23 (b); TEM observation of Al/SiC nanocomposite shows that the size of SiC particles was in the range of 25–100 nm as well.

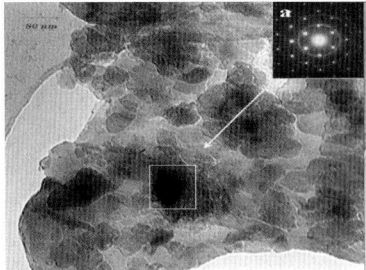 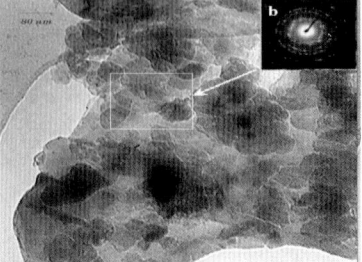

Fig. 8.23. TEM micrographs of Al/SiC nanocomposite (a) Selected area diffraction patterns of SiC nanoparticles and (b) Selected area diffraction patterns of nanocrystallites Al after 24 h of milling[44]

Based on the X-ray diffraction patterns of as-mixed and mechanically-milled Al/SiC after 6, 12, 18 and 24 h of milling, crystallite size and lattice strain were obtained by using the Williamson-Hall method. The XRD pattern indicated that no undesirable phase, such as Al_4Cl_3 was present in the interface of Al and SiC. Figure 8.24 shows the variations of the crystallite size and the lattice strain of Al as a function of milling time. During the first 12 h, the crystallite size decreased but by further milling up to 24 h it slightly increased [Fig. 8.24 (a)].

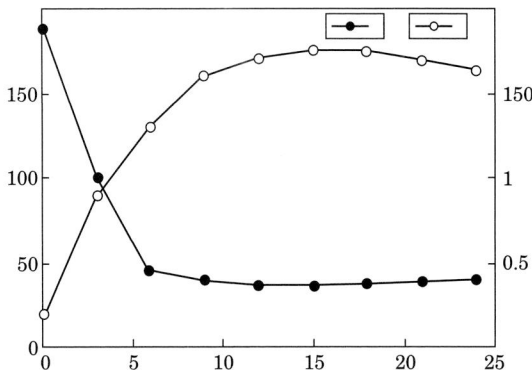

Fig. 8.24. The variations of (a) crystallite size, and (b) Lattice strain of nanocomposite as a function of milling time[44]

The mechanism of the formation of the nanostructure materials by mechanical alloying had described by altering of the crystallite size and the lattice strain. At the early stages of mechanical alloying, shear bands were observed due to the high deformation rates experienced during mechanical alloying. These shear bands, which contain a high dislocation density, have a typical width of approximately 0.5–1.0 μm. As was discussed by Suryanarayana, with continued milling, the average atomic level strain increases due to the increasing dislocation density, and at a certain dislocation density within these heavily strained regions, the crystal disintegrates into subgrains that are separated by low-angle grain boundary. These results in a decrease of the lattice strain. On further processing, deformation occurs in shear bands located in previously unstrained parts of the material. The grain size decreases steadily and the shear bands coalesce. The small angle grain boundaries are replaced by higher angle grain boundaries, implying grain rotation, as reflected by the absence of texture in the electron diffraction patterns and random orientation of the grain observed from the lattice fringes in the high resolution electron micrographs. Consequently, dislocation-free nanocrystalline grains are formed. According to the results of the present work, at the final stages of milling time (24 h), which is shown in Fig. 8.10 a little increasing of crystallite size and decreasing of lattice strain were occurred. The result can be interpreted due to competition between the plastic deformation via dislocation motion and restoration processes (recovery and recrystallization).

Figure 8.25 (a) shows cumulative distribution diagram of the particle size of mechanically-milled Al/SiC. X percent of the particle size of the nanocomposite that has less than D diameter is defined D_x. For example, D50 was taken to consideration as a mean diameter of the particle size. The variations of the particle mean diameter as a function of milling time can be seen in Fig. 8.13 (b). At about 6 h milling time, the particle size of the nanocomposite was decreased, further milling up to 24 h leads increasing of the particle size of nanocomposite. The mechanism of

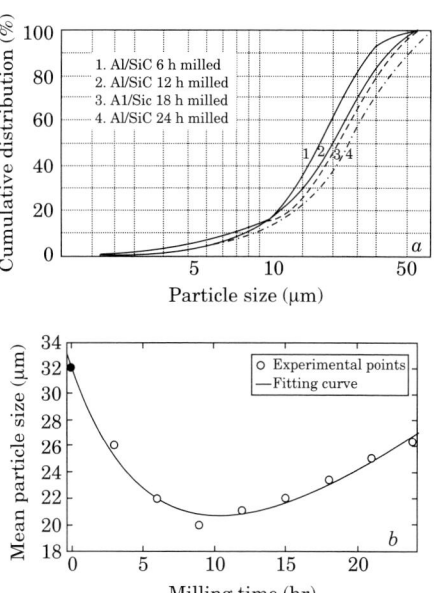

Fig. 8.25. (a) Cumulative distributions of Al/SiC nanocomposite after 6, 12, 18, and 24 h milled, and (b) The variations of mean diameter of particle size of Al/SiC mechanically mixed as a function of milling time[44]

phenomena happened during mechanical alloying can describe the variation of the particle size versus milling time. During mechanical alloying, the powder particles are repeatedly flattened, cold welded, fractured and rewelded. After milling for a certain length of time, steady-state equilibrium is attained when a balance is achieved between the rate of welding, which tends to increase the average particle size, and the rate of fracturing, which tend to decrease the average composite particle size. Al particles tend to agglomerate due to high energy produces by planetary mill, and ductility of Al phase itself. In the present research work, Stearic acid was added for decreasing of particle welding. Since the process control agent adsorbs on particle surfaces interfere with cold welding and lower the surface tension of particles under deformation. Thus, in the early stages of milling, decreasing of the nanocomposite particle size was observed, because the rate of fracturing is higher than the rate of welding of particles. From 12 to 24 h milling time, the Stearic acid gradually evaporates and the rate of welding becomes higher than the rate of fracturing. Consequently, this process leads to powder agglomeration phenomenon. Decreasing in the process control agent (Stearic acid) can be considered due to steady evaporation and oxidation. However, the sampling of powders in every stages of milling cannot be ignored. After 24 h milling time, agglomeration of powder particles occurred as shown in Fig. 8.26. The initial approximation using cumulative distribution of particle size shows that less than 10% of particles were agglomerated after 12 h milling time.

Fig. 8.26. SEM Micrograph of agglomerated nanocomposite powders after 24 h milling time

2.2. Sol-gel

Chemistry has always played an important role in development of new materials and technologies, from polymers and organic materials to inorganic materials with tailored properties. Now, introduction of nanochemistry, has broadened its applicability to much grater extent. Nanochemistry gives us the chance to produce nanoparticles of metals, oxides, intermetallics, semiconductors, quantum dots, and glasses. Broad as chemistry science itself, are chemical synthesis methods of nanostructured materials, among which sol-gel method has attracted so much attention and therefore, is going to be described in this section.

Sol–gel processing is a form of nanostructure processing which not only begins with a nanosized unit, but it undergoes reactions on the nanometer scale resulting in a material with nanometer features[45]. When choosing a classification for processes according to scale, there is none more appropriate for sol–gel processing than the nanometer scale[2].

Originally employed for the production of ceramic powders, sol-gel process could be used to synthesize nanocrystalline oxides, ceramics and nanocomposites. As a method for producing glass and ceramic compounds, sol-gel method is remarkable in several ways, including high purity and homogeneity of its products, low processing temperature and the possibility of controlling the structural features of the final products. There are, however, some drawbacks to this method; the sol-gel process cannot be easily calibrated and the physical properties of its products are hardly reproducible. This is partly due to the fact that the final state of products is highly dependent on the experiment conditions.

Before proceeding to description of sol-gel process, it's necessary to define some basic terms. A colloid is a suspension in which the dispersed phase is so small (~1–1000 nm) that the gravitational forces are negligible and interactions are dominated by short-range forces, such as van der Waals attraction and surface charges. A sol is a colloidal suspension of solid particles in a liquid. The term gel refers to the semirigid mass formed when the colloidal particles are linked to form a network or when the polymer molecules are cross-linked or interlinked. A ligand is an atom, molecule, ion or radical surrounding a central atom.

The sol-gel process involves formation of inorganic networks within a colloidal suspension (Sol), coagulation and formation of a continuous network in liquid. Starting materials include metals, metallic ions and sometimes, other elements surrounded by different ligands. Metal alkoxides ($M(OR')_n$), where M is the metal, O is the oxygen and R′ is the organic group) and alkoxysilanes are the most widely used, because they react readily with water (hydrolysis). Since the metal alkoxide and water are not soluble in each other, they are required to be dissolved in a common alcoholic solvent in order to carry out their reaction. The starting material must be able to dissolve in liquid and form the sol. removal of liquid leads to formation of the gel. The final size and shape of the particles are determined by the sol to gel transition. Generally, sol-gel process is accomplished in 4 steps:

- Hydrolysis
- Condensation and polymerization of monomers
- Growth

- Agglomeration of particles, development of a network in liquid and formation of the gel.

A schematic diagram, demonstrating sol-gel process and it's different types of processing routes and final products, is presented in Fig. 8.27. It can be seen that sol-gel process could be used to produce dense ceramics, thin films, nanoparticles or aerogel.

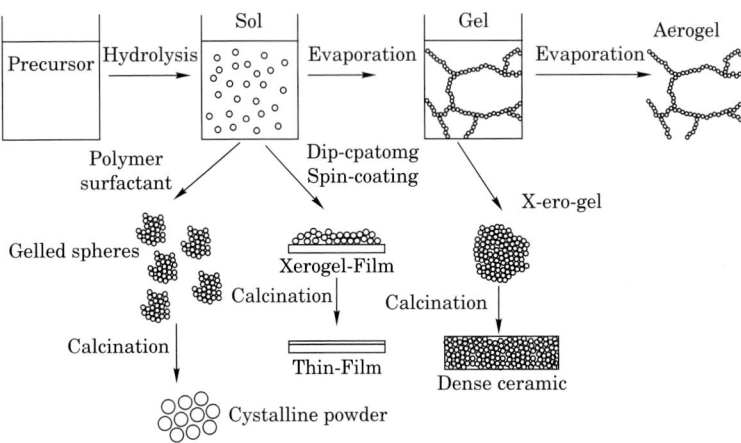

Fig. 8.27. A schematic diagram of sol-gel process

Depending upon the kind of material to be processed, the different steps of the sol-gel process could be different, which makes it impossible to give a thorough coverage of the subject; therefore our discussion will be limited to the case of Silica gel, which is indeed one of the most important and widely used materials produced by sol-gel.

2.2.1. *Silica gel*

Synthesis of silica gel involves two processes, hydrolysis and condensation[1]. The kinetics of hydrolysis and condensation reactions enables us to control the properties of the formed network within the gel; this could be achieved through a number factors including:

– PH

– Aging time and temperature

– Drying condition

– Reaction time and temperature

– Type and concentration of catalyst.

Between these factors, the first three play a more important role. At the very first stage of the process, Si-O-Si bonds are formed and all the isolated silicon atoms will be linked together by siloxane bridges, and the sol is formed. This is the sol which makes it possible to shape the material into the fibers or thin films, or to mold it into a bulk solid. At the second stage, by aid of a few chemical bonds, the colloids bond together and create a solid wide-meshed network, which would fill all the space in the container. This leads to formation of the gel. Drying the gel, and having the water or alcohol molecules removed from within the pore structure, condenses the network. Now, let's discuss the hydrolysis and condensation steps[1].

2.2.1.1. *Hydrolysis*

Hydrolysis is the reaction during which, hydroxy group (OR) is replaced by hydroxyl group (OH). This reaction, as its name suggests, is activated by addition of water. The next reaction makes silanol group (Si–OH) form the siloxane bonds (Si–O–Si), and water or alcohol are produced as byproducts. During hydrolysis, the oxygen from water attack silicon atoms; first water is transformed into H^+ and $(OH)^-$ ions, and alkoxy group is broke down into $(OR)^-$ and $Si(OR)^+_3$ ions; subsequently, $(OH)^-$ from hydrolysis of water is replaced by $(OR)^-$ from hydrolysis of alkoxy through the following reaction:

$$H_2O + ROSi(OR)_3 \rightarrow ROH + HOSi(OR)_3$$

The kinetics of this reaction could be improved by addition of catalysts such as ammines. Either an acidic catalyst (H^+) or basic catalyst (OH^-) could be used. While using an acidic catalyst, the underlying mechanisms of reaction would be slightly different. In this case, because of the presence of H^+ ions in environment, alkoxy groups rapidly attack H^+ protons, and the charge density of silicon atoms would be reduced, making them more prone to the attack of water molecules. In basic conditions, water hydroxyl ions replace OR groups.

2.2.1.2. *Condensation*

Two types of mechanism could occur during condensation, one forms water and the other forms alcohol. During condensation, the following products are formed in the order mentioned: monomers, dimers, linear trimers, cyclic trimers, linear tetramers, cyclic tetramers and cyclic molecules of higher order. The mechanism which leads to water formation and alcohol formation are as follows:

$$2HOSi(OR)_3 \rightarrow (OR)_3SiOSi(OR)_3 + H_2O$$
$$2HOSi(OR)_3 \rightarrow (OR)_2OHSiOSi(OR)_3 + HOR$$

During condensation, with increasing number of siloxane bonds, isolated molecules agglomerate and form the sol. then, the sol particles grow and collide to form a three dimensional network which fills the container and can stand stresses elastically; this is the so called sol-gel transition and the product is called the gel. When prepared, gel still undergoes some changes like rearranging itself by expelling the solvent; this process is called aging. Now it's time to dry out the large amount of liquid, entrapped in these fine interconnected channels. While drying seems like an easy step, it is totally critical to the properties of the final product and depends on the intended use of dried material. For production of ceramic powders, no special care is required to prevent fragmentation. For production of monoliths, though, controlling the capillary forces in the gel pores and the internal stresses associated with the volume changes are very important. Drying by means of thermal evaporation leads to formation of monoliths termed *xerogel*. But if the solvent is extracted under supercritical or near supercritical conditions, the product is an *aerogel*[46].

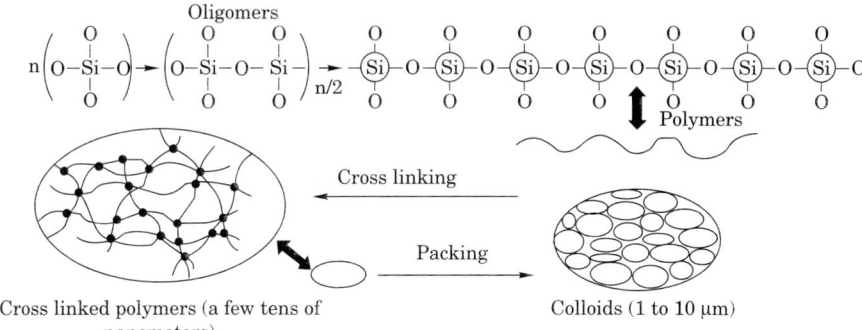

Fig. 8.28. Schematic diagram of the formation of oligomers which change into crosslinked polymers, then into micrometric colloids constituting the sol[46]

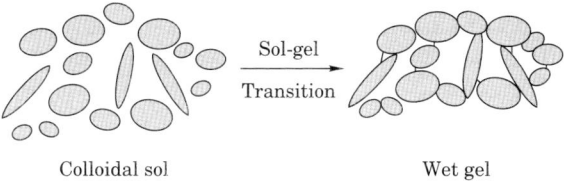

Fig. 8.29. The sol-gel transition[46]

It must have become clear till now that the finesse of the sol-gel process has a twofold effect, making it difficult to reproduce, but making it versatile enough to modify the final product to desired particle shape or size. Following sections review some examples of sol-gel process, demonstrating both its potential in production of various nanoparticles and its state of finesse.

2.2.2. Silica nanoparticles

Synthesis of silica particles via sol-gel was already discussed in a general way. As already indicated, while in sol-gel, the process and the final product is the same, but selection of reagents, catalysts and their concentrations, along with so many other process variables like time and temperature of aging and dehydration could affect the properties of the final product. Recently, silica nanoparticles were synthesized by sol-gel process using tetraethoxisilane (TEOS) in the solution of water and ethanol (C_2H_5OH), and in the presence of ammonia as a catalyst. Also, a systematic study of the effects of different molar ratio of reagents, on the features of final products was carried out[47–48].

Ethanol, Tetraethoxisilane (TEOS), and ammonia aqueous solution were used as provided from Merck Co., and the water used for the sample preparation was purified by both ion-exchange and distillation. Reagents were mixed into the two starting solutions of ethanol; solution (I) contained TEOS and C_2H_5OH, and solution (II) contained NH_3, H_2O and C_2H_5OH. The contents of the solutions (I) and (II) were adjusted in such a way that the concentrations of reagents would be at the prescribed concentrations when they are mixed with each other. The solutions (I) and (II) were mixed with each other at 298 K and the mixture was stirred vigorously. After the invisible hydrolysis reaction the condensation reaction could be easily recognized from an increase in opalescence of mixture 2–15 min after adding TEOS, depending on the molar ratio of reagents. Suspended white particles occurred regularly a few min later. For the synthesis of the first sample of colloidal silica nanoparticles, the molar ratio of reagents was chosen as shown in Table 8.6.

Table 8.6. Molar ratio of reagents in sample I[47]

Sample	TEOS	NH_3	H_2O	C_2H_5OH
I	0.1	0.2	0.5	2

Figure 8.30 shows the TEM images of the resultant colloidal particles. As can be seen in this figure, from the selected constituent concentrations, spherical silica particles were obtained. X-ray diffraction using CuK_α (Philips X'pert) was utilized to determine the crystalline structure of the silica particles, Fig. 8.31. This spectrum is typical and represents the nature of the all colloidal silica nanoparticles which have been synthesized in this study. It can be deduced from the results of this X-ray diffractometry that very few of the resultant silica particles have crystalline structure and the particles are generally amorphous. Moreover, the final products were analyzed by XRF. The result of this analysis is shown in Table 8.7.

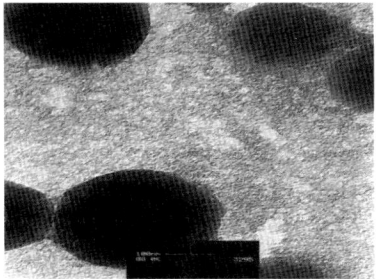
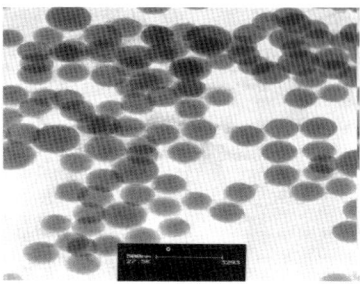

Fig. 8.30. Silica particles obtained from solution (I)[47]

Table 8.7. Result of XRF analysis[47]

	LOI	SiO$_2$	Cl	Zn	Br
Wt %	18.096	81.854	0.04	0.01	0.015

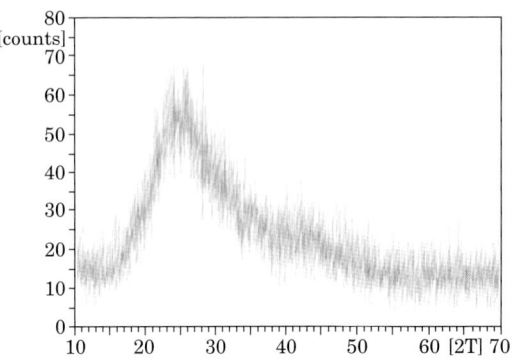

Fig. 8.31. XRD spectrum of colloidal silica particles[47]

According to result of this analysis the purity of final products is satisfactory. LOI could be consisted of water, CO_2, sulfur and etc. A kinetic study of silica particle structure evolution as a function of reaction time was carried out.

Figure 8.32 shows the SEM observations of this study. It seems that after the first 5 min, the small elementary silica particles were formed and subsequently, they aggregate to each other till they form final silica particles, which are considerably larger. Other SEM images showed that the size and shape of resultant particles do not change significantly after the first 30 min.

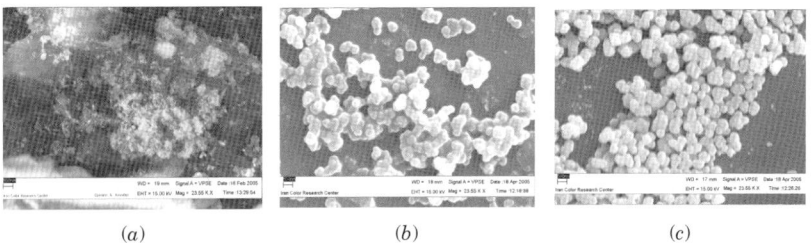

Fig. 8.32. SEM images of silica nanoparticles (a) 5, (b) 15, and (c) 30 min after mixing the starting solutions[47]

The BET surface area of the particles of solution (I) was measured in the conventional way with N_2 and was reported 153.12 (m^2/g). The density of the particles was measured three times and the average density was reported 2.0176 (g/cm^3). Average diameter of particles was estimated 19 nm from equation:

$$d = 6/\rho s \qquad \qquad ...(2)$$

In which d is the diameter of particles, ρ is the average of density (g/cm^3) and S is surface area (sq. cm/g). However, in Fig. 8.32 (c) the particles were estimated 100–200 nm. Figure 8.30 TEM shows that spherical silica particles have particulate structure. So, these results suggest that primary particles about 19 nm and lower diameter, aggregate in secondary particles about 100–200 nm. Therefore silica particles were formed via a multi-staged aggregation.

Finally, the influence of different molar ratio of reagents (Table 8.8) on the morphology and size of final silica particles has been investigated. The SEM images of resultant particles, obtained from samples II to VII, are shown in Figs. 8.33, 8.34 and 8.35.

Table 8.8. Molar ratios of reagents[47]

Sample	II	III	IV	V	VI	VII
TEOS	0.1	0.1	0.1	0.1	0.1	0.1
NH$_3$	0.3	0.3	0.2	1	1	1
H$_2$O	1.2	1.2	0.3	0.3	0.3	3
C$_2$H$_5$OH	0.5	3	0.5	0.5	5	5

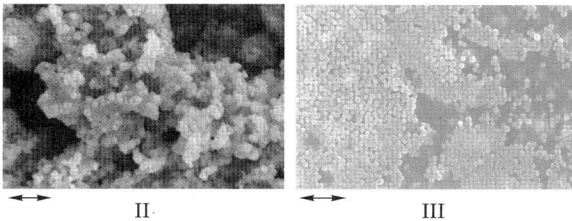

Fig. 8.33. Silica particles obtained from samples II and III[47]

Fig. 8.34. Silica particles obtained from samples IV and V[47]

Fig. 8.35. Silica particles obtained from samples VI and VII[47]

The molar ratios in Table 8.8 are selected in such a way to make it possible to study the probable effects of the changes in the concentration of reagents on the structural features of the obtained particles. By comparing SEM images of solutions II and III (Fig. 8.33), and subsequently solutions V and VI (Figs. 8.34, 8.35), it can be inferred that for equal proportions of TEOS, ammonia and water, higher concentrations of ethanol results in particles with smaller mean diameter. Also, by comparing solutions IV and V, It can be deduced that increasing the concentration of ammonia leads to an increase in the size of silica nanoparticles.

Furthermore, by comparing SEM images of solutions II, IV, and V, it can be understood that in solutions with lower proportions of solvents (water and ethanol), there are high chances that the final particles agglomerate to each other. This issue specifically holds true about ethanol. However, when narrow size distributions are required, small ratios of ethanol to water should be assumed. Finally, a comparison of solution VI and VII reveals that higher proportions of water can lead to larger silica particles.

2.2.3. ZnO nanoparticles

ZnO is an n-type semiconductor with a wide band gap of 3.37 eV and high exciton binding of 60 meV at room temperature, so it is very interesting for electronic and optical applications[49] such as photodetectors, varistors, gas sensors and solar cells. There are several methods available for synthesizing ZnO nanoparticles such as hydrothermal, solvothermal, microemulsion synthesis and direct precipitation. Chemical methods, among them sol-gel process, are more convenient and controllable. Several chemical routes have been used for synthesizing ZnO nanoparticles via sol-gel process. It will be very desirable if we could obtain finer ZnO nanoparticles using suitable surfactants. Low dimension ZnO nanostructures are expected to enhance sensing properties due to surface area increase and quantum confinement effect. Surfactants are known as important precursors in sol-gel processes which are mostly used to control the size and morphology of products. They also have an effective role in preventing agglomeration phenomena. In most of the related reports, Monoethanolamine (MEA) and Diethanolamine (DEA) were added to the solutions as surfactant and Zn^{2+} ions stabilizers. Shokuhfar et al.[50] used Triethanolamine (TEA) as a novel surfactant in order to see its effects on ZnO nanoparticles size and their morphology.

As it is shown in Table 8.9, three methanolic solutions of $ZnAc_2.2H_2O$ (100 ml, 0.5 mol) were prepared. TEA was added to each solution with the different weight ratios of TEA: $ZnAc_2.2H_2O$ = 1:2, 1:1 and 2:1. The obtained solutions were intensively stirred using magnetic stirrer for 30 min at the methanol boiling temperature (67°C). After that the solutions were kept in an oven at 50°C. After about 2 h a white porous wet gel (white cake) was formed. Permitting the hydrolysis and ploy condensation reactions complete the white cake was kept in the same condition for 24 h. After that for further purifications the product was dispersed in double distilled water and then filtered and dried at 50–60°C for 48 h. Finally obtained nanoparticles were calcinated at 500°C for 1 h.

Table 8.9. The synthesis conditions[50]

Sample	TEA: ZnAc$_2$.2H$_2$O	Stirring condition
I	1:2	30 min, 67°C
II	1:1	30 min, 67°C
III	2:1	30 min, 67°C

XRD patterns of obtained samples revealed that all samples consisted of pure ZnO and all patterns have same the form; this could be resulted from the same morphologies as shown in Fig. 8.37. The mean crystallite size can be determined by Debye-Sherrer method and is listed in Table 8.10 for all the samples and its variations by TEA:ZnAc ratio is shown in Fig 8.36.

Table 8.10. Crystalline size from XRD[50]

Sample	Crystallite size nm
I	26.4
II	24.3
III	23.1

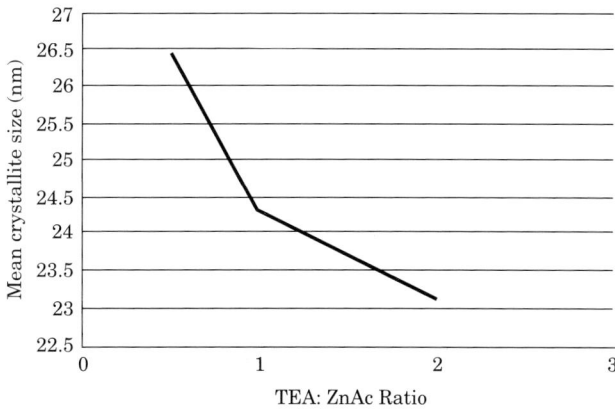

Fig. 8.36. Effect of surfactant ration on crystallite sizes[50]

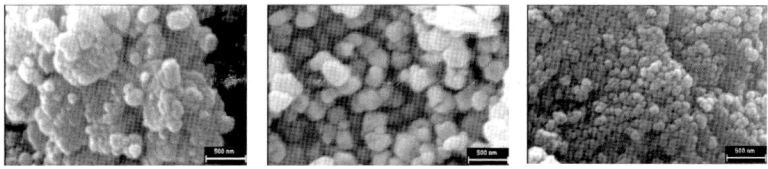

Fig. 8.37. Scanning electron microscopy images of sample (a) I, (b) II, (c) III[50]

Nonionic surfactants are produced mainly by alkoxylation technology, although amine oxides under alkaline conditions are also classified as nonionic like Triethanolamine. Figure 8.38 shows the histograms of samples I, II and III. From this figure, the effects of surfactant to precursor ratio on the particle size of ZnO can be investigated. According to Tables 8.9 and 8.11, the higher the amount of surfactant, the smaller the particle size. Also increase in amount of surfactant was resulted in decrease of particle size distribution ranges and enhancement of homogeneity of obtained samples as is shown in Fig. 8.40.

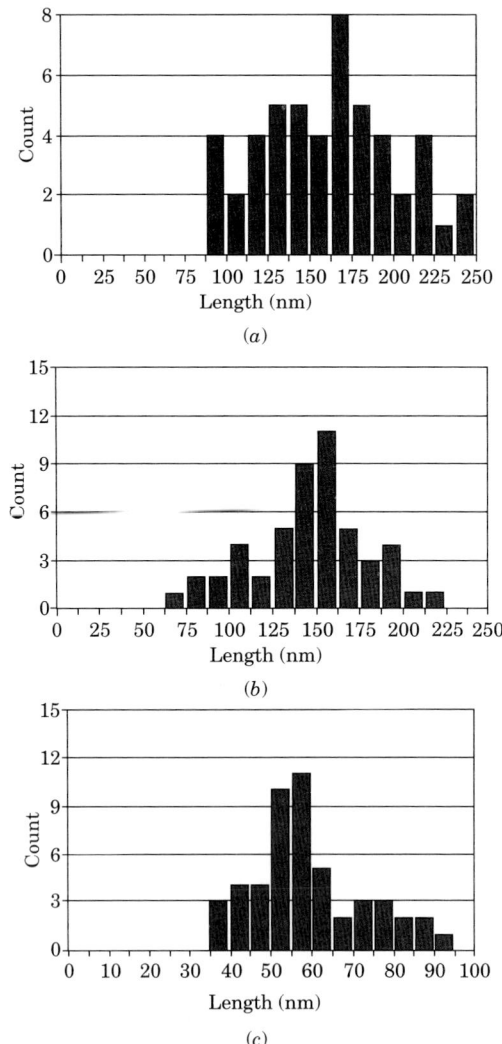

Fig. 8.38. Histograms of particle size of sample (a) I, (b) II, (c) III[50]

Figure 8.39 shows the effects of surfactant in the synthesis of ZnO via sol-gel process. This procedure is based on FTIR studies of Monoethanolamine and Diethanolamin. At first step, the ZnAc is hydrolyzed in absolute methanol as it is solved. The positive part connected to TEA in the second step. The first and second reactions offer an element of hydrolysis, dehydration and polycondensation. Finally the remained structures can join with each other and at last the sol is constructed by joining these structures together with hydrogen bands in the solvents.

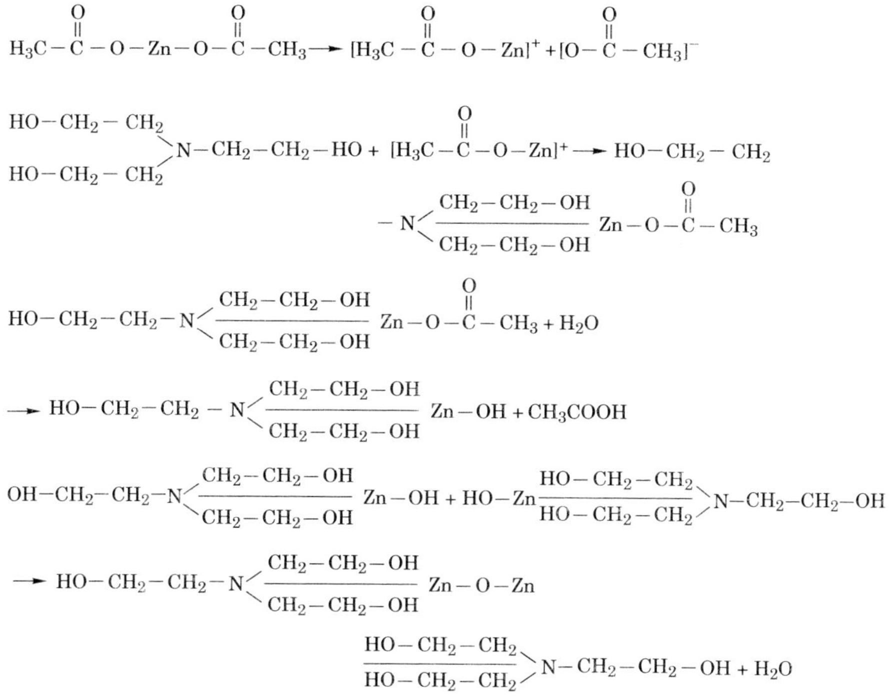

$$H_3C-\overset{\overset{O}{\|}}{C}-O-Zn-O-\overset{\overset{O}{\|}}{C}-CH_3 \rightarrow [H_3C-\overset{\overset{O}{\|}}{C}-O-Zn]^+ + [O-\overset{\overset{O}{\|}}{C}-CH_3]^-$$

$$\begin{array}{l} HO-CH_2-CH_2 \\ \diagdown \\ N-CH_2-CH_2-HO + [H_3C-\overset{\overset{O}{\|}}{C}-O-Zn]^+ \rightarrow HO-CH_2-CH_2 \\ \diagup \\ HO-CH_2-CH_2 \end{array}$$

$$-N\diagup^{CH_2-CH_2-OH}_{\diagdown CH_2-CH_2-OH}Zn-O-\overset{\overset{O}{\|}}{C}-CH_3$$

$$HO-CH_2-CH_2-N\diagup^{CH_2-CH_2-OH}_{\diagdown CH_2-CH_2-OH}Zn-O-\overset{\overset{O}{\|}}{C}-CH_3 + H_2O$$

$$\rightarrow HO-CH_2-CH_2-N\diagup^{CH_2-CH_2-OH}_{\diagdown CH_2-CH_2-OH}Zn-OH + CH_3COOH$$

$$OH-CH_2-CH_2-N\diagup^{CH_2-CH_2-OH}_{\diagdown CH_2-CH_2-OH}Zn-OH + HO-Zn^{HO-CH_2-CH_2}_{HO-CH_2-CH_2}\diagdown N-CH_2-CH_2-OH$$

$$\rightarrow HO-CH_2-CH_2-N\diagup^{CH_2-CH_2-OH}_{\diagdown CH_2-CH_2-OH}Zn-O-Zn$$

$$^{HO-CH_2-CH_2}_{HO-CH_2-CH_2}\diagdown N-CH_2-CH_2-OH + H_2O$$

Fig. 8.39. Effects of TEA in the synthesis of ZnO via sol-gel process according to FTIR studies[50]

Table 8.11. Statistical Data of the particle size of samples I, II and III[50]

Sample	Min particle size nm	Max particle size nm	Mean particle size nm
I	87.8	249.8	161.5
II	69.9	214.9	146.8
III	36.7	94.5	59.2

According to the above illustrations, one can say that increase in concentration of the surfactant to a certain amount will help on compellation of the reaction. Furthermore, by increasing of the amount of surfactant the surface tension decreases and nuclei form farer to each other.

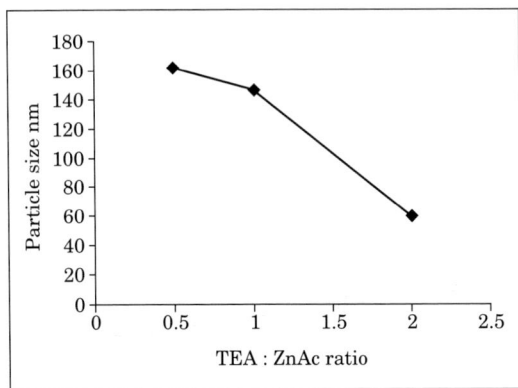

Fig. 8.40. Effect of amount of TEA on the spherical particle's diameters[50]

2.3. Sonochemistry

As its name suggests, Sonochemistry employs sound energy to modify chemistry. Chemical engineers usually use external variables like pressure, heat or light to modify the reactivity of the system. Though it's quite a long time that sound waves are known to have some effects on reactivity of chemical reactions, but it's not more than two decades that the term sonochemistry has come into existence; however, during this almost short period, it has proved to be very promising in the enhancement of chemical reactivity and reduction of particle size. Until now, a fair number of nanoparticles have been produced, taking advantage of sonochemistry.

Sonochemistry is the research area in which molecules undergo a chemical reaction due to the application of powerful ultrasound radiation (20 kHz–10 MHz)[51]. The magical effect of ultrasound is mostly attributed to cavitational bubbles. The generation and collapse of these bubbles can break down chemical bonds. There is also a hot spot mechanism, which is supposed to be responsible for production of nanoparticles. In the following sections, first, the principles of cavitation and the parameters controlling the sonochemical reactions are discussed. Then the hot spot mechanism for formation of nanoparticles would be explained. Finally, the advantages of sonochemistry in materials science would be reviewed.

2.3.1. Cavitation

Sonochemistry takes advantage of a transducer (a piezoelectric crystal) to convert an alternating electric field into vibration; the probe is either in the form of a horn which will be placed in the liquid medium or the bottom of the bath. The vibration is then transmitted through the liquid medium, oscillating every molecule on its way about a mean position. The average distance between molecules which are bonded together by weak van-der waals increase and decrease during the rarefaction and compression cycles of the sound wave, respectively. When the ultrasound energy is large enough to increase the intermolecular spacing beyond a certain threshold where the bonding is broken down, then a bubble will be created. This phenomenon is called cavitation and the bubbles created have a major role in sonochemistry.

When bubbles are created during the rarefaction cycle of the ultrasound wave, they grow until the point when the compression cycle starts; now, they tend to decrease in volume and ultimately, some would collapse. The collapse of these cavitational bubbles dissipates considerable amounts of energy into the neighboring liquid medium; this energy then alters chemical reactivity of the system, which consists of a liquid solution.

Process variables in sonochemistry include those of the ultrasound source such as ultrasound energy and frequency, and those of chemical reaction like the choice of solvent, temperature and pressure. The frequency controls the lifespan of the bubbles. Typical frequencies used in sonochemistry range between 20 to 50 KHz, which provides enough time for the bubble to grow and collapse so that it produces sufficient energy for breaking the chemical bonds in the liquid. This also applies to the ultrasound energy, which must be sufficiently high to disrupt the liquid. Increasing the ultrasound energy will expand the region affected by cavitation until a critical point where the number of bubbles becomes so high that the energy would not be effectively transmitted into the liquid.

While selection of the proper ultrasonic power and energy is an essential step in design of sonochemical reaction systems, but these are the chemical variables which give us much possibility to control the reactions and synthesize materials with desired size, shape and other properties. For the solvent, viscosity and evaporating temperature indicate the intermolecular bonding energy. If the solvent has a low evaporating temperature or viscosity, then the bonding between molecules may break down in much lower ultrasound energies, and at the same time, upon their collapse, they don't produce sufficient energies for altering chemical reactions. In these cases, we may therefore lower the temperature, so that the viscosity would go up. This simply shows that selection of the solvent and the bath temperature must be done simultaneously. The choice of the gas type and the pressure applied to the liquid medium is also central to the design of a successful sonochemical experiment.

The sonochemical effect is known to vary in different liquid systems. In all the liquid systems, when bubbles are formed, they most probably contain some of the liquid medium or dissolved volatile reagents in the form of vapor. The high temperatures and pressures within the bubble decompose trapped molecules into highly reactive radicals which may then be released into the liquid and react with dissolved molecules. Also, the same process may occur in the neighboring medium, but to a much less extent. The other phenomenon is the inrush of the liquid to fill the void, when the bubble collapses. The liquid can move so rapidly toward the void that it can induce shear forces in the liquid, large enough to break chemical bonds in polymeric materials, if present in the liquid.

Bubbles formed next to phase interfaces are deformed and upon their collapse a liquid jet is generated toward the interface. In liquid-liquid interfaces, these liquid jets cause droplets of the liquids to enter the other one and an emulsion is formed. At a solid-liquid interface, the jets strike the solid surface with high speeds and cause erosion, as it's the case in cavitational erosion in propellers. Finally, when solid particles are present in the liquid, they accelerate when hit by the jet, and if they collide with other particles may cause abrasion or in some cases particle fusion. The readers are referred to the book by Mason and Lorimer[52] for a comprehensive and quantative study of cavitation and its effects in different liquid systems.

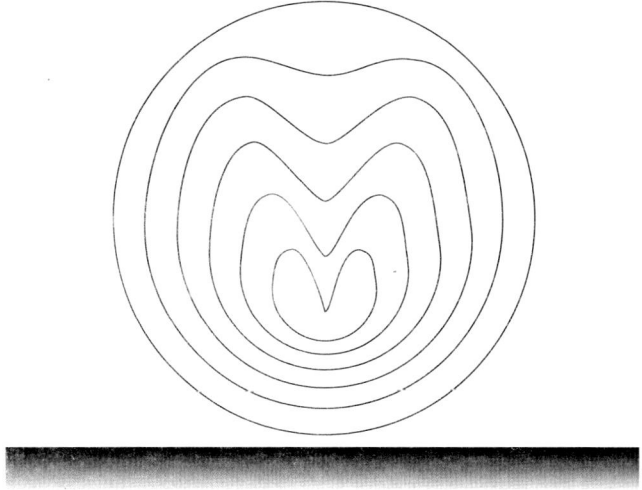

Fig. 8.41. Formation of a liquid microjet during bubble collapse next to a solid interface[53]

Simplicity, low production costs and high yielding rates in sonochemistry have attracted much attention in preparation of nanoparticles of different size and shapes. The mechanism responsible for synthesis of nanoparticles is known as hot spot mechanism. According to this theory, high temperatures are attained upon the collapse of the bubbles in a fraction of nanosecond, providing cooling rates of 10^9 K/s, which is considerably higher than those attained in rapid solidification techniques. According to Gedanken[54], in cases with volatile precursors, gas phase reactions are predominant; high cooling rates does not allow crystallization of the products and amorphous nanoparticles are synthesized. However, it's not completely clear what is responsible for the formation of nanostructured particles, but there are theories proposing that these particles are composed of a number of nuclei which have not had enough time to grow during short collapse times. When a non-volatile precursor is used, the reaction takes place in a ring adjacent to the collapsing bubble (Fig. 8.42). Depending upon the temperatures achieved in the region, either nanoamorphous or nanocrystalline particles may be obtained. The synthesized nanoparticles by sonochemistry can take different size and shapes. Although a lot of nanoparticles with different size and shapes have been prepared by sonochemistry, but yet our knowledge of the controlling variables is limited and much is still to be discovered. Sonochemistry can be used in other areas as well; Gedanken[54] has summarized the applications of sonochemistry which are of special interest to materials scientists:

- Preparation of amorphous products. This may also be achieved by rapid solidification techniques, however in case of materials with high melting temperatures like metal oxides, sonochemistry is much more effective since it does not require high amounts of energy for melting and at the same time, it can yield nanoamorphous particles.

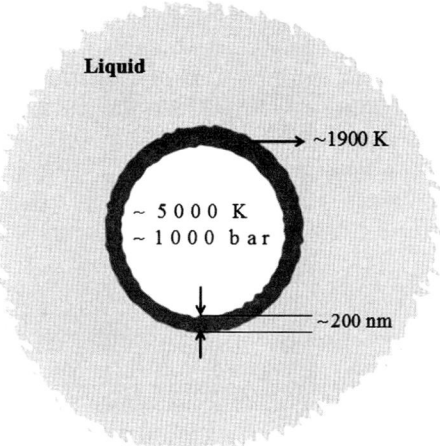

Fig. 8.42. Schematic representation of the temperatures and pressure attained upon the collapse of cavitational bubble (estimated by Suslick[53])

- Insertion of nanomaterials into mesoporous materials. Sonochemistry has proved to be an effective method for deposition of nanoparticles on inner mesopores walls.
- Deposition of nanoparticles on ceramic and polymeric surfaces. Nanoparticles may bond chemically to the ceramic and polymeric surfaces and form a coating layer upon sonication.
- The formation of proteinaceous micro- and nanospheres. Proteins can be converted into sphere upon sonication. Drugs may then be encapsulated in these spheres.

A large number of metal oxides, chalcogenides, sulfides and other metallic compounds and alloys of different shapes and size have been synthesized by sonochemistry. The readers are referred to the review by Gedanken[54] for a detailed study of the nanostructured materials prepared by sonochemistry.

2.3.2. SnO/ZnO Nano-Composite

Tin oxides and zinc oxides have attracted scientists' attention for many years. Recently, Shokuhfar et al.[55] synthesized ZnO/SnO nanocomposites by sonochemical methods starting from $ZnCl_2$ and $SnCl_2$ as the source of Zn and Sn respectively, reacting with a solution of NaOH as the source of oxygen. Ultrasonic waves were applied by using a MISONIX sonicator 3000 with a 0.5 inch horn made up of titanium alloy producing 20 KHz ultrasonic frequencies. Two aqueous $ZnCl_2$ and $SnCl_2$ solutions (1 M, 100 ml) were added dropwise to an aqueous NaOH solution (4 M, 100 ml) within about 30 min in a beaker placed in a thermostatic cooling-heating water bath. The resulting solution was kept at different constant temperatures under constant sonication output power of 21 W for 2 hours. Different samples were prepared by this simple procedure under different applied conditions as listed in Table 8.12. For samples I, II and III the magnetic stirrer was used for stirring the beakers' solutions. Molar ratios of initial solutions were selected according to below reactions.

$$ZnCl_2 + 2\ NaOH \rightarrow Zn\,(OH)_2 + 2\ NaCl \qquad ...(3)$$
$$SnCl_2 + 2\ NaOH \rightarrow Sn\,(OH)_2 + 2\ NaCl \qquad ...(4)$$

Table 8.12. Synthesis Conditions[55]

Sample	Temperature [°C]	Time [hr]	Sonication output power [W]
I	20–30	0.5	0
II	50–60	0.5	0
III	70–80	0.5	0
IV	20–30	0.5	21
V	50–60	0.5	21
VI	70–80	0.5	21

Finally, all of the samples were centrifuged and washed with distilled water and pure methanol several times and dried at 50°C.

Effect of temperature without applying sonication has been studied to investigate the effect of synthesis temperature on particle size and morphology of the samples for comparison with the samples which synthesized under sonication. Samples I, II and III synthesized without sonication at 25, 50 and 70°C respectively. The XRD patterns of these samples are shown in Fig. 8.43. It shows that SnO and ZnO are produced in sample II and III but for sample I no peaks corresponding to SnO particles can be found. Also as shown in Table 8.13, the XRF analyze of this sample indicate the very low quantity of SnO phase, so it can be concluded that SnO production in this synthesis temperature would be very low.

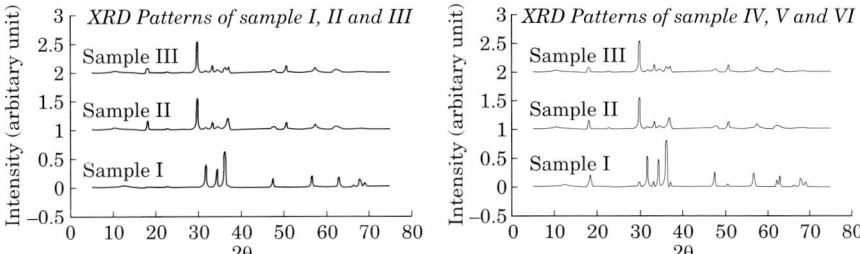

Fig. 8.43. XRD of Samples I, II, III, IV, V and VI[55]

Table 8.13. XRF analysis of Sample I[55]

Sample	Percentage of products (% wt)		
	ZnO	SnO	Other products (mainly NaCl)
1	81.3	7.8	7.9

Also from XRD patterns it can be seen that the intensity of ZnO and SnO peaks increases with temperature but for SnO this increase is much more considerable. This can be related to increase in crystallinity of both ZnO and SnO nanoparticles. The SEM images of synthesized samples are shown in Fig. 8.44. For sample I, II and III change in particle size and morphology with increase in temperature is obvious. It can be seen that by increase in synthesis temperature the particle size of samples decreases and their morphology changes from plate like to semispherical particles.

By inducting ultrasonic power to the samples, as cited in introduction, the synthesis conditions would be different. This phenomenon can be detected in XRD patterns of samples. In our synthesis the main difference was the increase in peaks intensities (Fig. 8.43). This phenomenon influences XRD peaks of ZnO rather than SnO. But as a result, the whole intensities of ZnO and SnO peaks are increased. This increase in sample IV is resulted to nucleation of SnO which could not be detected in sample I. This could be deduced from insufficient energy for crystallization of SnO in presence of ZnO in ambient temperature which can be detected in XRD pattern of sample I. Due to 2θ shifts of ZnO in its XRD pattern with regard to its standard patterns, it can be concluded that the remained Sn in XRF analyses acts as a dopant in the ZnO structure. Also in the presence of the sonication, as an exterior energy source, tiny peaks of SnO can be detected. This could be due to enhancement of whole reaction media energies because of existence of sonication. On the other hand, according to the increase of

XRD peaks, the crystallinity of obtained samples was increased by using ultrasound source in the synthesis.

Fig. 8.44. SEM Images of samples I, II and III[55]

Fig. 8.45 shows the SEM images of samples IV, V and VI. From this images most of the flakes which were produced in sample I, were eliminated in sample IV by using sonication in synthesis. From Fig. 8.45, the samples IV, V and VI which were synthesized at different temperature with presence of sonication, *i.e.*, synthesis via sonochemical route, have finer particles in comparison with the samples I, II and III which were synthesized via Chemical Bath Deposition (CBD). The histograms of median particle sizes of samples were illustrated in Fig 8.46. From this figure the behavior of particle size with temperature and sonication can be investigated.

Fig. 8.45. SEM Images of Samples IV, V, VI[55]

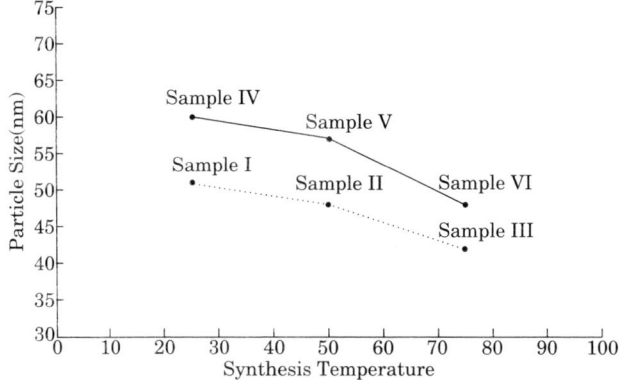

Fig. 8.46. Median particle size of Samples vs. Synthesis temperature[55]

2.4. Chemical Vapor Deposition

The desire to gain more control on the environment, has led humans to develop methods which enable them to control very fine details. This also applies to nanotechnology, where methods like Chemical Vapor Deposition (CVD) which is capable of controlling the final shape, size and chemical composition of products, attracts increasing attention of researchers[1].

CVD is a bottom-up process that involves deposition of gas particles. It is one of the vapor-transfer processes which is atomistic in nature, *i.e.*, involves transfer of atoms and molecules. Basically, CVD is used to deposit solid thin film coatings on surfaces, but it may also be used to produce bulk materials and also composites. Talking about nanostructured materials, CVD is capable of producing nanofilms and coatings, as well as nanopowders. There are a number of enhanced CVD variants which take advantage of plasmas, ions, lasers or combustion reactions to improve deposition rate and/or decrease the deposition temperature. Since there is an extensive literature on the subject, here the fundamentals of the CVD process for production of nanostructured materials will be discussed.

As a method of producing nanopowders, CVD has a number of advantages like the ability to control the shape, size, crystallization and chemical composition of particles, as well as the high purity of products and the capability of producing multi-component systems which are also applied to CVD of thin films. However, CVD has several disadvantages too. Compared to many other methods of producing nanoparticles or nanolayers, CVD is costly. Not all elements could be used as a substrate, mainly due to their instability at high temperatures, though the development of plasma-CVD and metal-organic CVD has partially compensated for this difficulty. Also, the by-products of CVD reactions may be toxic, explosive or corrosive.

2.4.1. *CVD for production of nanoparticles*

CVD was first used in industrial production of carbonyl iron powder and carbon black. Thermal CVD is probably one of the most efficient gas phase processes for synthesis of nanometric powders. The principle is quite simple: gaseous precursors are directly injected into an empty cylindrical tube conveniently heated by electrical furnaces. Homogeneous nucleation leads to formation of seeds in the gas phase, which are then able to grow by classical heterogeneous CVD or coagulation mechanisms. Their final size will depend on their residence time in the reactive zone and on the local operating conditions prevailing in the reactor chamber, such as temperature, pressure, gas velocity or precursor mass fractions[56].

In order to limit agglomeration and even to reduce average particle size, unipolar charges have been applied in the CVD chamber using a corona discharge ionizer. Electrospray (ES) assisted CVD has also been developed for that purpose, consisting in unipolarly charging a spray of precursor droplets then evaporating them to form charged non-agglomerated nanoparticles. The mechanisms involved in the two routes: (*a*) homogeneous thermal CVD and (*b*) ES-CVD according to Nakaso *et al.*[57] are schematically presented in Fig. 8.47. Unipolar ions are produced during evaporation of charged droplets in the ES, and on one hand these ions probably act as seed nuclei, increasing nucleation, and on the other hand, they are attached to the nanoparticles produced, thus lowering agglomeration through repulsive forces[56].

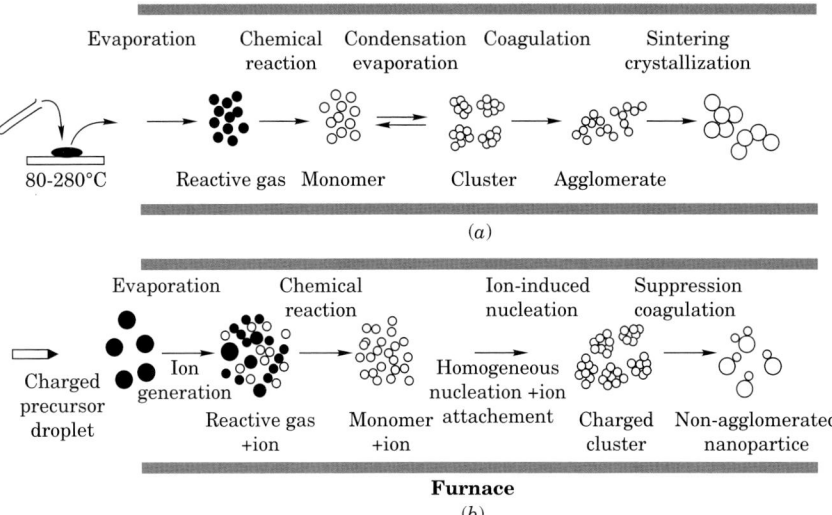

Fig. 8.47. Particle formation mechanisms in (*a*) homogeneous thermal CVD and (*b*) ES-CVD[57]

2.4.2. *CVD for production of nanocoatings*

Almost every CVD process involves the following order[58], sketched in Fig. 8.48:

- Convective and diffusive transport of reactants from the gas inlets to the reaction zone
- Chemical reactions in the gas phase to produce new reactive species and by-products
- Transport of the initial reactants and their products to the substrate surface
- Adsorption (chemical and physical) and diffusion of these species on the substrate surface
- Heterogeneous reactions catalyzed by the surface leading to film formation
- Desorption of the volatile by-products of surface reactions

- Convective and diffusive transport of the reaction by-products away from the reaction zone.

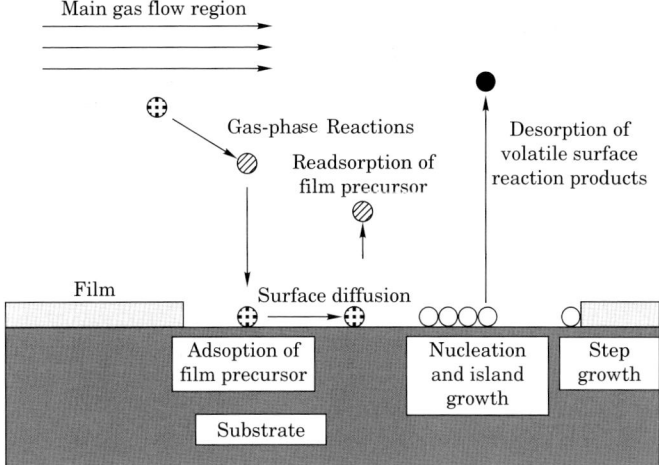

Fig. 8.48. Sequence of gas transport and reaction processes contributing to CVD film growth[58]

The first deposited layer of atoms or molecules may interact with precursor surface and will provide a pattern for growth of the next layers. Since neighboring molecules affect each other, the pattern formed on the surface will be an ordered one. Therefore, if the surface to be deposited is totally smooth, then the deposition will have its maximum efficiency. During deposition, one position may be thermodynamically more favorable for the next atom or molecule to be deposited on. Therefore, it's possible to produce a pattern for deposition of the next layers and produce an anisotropic material with different properties in direction of deposition. Today, electronic etching and lithography as well as laser and other tools, provide us with the means to produce desired patterns on the precursor and produce the material with our desired properties.

It has become obvious till now that the CVD process is a multidisciplinary concept dealing with thermodynamics, fluid flow, plasma physics, kinetics and above all chemistry. Therefore, in the following sections, the various types of reactions that have been employed to deposit nano films and coatings will be discussed.

2.4.2.1. Reaction types

A large number of materials can be deposited using CVD technique via a wide range of precursors and reactions. The choice of a particular process is based on the purity, morphology and cost requirements for the product films, plus compatibility with the substrate material[59]. To provide the reader with a general overview of CVD chemistry, overall chemical reactions will be discussed through different categories[58].

2.4.2.1.1. Pyrolysis: Pyrolysis or thermal decomposition is the simplest of all CVD reactions in which gaseous species like hydride, carbonyl or organometallic compound molecules are decomposed into their constitutive atoms or molecules on the hot substrate. Some typical examples of Pyrolysis are:

$$SiH_4(g) \rightarrow Si(s) + 2H_2(g) \text{ (Hydride Decomposition)}$$
$$Ni(CO)_4(g) \rightarrow Ni(s) + 4CO(g) \text{ (Carbonyl Decomposition)}$$
$$TiI_4(g) \rightarrow Ti(s) + 2I_2(g) \text{ (Halide Decomposition)}$$
$$WF_6(g) \rightarrow W(s) + 3F_6(g) \text{ (Halide Decomposition)}$$
$$CH_4(g) \rightarrow C(s) + 2H_2(g) \text{ (Hydrocarbon Decomposition)}$$

2.4.2.1.2. Reduction: Reduction reactions in CVD usually take advantage of hydrogen carrier gas to effect the reduction of gaseous species like halides, carbonyl halides, oxyhalides or other oxygen-containing compounds, as in the cases of tungsten deposition from the hexafluoride:

$$WF_6(g) + 3H_2(g) \rightarrow W(s) + 6HF(g)$$

However, sometimes a separate reducing agent may be employed, as in the case of tungsten hexafluoride reduction by silane:

$$2WF_6(g) + 3SiH_4(g) \rightarrow 2W(s) + 3SiF_4(g) + 6H_2(g)$$

Metal vapors may also be used in certain processes, such as the reduction of titanium tetrachloride by magnesium:

$$TiCl_4(g) + 2Mg(s) \rightarrow Ti(s) + 2MgCl_2(g)$$

2.4.2.1.3. Oxidation and Hydrolysis Reactions: Oxidation and hydrolysis are two important reactions in CVD technique which lead to formation of metal oxides. Oxygen sources including oxygen itself, water, CO_2 and in some cases Ozone, will oxidize the main precursor, removing hydrogen or halides from the elements of interest. Typical examples of these kind of reactions are:

$$SiH_4(g) + O_2(g) \rightarrow SiO_2(s) + 2H_2(g)$$
$$SiCl_4(g) + 2CO_2(g) + 2H2(g) \rightarrow SiO_2(s) + 4HCl(g) + 2CO(g)$$
$$2AlCl_3(g) + 3H_2O(g) \rightarrow Al_2O_3(s) + 6HCl(g).$$

3. PROPERTIES OF NANOSTRUCTURED MATERIALS

Unique properties of nanostructured materials, compared with their conventional counterparts, have fascinated the boundless curiosity of scientists all over the world. This interest is because of both their unusually superior properties and the unknown mechanisms leading to them. Though it's clear that the rise in the amount of surfaces and interfaces due to the miniaturization of the constitutive parts of the system plays an important role in the improved properties of nanostructured materials, but the lack of powerful characterization techniques, capable of investigating very fine details of these materials, doesn't allow scientists to come to a thorough understanding of the underlying mechanisms.

3.1. Diffusion in Nanostructured Materials

With the ever-increasing attitude toward words starting with the "nano" prefix, one could expect the future dictionary of nanotechnology to have just an N section; well, nanostructured materials often exhibit improved properties, yet there are times that their differences with their conventional counterparts go beyond a level that it wouldn't be possible to use the same word for expressing the same quality. This is the case for diffusion in nanomaterials which is almost being substituted by the word nanodiffusion. Having a very basic knowledge of the field, one would expect an improved diffusion in nanostructured materials, since both diffusion and nanomaterials share the common feature of surface. However studies have shown that diffusion in nanostructured materials have got much more than a simple improved interfacial diffusion.

Diffusion is a key property of nanostructured materials, influencing the applications that involve atomic transfer as well as many physical properties of materials like creep. The diffusion coefficient in nanostructured materials is influenced by both the size effects and the specific interfacial structure. It is known that the diffusion coefficient (D) and activation enthalpies (ΔH) vary in bulk material, surface, grain boundaries and defects like dislocations and experiments have proved that they have the following relationship:

$$D \ll D_d \leq D_{gh} \leq D_s$$
$$\Delta H > \Delta H_d \geq \Delta H_{gh} > \Delta H_s$$

where D_d, D_{gb} and D_s denote diffusion coefficient of defects, grain-boundaries and surfaces, respectively. It is quite clear that higher diffusion coefficients and lower activation enthalpies in surfaces and grain boundaries are responsible for improved diffusion rates in polycrystals compared with single crystals. Until now, we have had enough clues that diffusion in nanostructured materials must be much higher than conventional materials, since they possess a higher surface to volume ratio and/or much higher volumes fractions of grain boundaries. However, this simple conclusion is further completed by studying the diffusion kinetics in nanomaterials.

Harrison's classification of kinetic regimes[60] is shown in Fig. 8.49. Regime C occurs when diffusion takes place only inside the grain boundaries, while in regime B, some leakage outside the grain boundaries is observed and finally, in regime A the leakage from adjacent boundaries overlap. This classification was then replaced by a more sophisticated classification by Kaur, Mishin and Gaust[61]. In this new classification, new subregimes were introduced which corresponded to different characteristic lengths, as shown in Fig. 8.50 and Table 8.14.

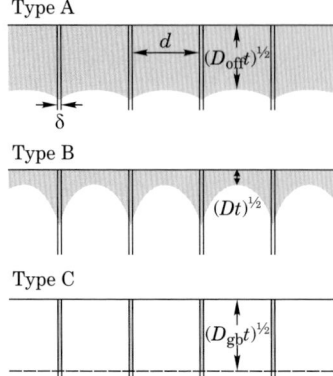

Fig. 8.49. Harrisons classification of diffusion kinetics regimes[62]

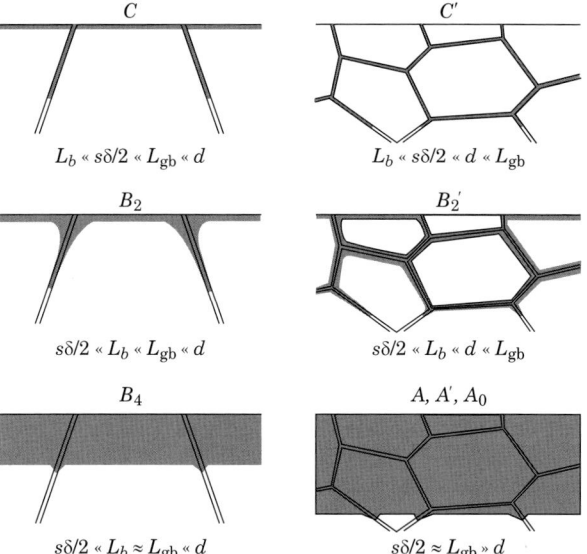

Fig. 8.50. Kinetic regimes and subregimes of diffusion in polycrystals according to Kaur, Mishin and Gust[62]

Table 8.14. Characteristic length scales for diffusion in polycrystals[62]

Lattice (or bulk) diffusion length	$L_b = \sqrt{Dt}$
Average grain size	d
Grain-boundary width	δ
Effective grain-boundary width	$s\delta$
Diffusion length inside boundaries (C regime)	$L_{gb}^C = \sqrt{D_{gb}t}$
Effective grain-boundary diffusion length (B regime)	$L_{gb}^C = \dfrac{\sqrt{sD_{gb}\delta}}{(4D)^{1/4}}\, t^{1/4}$

s: segregation factor of the diffuser

In coarse-grained polycrystals, diffusion starts in the C regime, where diffusion takes place only inside the grain boundary, until the time when lattice diffusion becomes larger than the effective grain-boundary width, i.e., $L_b \gg s\delta/2$ and diffusion enters the B_2 regime. The characteristic time and characteristic length of the transition are:

$$t' = \frac{s^2 \delta^2}{4D} \quad \text{and} \quad L' = \frac{s\delta}{2} \sqrt{\frac{D_{gb}}{D}} \quad \ldots(5)$$

It's known that $L_{gb}^C \propto t^{1/2}$ and $L_{gb}^B \propto t^{1/4}$, and when L_b reaches to L_{gb}^B then diffusion proceeds with a nearly planar front and diffusion enters region B_4. The characteristic time and length of this transition are:

$$t'' = \frac{(s\delta D_{gb})^2}{4D^3} \quad \text{and} \quad L'' = \frac{s\delta D_{gb}}{2D} \quad \ldots(6)$$

By coarse-grain in here, we mean $d \gg L''$ and therefore it takes quite a long time (not in sufficiently high temperatures) until L_b reaches the magnitude of d and exceeds it and this is the moment where diffusion enters the A regime.

In ultra-fine grained polycrystals where $d < L'$, diffusion starts in the C regime ($L_b \ll s\delta/2$) and enters the C' regime in $t_0 = (d/L')^2 t'$. Since $d \ll L'$, $t_0 \ll t'$. Later, diffusion enters the B'_2 regime in which $L_{gb}^C \gg d \gg L_b \gg s\delta/2$. It is not until region A where the out-diffusion from grain boundaries overlap, even though they exist in B'_2 regime. If $d > L'$ then the A regime starts before the B'_2 regime. A more detailed description of the subject along with calculations is given in[62].

For having a clearer picture of diffusion times in ultra-fine grained polycrystals in comparison with coarse-grained materials, the schematic evolution of these regions by time are depicted in Fig. 8.51. It is clear that diffusion continues in the A regime with a planar interface with an effective diffusitivity at much lower time and/or temperature.

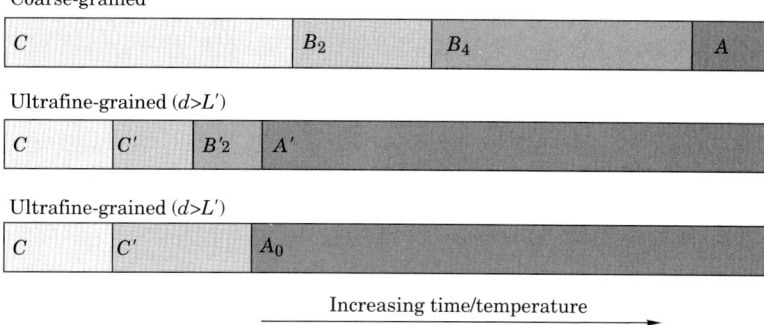

Fig. 8.51. Schematic representation of the evolution of kinetic regimes for coarse-grained polycrystals and ultrafine-grained material.

Apart from the direct effects of nanostructure on diffusion characteristics of material, the processing route may give rise to an increase in diffusion rate in nanostructured materials. For example, mechanical processes which are accompanied by cold work, such as mechanical alloying or severe plastic deformation lead to an increase in defect density. On the other hand, the higher defect density decreases the activation energy for diffusion according to the following equation:

$$D = D_0 \exp(-\Delta Q/RT) \qquad \ldots(7)$$

where D is the diffusion coefficient, D_0 is the material constant, ΔQ is the activation energy. We[63–64] investigated the mutual diffusion of Al and Cu during synthesis of Al-4.5 wt% Cu alloy with and without presence of SiC particles. X-Ray diffraction patterns of the monolithic and composite samples imply the fact that a higher level of mutual diffusion of constituents, Al and Cu, happened in the matrix in the presence of SiC particles. This effect of the reinforcing particles can be attributed to the augmented density of dislocations and vacancies introduced by the presence of SiC particles within the matrix, which give rise to an increase in micro-strain, lattice parameter and a decrease in crystallite size.

In conclusion, it seems reasonable to use the word nanodiffusion to identify it from diffusion in conventional polycrystals. However, difficulties in evaluating the diffusivity in ultra-fine grained polycrystals has made it impossible, to date, to give a clear picture of the differences between diffusion in nano- and conventional crystalline materials. The same problem applies to nanoamorphous materials where there is still much debate on how to explain diffusion in amorphous materials on nanoscale; anyhow, it is known that nanodiffusion is much higher than conventional diffusion in coarse-grained materials, at a constant temperature.

3.2. Mechanical Properties of Nanostructured Materials

Mechanical behavior of nanostructured materials present one of the most challenging problems to materials scientists, since there is no single rule which applies to all the materials when their length scale enters the nanometer range. Nanostructured materials behave much different from their conventional counterparts, showing high degrees of strength along with a reasonable ductility, an inverse Hall-Petch effect when grain size decreases to less than 100 nm and a remarkable superplastic behavior at not very high temperatures. Since now a number of theories have been proposed trying to link these unusual behaviors to the nanostructure, but there are always some examples violating the rules.

3.2.1 *Hardness and Strength*

Grain refinement has always been a method of improving the hardness and yield strength of metals and alloys. The well-known Hall-Petch equation describes how yield stress σ_y changes with average grain size d:

$$\sigma_y = \sigma_0 + kd^{-1/2} \qquad \ldots(8)$$

where σ_0 is the friction stress and k is a constant. This equation may also be used for hardness. According to this relationship, if grain size was reduced from 10 μm to 10 nm then the strength must have increased by a factor of about 30. Indeed, nanocrystalline materials have much higher yield stress and hardness than coarse-grained polycrystals, but not as much as that predicted by Hall-Petch equation. Hall-Petch equation is derived based on the hypothesis that grain boundaries act as barriers to dislocation movements. Dislocations are generated by Frank-Read sources in the center of the grain and are piled up before the grain boundary. Pile-ups at grain boundaries induce shear concentrations that can either push the lead dislocation through the boundary or activate a dislocation source in the adjacent grains. The violation from Hall-Petch equation is explained based on pile-up length. Decreasing the grain size until the length of the pile-ups increases yield stress and hardness, but any decrease beyond this point results in a softening behavior (Fig. 8.52).

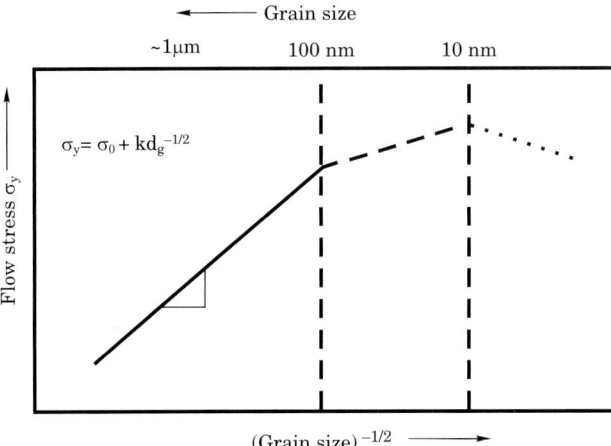

Fig. 8.52. Schematic of the variation of yield stress as a function of grain size

Several models have been proposed for describing this violation. In the first model, the material is assumed as a composite of crystal interior, grain boundaries, triple junctions and quadruple nodes, each having its own strength. Therefore material strength can be calculated by rule of mixtures. The softening behavior is then explained by higher volume fractions and lower strengths of grain boundaries, triple junctions and quadruples in nanocrystalline materials. This model does not have physical grounds[65]. The other model proposed by Scattergood and Koch employed dislocation network model which determines yield stress by the stress required to glide a dislocation through a dislocation network. The yield stress is calculated by the dislocation cut mechanism of deformation in materials with larger grain sizes and by Orowan bypassing of network dislocations in smaller grain sizes; the critical grain size would then be calculated. Other models and their pros and cons are reviewed in[8].

3.2.2. Ductility

While strength and hardness are important mechanical properties, but when it comes to manufacturing and performance, ductility plays a key role. Ductility is defined as the ability of a material to undergo plastic deformation and is measured by percentage elongation or percentage reduction in area. Yield stress and ductility usually follow a different trend when grain sizes are reduced. Nanostructured metals and alloys prepared by different methods usually have very low ductility and a lower ductile to brittle transition temperature. The low ductility of nanostructured materials is mostly due to the fact that they usually have much higher intrinsic and extrinsic defects. For processing routes that start with preparation of nanoparticles and follows a consolidation step, the porosity and inclusions are the main reason behind the low ductility of the products. However, in processing routes that no such defects are present, there are usually higher defect densities, which eventually lead to much lower cold work capacity. For instance, SPD techniques like ARB and ECAP are one-step processes, but they induce very high dislocation densities in the materials which considerably reduce the work hardening capacity of the materials and homogeneous flow before necking becomes very limited. Annealing at low temperatures can be helpful, reducing strength for the sake of ductility, but to a limited degree.

The combination of high ductility and strength has been observed in some nanocrystalline materials. Nanostructured Cobalt, Copper and Titanium had been prepared that exhibited unique combination of high strength and ductility. The reason behind this unusual behavior is not very clear, but a number of theories have been proposed relying on the role of twin boundaries, the change in deformation mechanism by grain boundaries and more[8]. Another approach to this combination is introducing a bi-modal grain size distribution; it has been said that the larger grains would strain harden with conventional strain-hardening mechanisms, while nanoscale grains would provide strength and hardness.

3.2.3. Superplasticity

Superplasticity is the capability of some materials to undergo homogeneous plastic deformations in tension, usually in excess of 1000%. This behavior is observed in conventional coarse-grained metals and alloys and also ceramics at above half the melting temperature of the material and/or low strain rates. Nanostructured materials exhibit superplastic behavior at high strain rates and/or much lower temperatures required in coarse-grained materials. The main mechanism for superplastic behavior is grain boundary sliding. The general constitutive equation for Superplasticity at high temperatures is:

$$\dot{\varepsilon} = A \frac{DGb}{kT} \left(\frac{b}{d}\right)^p \left(\frac{\sigma}{G}\right)^n \qquad \ldots(9)$$

where $\dot{\varepsilon}$ is the steady-state strain rate, D is the appropriate diffusivity (lattice or grain boundary), G is the shear modulus, b is the Burgers vector, k is the Boltzmann constant, T is the test temperature, d is the grain size, p is the grain-size exponent, σ is the applied stress, n is the stress exponent, and A is a constant. The equation shows that Superplasticity will increase by decreasing grain size. Superplastic behavior is of considerable interest to ceramics where high temperatures are required for forming them and thus nanostructured ceramics are very promising in this respect.

4. APPLICATIONS

The broad range of potential applications of nanostructured materials makes it impossible to summarize them all in some pages. Considering the fact that new applications are being developed and are going to be studied, makes it even more difficult to give an overview of the subject. Hence, the following pages discuss some of the applications of nanostructured materials in different fields, demonstrating how these materials impact these fields of research.

4.1. Biomedical Applications

Biomedical engineering is amongst the most promising fields of engineering for application of nanostructured materials. Development of synthetic organs and prosthesis has long been studied but has almost revolutionized by nanostructured materials. Site-specific drug delivery, engineered scaffolds in tissue engineering, non-viral gene carriers, biosensors and bio-actuators, medical diagnostic and cancer therapy are a few examples of the applications of nanostructured materials in biomedical engineering. The reason behind this increasing popularity lies mainly on the fact that most of our body parts have a nanostructure and our genius cells do understand and act correspondingly to this nanostructure. Here, we review some of the exciting advantages nanotechnology has brought to tissue engineering and examine bone substitutes as an example.

Tissue engineering has long been investigated and scientists have been trying hard to produce artificial organs which could replace non functional tissues and organs. Current solutions in tissue engineering include synthetic implants and transplantation of allografts. Implants made from conventional synthetic materials have a limited lifetime, while allografts and autografts (tissue or organs transplanted from one individual to another of the same species, or in the same person) have the problem of limited availability and the risk of pathogen transfer. While synthetic implants were bio-inert (not provoking biological response) in the past, like titanium or stainless-steel implants, but now scientists are more interested in bioactive materials which can integrate with biological molecules or cells to regenerate tissues[66]. Design of scaffolds which would stimulate cells }to regenerate the tissue and degrade in a timely manner is essential in tissue engineering. The scaffold may then be cultured with harvested cells from the patient or a donor (*in vitro*) or be transplanted into the body (*in vivo*).

Biomaterials for tissue engineering must have a high surface area and an appropriate surface roughness, since these properties influence cell adhesion and cell density. It has been hypothesized that cell–matrix interactions are governed by surface properties and their imperative interactions found to occur within 1 nm of the implant surface[67]. Bone tissue has a hierarchical organization over length scales that span several orders of magnitude from the macro- (centimeter) scale to the nanostructured (extracellular matrix or ECM) components. Bone ECM comprises both a nonmineralized organic component (predominantly type-1 collagen) and a mineralized inorganic component (composed of 4 nm thick plate-like carbonated apatite mineralites)[66]. Biomaterials used for bone substitute applications include ceramics, polymers, metals and composites, each having its own advantages and disadvantages. A number of hydroxyapatite nanoparticles and polymeric nanofibers like poly(lactic-co-glycolic acid) (PLGA) have been synthesized and tested for biological interactions. The enhanced adhesion, proliferation and differentiation of osteoblasts (bone-forming cells) has been observed during *in vitro* experiments by measuring intracellular and extracellular matrix protein synthesis such as alkaline phosphatase and calcium deposition. P. Li[68] studied bone formation by osteoblasts on nano-apatite coated titanium and non-coated titanium and found out that bone formation was considerably higher on titanium coated with nano-apatite, as shown in Fig. 8.53.

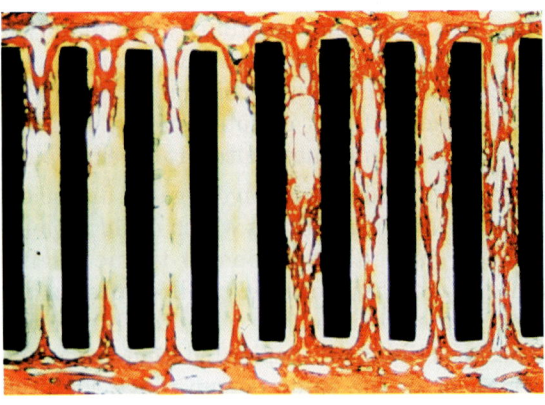

Ti   Nano-apatite coating

Fig. 8.53. Increased bone regeneration through the titanium channels coated with nano-apatite[68]

Apart from cell response to nanofeatures of the surface, nanobiomaterials have usually much better mechanical properties. Again, in bone tissue engineering where nanocomposites provide the best candidates for bone substitutes and scaffolds, mechanical properties of the scaffold must match that of the natural bone. The scaffolds must have a compressive and bending strength and elastic modulus close to that of natural bone in order to prevent stress shield effect resulting in bone resorption and loosening of the implant. Nanocomposites give us the opportunity to adjust these mechanical properties by changing the volume fractions of the constituents such as softer biopolymer matrix and harder ceramic nanoparticles. For a comprehensive study of the applications of nanotechnology in life sciences and biomedical engineering, readers are referred to the series on *Nanotechnologies for Life Sciences* by C. Kumar[69].

4.2. Catalysis by Gold Nanoparticles

Bulk gold has been considered chemically inert and consequently, not a candidate for catalysis. However, gold nanoparticles have been found very active in the oxidation of carbon monoxide, when deposited on reducible oxides such as TiO_2, Fe_2O_3 and NiO. Au nanoparticles also exhibit high activity for partial oxidation of hydrocarbons, hydrogenation of unsaturated hydrocarbons and reduction of nitrogen oxides. These reaction systems are of extreme importance in pollution control and fuel cell technology.

Catalytic activity of gold nanoparticles depends mostly on the size of particles, however, nature of the support material, preparation method and activation procedure are also of great importance. Apart from the size effects, gold atoms have unusual properties which are attributed to the relativistic effect. Relativistic effect in gold accounts for its resistance to oxidation and its higher extent of oxidation when oxidized. In heavier elements where atomic nuclear charge is high, the velocity of the **s** and **p** (with less extent compared to **s**) electrons increases to 60% velocity of light, and they become more bound to the nucleus. In contrast, the **d** and **f** orbitals become less bound and expand when relativistic effects are present[70]. Most properties of gold like its higher melting temperature, its yellow color and its smaller atomic size are described by large relativistic effects in gold and consequently, the higher energy and reactivity of **5d** electrons.

In order to make gold nanoparticles with high oxidative activity, they must be prepared by a method which yields particles of less than 5 nm combined with an oxidic catalyst (MO_x). In such small sizes, most of the atoms are located at the surface, and since these atoms have much weaker band structure, they behave more like individual atoms and more atoms are in contact with the support. The higher oxidative reactivity of gold nanoparticles makes them extremely useful as a catalyst in CO and hydrocarbon oxidation in lower temperatures.

The other important catalytic application of gold nanoparticles is in polymer electrolyte membrane (PEM) fuel cells, which are one of the potential candidates of clean energy sources. In PEM fuel cells, hydrogen must be generated by hydrogenation of methanol, natural gas and other hydrocarbons. The major problem is the production of CO in hydrogen product steam which can poison the Pt-based electrodes and reduce the efficiency of the fuel cell. The best solution to this problem is selective catalytic oxidation of CO. The catalyst must have great tendency toward oxidation of CO and production of CO_2 at working temperature of the fuel cell and with a rate comparable to that of oxygen production in hydrogenation. Again gold nanoparticles are very promising, since they have high oxidation activity in low working temperature range of fuel cells and also, the rate of CO oxidation is higher than rate of hydrogen oxidation[71].

4.2. Gas Sensors

With the growing environmental concerns in modern technology, gas sensors capable of monitoring the smallest changes in gas release are of considerable interest. Toxic gas release from combustion engine automobiles to big automotive plants include CO, SO_x, NO_x, and also CH_4 and CO_2 which are the main contributors to the greenhouse effect, have to be detected in ppm and ppb ranges. Solid state gas sensors are the best candidates for commercial applications, due to the low production costs, high sensitivity and the ability to detect wide range of gaseous molecules. However, they have limited measurement accuracy and short effective lifetimes. One can summarize the most important characteristics of a gas sensor in sensitivity, selectivity and stability; the other important criteria are short response time, easy processing long life time and low cost[72]. Different types of gas sensors such as potentiometric, resistive and capacitive gas sensors have been developed, but the most important class of gas sensors are semiconducting materials, which will be discussed in more details here.

Gas detection is achieved through the reaction of gas molecules on sensor surface. The oxidation or reduction reaction on the sensor surface takes or leaves electrons on the sensor surface; this change in conductivity is then related to the gas concentration in the atmosphere. Nanostructured semiconducting oxides are therefore very promising, since much greater numbers of atoms are located at surface and exposed to gas. Semiconducting oxides in the form of nanowires, nanobelts, nanofibers, nanoribbons and nanorods are of particular interest. The advantages of nanostructured materials for gas sensing applications as summarized by Traversa et al.[72], include higher surface area exposed to gas which increases sensitivity of the gas sensor, faster response due to enhanced diffusion, expansion of the charge space thickness to grain size which depletes charge carriers in grain and improves sensitivity, and finally lower processing temperatures which can yield metastable phases with better selectivity. Grain coarsening at moderate temperatures may however limit the stability of nanostructured materials and reduce their lifetime as gas sensors.

Since now, numerous nanostructured materials have been synthesized for gas sensing applications. ZnO nanobelts for detection of H_2 and NO_2[73], In_2O_3 nanowires for NO_2 detection[74], SnO_2 nanobelts for CO and NO_2 sensing[75], carbon nanotubes for H_2 detection[76], ZnO nanowires for detection of H_2S and NO[77], and many other examples have been investigated in literature. A good text on the advantages of nanostructured materials over microcrystalline materials is written by Traversa et al.[72].

5. CONCLUSIONS

The attention paid to nanostructured materials in the last decades has helped in both introduction of new synthesis and preparation methods, and development of the conventional ones. On the other hand, by the aid of new and improved characterization techniques, a better understanding of the subject has been achieved. It has been well established that properties of nanocrystals depend not only on the average grain size but also on the preparation routes. Some of these preparation techniques have remained an academic interest, and they need to be modified to become industrial.

Large amounts of nonequilibrium grain boundaries, high density of defects and internal stresses are all sources of enhanced energy, making them key structural features of nanostructured materials; modifying these factors enables us to enhance the ultimate properties of the material. However, this requires a much better understanding of the underlying mechanisms and the structure-property relationships for nanocrystals, which are known only in general at present. Though just diffusional and mechanical properties were described in this chapter, significant improvements in many other properties have been observed and nanostructured materials with enhanced electrical, thermal, magnetic, photonic, etc. properties have been synthesized and studied. Nanotechnology has proved to be more efficient and often the unique way to obtain materials with very high strengths and with the ability to deform superplastically at lower temperatures and/or higher strain rates.

Nanostructured materials have found different applications in many fields, from electronics and photonics to biomedical and pharmaceuticals and so many others. There is no doubt that nanostructured materials are the materials of tomorrow and that the nanotechnology is going to change our standard point of view, helping us to make our world from the very basic elements, atom by atom.

REFERENCES

1. Shokuhfar, A. and Momeni, K. 2005. An Introduction to Nanotechnology (In Persian). Nashr Gostar.
2. Edelstein, A.S. and Cammarata, R.C. (Eds.) 1996. Nanomaterials: Synthesis, Properties and Applications. Bristol: Institute of Physics Publishing.
3. Koch, C.C. 2002. Nanostructured Materials. Norwich. William Andrew Publishing.
4. Nalwa, H.S. 2002. Nanostructured Materials and Nanomaterials. Academic Press.

5. Philippe Knauth, J.S. 2004. Nanostructured Materials. New York: Kluwer Academic Publishers.
6. Cao, G. 2004. Nanostructures and Nanomaterials. Imperial College Press.
7. Müller, A.K. 2004. The Chemistry of Nanomaterials. Weinheim: Wiley-VCH Verlag GmbH & Co. KGaA.
8. Koch, C.C. 2007. Structural Nanocrystalline Materials. Cambridge: Cambridge University Press.
9. Whitesides, G.M. 1996. Self assembly and nanotechnology. The Fourth Foresight Conference on Molecular Nanotechnology, November 9–11, Palo Alto, California.
10. Benjamin, J.S. 1970. Dispersion strengthened super alloys by mechanical alloying, *Metall. Trans.* **1(10)**: 2943–51.
11. Suryanarayana, C. 2001. Mechanical alloying and milling, *Progr. Mater. Sci.* **46 (1–2)**: 1–184.
12. Adamop, O. 2007. Nanostructured bioceramics for maxillofacial applications, *J. Mat. Sci. Mat. in Med.* **18**: 1587–97.
13. Eckert, J., Holzer, J.C., Krill III, C.E. and Johnson, W.L. 1992. Reversible grain size changes in ball-milled nanocrystalline Fe-Cu alloys, *J. Mater. Res.* **7(8)**: 1980–83.
14. Koch, C.C. 1993. The synthesis and structure of nanocrystalline materials produced by mechanical attrition: A review, *Nanostructured Mater.* **2(2)**: 109–29.
15. Yuan, Q., Yong, Z. and Haijun, Y. 2009. Synthesis of nanocrystalline Ti(C,N) powders by mechanical alloying and influence of alloying elements on the reaction, *Int. J. Refract. Metals & Hard Materials* **27(1)**: 121–5.
16. Zhou, Y.-L., Niinomi, M. and Akahori, T. 2008. Changes in mechanical properties of Ti alloys in relation to alloying additions of Ta and Hf, *Mater. Sci. Eng. A.* **483–484**: 153–6.
17. Lee, Y.-S. and Lee, S.-M. 2007. Phase formation during mechanical alloying in the Ti–Si system, *Mater. Sci. Eng. A* **449–451**: 1099–1101.
18. Liu, X., Liu, Y., Ran, X., An, J. and Cao, Z. 2007. Fabrication of the supersaturated solid solution of carbon in copper by mechanical alloying, *Mater. Charact.* **58(6)**: 504–8.
19. Hotta, H., Abe, M., Kuji, T. and Uchida, H. 2007. Synthesis of Ti–Fe alloys by mechanical alloying, *J. Alloy. Compd.* **439(1–2)**: 221–6.
20. Wilkes, D.M.J., Goodwin, P.S., Ward-Close, C.M., Bagnall, K. and Steeds, J. 1996. Solid solution of Mg in Ti by mechanical alloying, *Mater. Lett.* **27(1)**: 47–52.
21. Tavoosi, M., Enayati, M.H. and Karimzadeh, F. 2008. Softening behaviour of nanostructured Al–14 wt% Zn alloy during mechanical alloying, *J. Alloy. Compd.* **464(1–2)**: 107–10.
22. Scudino, S., Sakaliyska, M., Surreddi, K.B. and Eckert, J. 2008. Mechanical alloying and milling of Al–Mg alloys, *J. Alloy. Compd.* **483(1–2)**: 2–7.
23. Sasaki, T.T., Mukai, T. and Hono, K. 2007. A high-strength bulk nanocrystalline Al–Fe alloy processed by mechanical alloying and spark plasma sintering, *Scripta Mater.* **57(3)**: 189–92.
24. Fogagnolo, J.B., Amador, D., Ruiz-Navas, E.M. and Torralba, J.M. 2006. Solid solution in Al–4.5 wt% Cu produced by mechanical alloying, *Mater. Sci. Eng. A* **433(1–2)**: 45–49.
25. Mostaed, A.S., Shokuhfar, A., Mostaed, E., Saghafian, H. and Rezaie, H.R. 2009. Investigation on nanostructured AL-4.5 wt% Cu alloy synthesized via mechanical alloying process, *J. Alloy. Compd.* (in Press).

26. Martínez, V.P., Aguilar, C., Marín, J., Ordoñez, S. and Castro, F. 2007. Mechanical alloying of Cu–Mo powder mixtures and thermodynamic study of solubility, *Mater. Lett.* **61(4–5)**: 929–33.
27. Rojas, P.A., Penaloza, A., Worner, C.H., Fernandez, R. and Zuniga, A. 2006. Supersaturated Cu–Li solid solutions produced by mechanical alloying, *J. Alloy. Compd.* **425(1–2)**: 334–8.
28. Botcharova, E., Heilmaier, M., Freudenberger, J., Drew, G., Kudashow, D., Martin, U. and Schultz, L. 2003. Supersaturated solid solution of niobium in copper by mechanical alloying, *J. Alloy. Compd.* **351(1)**: 119–25.
29. Enayati, M.H. and Bafandeh, M.R. 2008. Phase transitions in nanostructured Fe–Cr–Ni alloys prepared by mechanical alloying, *J. Alloy. Compd.* **454(1–2)**: 228–32.
30. Suñol, J.J., Güell, J.M, Bonastre, J. and Alleg, S. 2009. Structural study of nanocrystalline Fe–Co–Nb–B alloys prepared by mechanical alloying, *J. Alloy. Compd.* **483(1–2)**: 604–7.
31. Valderruten, J.F., Perez Alcazar, G.A. and Greneche, J.M. 2006. Study of Fe–Ni alloys produced by mechanical alloying. *Physica B* **384(1–2)**: 316–8.
32. Liu, Z.G., Fecht, H.J. and Umemoto, M. 2004. Microstructural evolution and nanocrystal formation during deformation of Fe–C alloys, *Mat. Sci. Eng. A* **375-7**: 839–43.
33. Yelsukov, E.P., Ulyanov, A.L. and Dorofeev, G.A. 2004. Comparative analysis of mechanisms and kinetics of mechanical alloying in Fe–Al and Fe–Si systems, *Acta Mater.* **52(14)**: 4251–7.
34. Hightower, A., Fultz, B. and Bowman, R.C. 1997. Mechanical alloying of Fe and Mg, *J. Alloy. Compd.* **252(1)**: 238–44.
35. Chen, Y.-L., Hu, Y.-H., Tsai, C.-W., Hsieh, C.-A., Kao, S.-W., Yeh, J.-W., Chin, T.-S. and Chen, S.-K. 2009. Alloying behavior of binary to octanary alloys based on Cu–Ni–Al–Co–Cr–Fe–Ti–Mo during mechanical alloying, *J. Alloy. Compd.* **477**: 696–705.
36. Chen, Y.-L., Hu, Y.-H., Hsieh, C.-A., Yeh, J.-W. and Chen, S.-K. 2009. Competition between elements during mechanical alloying in an octonary multi-principal-element alloy system, *J. Alloy. Compd.* **481(1–2)**: 768–75.
37. Samani, M.N., Shokuhfar, A., Kamali, A.R. and Hadi, M. 2009. Production of a nanocrystalline Ni3Al-based alloy using mmechanical alloying, *J. Alloy. Compd.* (in Press).
38. Dashtbayazi, M.R., Shokuhfar, A. and Simchi, A. 2007. Artificial neural network modeling of mechanical alloying process for synthesizing of metal matrix nanocomposite powders. *Mat. Sci. Eng. A* **466**: 274–83.
39. Dashtbayazi, M.R. and Shokuhfar, A. 2007. Statistical modeling of the mechanical alloying process for producing of Al/SiC nanocomposite powders, *Comput. Mat. Sci.* **400**: 466–79.
40. Mostaed, E., Saghaphiyan, H., Mostaed, A., Shokuhfar, A. and Rezaie, H.R. 2009. Fabrication of nanostructured Al-4.5wt%CU Alloy and Al-4.5 wt% Cu/SiC composite via mechanical alloying. (in Press).
41. Shokuhfar, A. and Nasiri-Tabrizi, B. 2009. Nanostructured materials prepared by mechanical alloying and mechanochemical process, *Defect Diff. Forum* **283–286**: 90–97.
42. Shokuhfar, A., Nasiri-Tabrizi, B., Gashti, O. and Ebrahimi-Kahrizsangi, R. 2009. Effect of polymeric milling media on the mechanosynthesis and structural

properties of nanocrystalline hydroxyapatite, *Defect Diff. Forum* **283–286**: 98–105.
43. Nasiri-Tabrizi, B., Shokuhfar, A. and Ebrahimi-Kahrizsangi, R. 2009. Synthesis and structural evaluation of nanocrystalline hydroxyapatite obtained by mechanochemical treatment in polyamide6 vials, *J. Nano Res.* **7**: 51–57.
44. Shokuhfar, A., Dashtbayazi, M.R., Alinejad, M.R. and Shokuhfar, T. 2007. Charecterization of Al/SiC nanocomposite prepared by mechanical alloying method, *Mater. Sci. Forum* **535**: 257–65.
45. Kumar, K.-N.P., Keizer, K., Burggraaf, A.J., Okubo, T., Nagamoto, H. and Morooka, S. 1992. Densification of nanostructured titania assisted by a phase transformation, *Nature* **358(6381)**: 48–50.
46. Corriu, R. and Anh, N.T. 2009. Nanomaterials, Molecular Chemistry of Sol-Gel Derived. John Wiley & Sons, Ltd.
47. Shokuhfar, A., Shokuhfar, T., Ghazinejad, M., Babazade, R. and Tabatabae, S. 2007. An Experimental Study on Synthesis and Characterization of Silica Nanoparticles prepared by Sol-Gel Method, *Mat. Sci. Forum* **553**: 245–51.
48. Aghababazadeh, R., Tabatabae, S., Shokuhfar, A. and Mirhabibi, A.R. 2007. Synthesis and characterization of silica nanoparticles, *Solid St. Phenom.* **121–123(1)**: 49–52.
49. Kandjani, A.E., Shokuhfar, A., Tabriz, M.F., Arefian, N.A. and Vaezi, M.R. 2009. Optical properties of Sol-Gel prepared nano ZnO, The effects of aging period and synthesis temperature, *Optoelectron. Adv. Mat.* **11(3)**: 289–95.
50. Shokuhfar, A., Sameei, J., Kandjani, A.E. and Vaezi, M.R. 2008. Synthesis of ZnO Nanoparticles via Sol-Gel Process Using Triethanolamine as a Novel Surfactant, *Def. Dif. Forum* **273–276**: 626–31.
51. Suslick, K.S., Choe, S.-B., Cichowlas, A.A. and Grinstaff, M.W. 1991. Sonochemical synthesis of amorphous iron. *Nature* **353(6343)**: 414–6.
52. Mason, T.J. and Lorimer J.P. 2002. Applied Sonochemistry, uses of power ultrasound in chemistry and processing. Wiley-VCH Verlag GmbH & Co.
53. Suslick, K.S., and Price G.J. 1999. Applications of ultrasound to materials chemistry. *Annu. Rev. Mater. Sci.* **29**: 295–326.
54. Gedanken, A. 2004. Using sonochemistry for the fabrication of nanomaterials. *Ultrason. Sonochem.* **11(2)**: 47–55.
55. Arefian, N.A., Shokuhfar A., Vaezi M.R., Kandjani A.E. and Tabriz M.F. 2008. Sonochemical Synthesis of SnO/ZnO Nano-Composite: The Effects of Temperature and Sonication Power. *Def. Dif. Forum* **273–76**: 34–39.
56. Vahlas, C., Caussat B., Serp P. and Angelopoulos G.N. 2006. Principles and applications of CVD powder technology. Mater. Sci. Eng. R 53 **(1–2)**: 1–72.
57. Nakaso, K., Han B., Ahn K.H., Choi M. and Okuyama K. 2003. Synthesis of non-agglomerated nanoparticles by an electrospray assisted chemical vapor deposition (ES-CVD) method. *J. Aerosol Sci.* **34(7)**: 869–81.
58. Ohring, M. 2001. Materials Science of Thin Films. Second Edition, Elsevier Inc.
59. Wang, Z.L., Liu, Y. and Zhang, Z. 2002. Handbook of nanophase and nanostructured materials. Vol. 1: Synthesis, Springer.
60. Harrison, L.G. 1961. Trans. Faraday. Soc. **57**: 1191.
61. Kaur, I., Mishin, Y. and Gust, W. 1995. Fundamentals of grain and interphase boundary diffusion. John Wiley & Sons, Ltd.
62. Mehrer, H. 2002. Diffusion in solids. Springer.
63. Mostaed, A., Mostaed, E., Shokuhfar, A., Saghafian, H. and Rezaie, H.R. 2009. The Influence of Milling Time and Impact Force on the Mutual Diffusion of Al and Cu During Synthesis of Al-4.5 wt% Cu Alloy via Mechanical Alloying. *Def. Dif. Forum* **283–286**: 494–8.

64. Mostaed, E., Mostaed, A., Saghafian, H., Shokuhfar, A. and Rezaie, H.R. 2009. Effect of SiC Particles Volume Fraction on the Mutual Diffusion of Al and Cu During Fabrication of Al-4.5 wt% Cu/SiC via Mechanical Alloying. *Def. Dif. Forum* **283–286**: 499–503.
65. Takeuchi, S. 2001. The mechanism of the inverse Hall-Petch relation of nanocrystals. *Scripta. Mater.* **44**: 1483–7.
66. Stevens, M.M. 2998. Biomaterials for bone tissue engineering. Materials Today **11(5)**: 18–25.
67. kesemo, B., and Lausmaa J. 1986. Surface science aspects on inorganic biomaterials. *CRC Crit. Rev. Biocomp.* **2**: 335–80.
68. Li, P. 2002. Biomimetic nano-apatite coating capable of promoting bone ingrowth. *J. Biomed. Mater. Res.* **66A**: 79–85.
69. Kumar, C., 2006. Tissue, cell and organ engineering. Weinheim: Wiley-VCHVerlag GmbH & Co. KGaA.
70. Bartlett, N. 1998. Relativistic effects and the chemistry of gold. *Gold Bull.* **31**: 22–25.
71. Grisel, R., Weststrate, K-J., Gluhoi, A. and Nieuwenhuys, B.E. 2002. Catalysis by gold nanoparticles. *Gold Bull.* **35**: 39–45.
72. Traversa, E., Schäf, O., Di Bartolomeo, E. and Knauth, P. 2004. Nanocrystalline oxides for gas sensing. *In:* Nanocrystalline Metals and Oxides, Selected Properties and Applications. *Editor* Philippe, J.S. Knauth, Kluwer Academic Publishers.
73. Sadek, A.Z., Wlodarski W., Kalantar-zadeh K. and Choopun S. 2005. ZnO nanobelt based conductometric H_2 and NO_2 gas sensors. IEEE: 1326–9.
74. Zhang, D., Liu, Z., Li, C., Tang, T., Liu, X., Han, S., Lei B. and Zhou, C. 2004. Detection of NO2 down to ppb levels using individual and multiple In2O3 nanowire devices. *Nanoletters* **4(10)**: 1919–24.
75. Comini, E., Faglia, G., Sberveglieri, G., Pan, Z. and Wang, L.Z. 2002. Stable and highly sensitive gas sensors based on semiconducting oxide nanobelts. *Appl. Phys. Lett.* **81**: 1869–71.
76. Wong, Y.M., Kang, W.P., Davidson, L.J., Wisitsora, A. and Soh, K.L. 2003. A novel microelectronic gas sensor utilizing carbon naotubes for hydrogen gas detection. Sensors and Actuators B **93**: 327–32.
77. Gupta, S.K., Joshi A. and Kaur A. 2010. Development of gas sensors using ZnO nanostructures. *J. Chem. Sci.* **122(1)**: 57–62.

9

Modelling and Simulation of Interfacial Heat Transfer

M.J.S. DE LEMOS[*] AND M.B. SAITO

ABSTRACT

In this chapter a computational procedure for determining the convective coefficient of heat exchange between the porous substrate and the working fluid for a porous medium was detailed. Macroscopically uniform laminar and turbulent flows through a periodic cell formed by square and elliptic rods were computed. Quantitative agreement was obtained when comparing laminar results herein with simulations of other authors. For turbulent flows, Low and High Reynolds turbulence models were employed in order to obtain the interfacial heat transfer coefficient. Correlations for determining the such were compared. Macroscopic results in a composite channel were also presented.

1. INTRODUCTION

The transport of heat inside highly permeable media has attracted the attention of scientists and engineers due to its many engineering applications. Such applications can be found in solar energy receiver devices, heat exchangers, porous combustors, grain drying equipment, heat sink units, energy recovery systems, etc. In many of these modern engineering systems the use of cellular and metallic porous foams brings the advantages of having a large specific heat transfer area, or say, the interfacial transport area per unit volume is large when compared with other heat capturing devices. More realistic modeling of transport processes in such media is then essential for reliable design and analysis of high efficiency engineering systems.

Motivated by the wide spectrum of practical engineering applications, macroscopic transport modeling of incompressible flows in porous media has been developed over the last decades, mostly based on the volume-average methodology for either heat (Hsu and Cheng, 1990) or mass transfer (Whitaker, 1966; Whitaker, 1967). Classical books by Bear (1972), Nield

---
[*] Departamento de Energia–IEME, Instituto Tecnológico de Aeronáutica-ITA, 12228-900-São José dos Campos-SP-BRAZIL
[*] Corresponding author : E-mail : delemos@ita.br

and Bejan (1992) and Ingham and Pop (1998), to mention a few, also document forced convection and related models for heat transport in porous media.

From the point of view of energy transfer between phases, namely the cellular material phase and the working fluid, there are basically two different models commonly found in the literature: (a) a *"local thermal equilibrium"* model and (b) a *"two-energy equation"* or *"thermal non-equilibrium"* model. The first one assumes that the lumped solid temperature does not differ much from the average value of the fluid temperature, thus local thermal equilibrium between the fluid and the solid phase is assumed. This model greatly simplifies theoretical and numerical research but the assumption of local thermal equilibrium between the fluid and the solid is inadequate for a number of practical problems (Schumann, 1929; Quintard, 1998; Kaviany, 1995). As a result, in recent years more attention has been paid to the local thermal non-equilibrium model, both theoretically and numerically (Quintard *et al.*, 1997; Ochoa-Tapia and Whitaker, 1997).

Accordingly, **two-energy equation models** have been investigated by a number of researchers. Kuwahara *et al.* (2001) proposed a numerical procedure to determine macroscopic transport coefficients from a theoretical basis without any empiricism. They used a single unit cell and determined the interfacial heat transfer coefficient for the asymptotic case of infinite conductivity of the solid phase. Nakayama *et al.* (2001) extended the conduction model of Hsu (1999) for treating also convection in porous media. Having established the macroscopic energy equations for both phases, useful exact solutions were obtained for two fundamental heat transfer processes associated with porous media, namely, steady conduction in a porous slab with internal heat generation within the solid, and also, thermally developing flow through a semi-infinite porous medium. Saito and de Lemos (2005) considered the distribution of flow variables in a cell unit and, after volume-averaging them, the interfacial heat transfer coefficient for laminar flow was obtained.

In all of the above, only laminar flow has been considered. When treating turbulent flow in porous media, however, difficulties arise due to the fact that the flow fluctuates with time and a volumetric average is applied (Gray and Lee, 1977). For handling such situations, a new concept called *double decomposition* has been proposed for developing macroscopic models. (Pedras and de Lemos, 2000; Pedras and de Lemos, 2001a; Pedras and de Lemos, 2001b; Pedras and de Lemos, 2001c; Pedras and de Lemos, 2003; de Lemos, 2005a). This methodology has been extended to non-buoyant heat transfer (Rocamora and de Lemos, 2000), buoyant flows (de Lemos and Rocamora, 2002; de Lemos and Braga, 2003; Braga and de Lemos, 2004; Braga and de Lemos, 2005a; Braga and de Lemos, 2005b; Braga and de Lemos, 2006a; Braga and de Lemos, 2006b), mass transfer (de Lemos and Mesquita, 2003) and double diffusion (de Lemos and Tofaneli, 2004). In addition, a general classification of models has been published (de Lemos and Pedras, 2001). Further, the problem of treating interfaces between a porous medium and a clear region, considering a diffusion-jump condition for laminar (Silva and de Lemos, 2003a) and turbulent fields (Silva and de Lemos, 2003b; de Lemos, 2005b; de Lemos and Silva, 2006), have also been investigated under the concept first proposed by Pedras and de Lemos (2000), Pedras and de Lemos (2001a), Pedras and de Lemos (2001b), Pedras and de Lemos (2001c), Pedras and de Lemos (2003) and de Lemos (2005a). Furthermore, Saito and de Lemos (2006) proposed a new correlation for obtaining the interfacial heat transfer coefficient for turbulent flow in a packed bed, which was modeled as an infinite staggered array of square rods. Recently, a book has been published on the subject of turbulence modeling in porous media (de Lemos, 2006).

Motivated by the foregoing, this work focuses on laminar and turbulent flow through a packed bed, which represents an important configuration for efficient heat and mass transfer and suggests the use of equations governing thermal non-equilibrium involving distinct energy balances for both the solid and fluid phases. Hence, the use of such a two-energy equation model requires an extra parameter to be determined, namely the heat transfer coefficient between the fluid and the solid. This chapter reviews recent efforts in proposing correlations for obtaining the interfacial heat transfer coefficient for laminar and turbulent flow in porous material. The medium is here modeled as an infinite array of rods of distinct shapes, over which the range of Reynolds number, based on a characteristic size of the rod, is extended up to 10^7.

The next sections detail the basic mathematical model, including the mean and statistical fields for turbulent flows. Although the discussion of turbulent motion in porous media is not presented in this work, the definitions and concepts to calculate the interfacial heat transfer coefficient for macroscopic flows are presented here.

2. GOVERNING EQUATIONS

In a number of previous publications a mathematical model for turbulent transport in porous media has been detailed, in which fluctuation in time and deviations in space of all flow variables have been considered. As such, there is no need to repeat those derivations of macroscopic equations and to the interested reader the aforementioned references are suggested for further reading. Here, the equations will be simply presented.

2.1. Macroscopic Flow and Energy Equations

When the time and volume average operators are simultaneously applied over local instantaneous equations for turbulent flow, a macroscopic model is obtained. Volume integration is performed over a Representative Elementary Volume (REV) (Gray and Lee, 1977) and Slattery (1967), resulting in:

Continuity: $\nabla \cdot \bar{\mathbf{u}}_D = 0$ ...(1)

where, $\bar{\mathbf{u}}_D = \phi \langle \bar{\mathbf{u}} \rangle^i$ and $\langle \bar{\mathbf{u}} \rangle^i$ identifies the intrinsic (liquid) average of the time-averaged velocity vector $\bar{\mathbf{u}}$.

Momentum:
$$\rho \left[\frac{\partial \bar{\mathbf{u}}_D}{\partial t} + \nabla \cdot \left(\frac{\bar{\mathbf{u}}_D \bar{\mathbf{u}}_D}{\phi} \right) \right]$$
$$= -\nabla (\phi \langle \bar{p} \rangle^i) + \mu \nabla^2 \bar{\mathbf{u}}_D - \nabla \cdot (\rho \phi \langle \overline{\mathbf{u}' \mathbf{u}'} \rangle^i)$$
$$- \left[\frac{\mu \phi}{K} \bar{\mathbf{u}}_D + \frac{c_F \phi \rho |\bar{\mathbf{u}}_D| \bar{\mathbf{u}}_D}{\sqrt{K}} \right] \quad \ldots(2)$$

where the last two terms in Eq. (2) represent the Darcy and Forchheimer (1901) contributions. The symbol K is the porous medium permeability, c_F is the form drag or Forchheimer coefficient, $\langle \bar{p} \rangle^i$ is the intrinsic average pressure of the fluid and ϕ is the porosity.

The macroscopic Reynolds stress, $-\rho \phi \langle \overline{\mathbf{u}' \mathbf{u}'} \rangle^i$, appearing in Eq. (2) is given as:

$$-\rho \phi \langle \overline{\mathbf{u}' \mathbf{u}'} \rangle^i = \mu_{t_\phi} 2 \langle \bar{\mathbf{D}} \rangle^v - \frac{2}{3} \phi \rho \langle k \rangle^i \mathbf{I} \quad \ldots(3)$$

where
$$\langle \bar{\mathbf{D}} \rangle^v = \frac{1}{2} [\nabla (\phi \langle \bar{\mathbf{u}} \rangle^i) + [\nabla (\phi \langle \bar{\mathbf{u}} \rangle^i)]^T] \quad \ldots(4)$$

is the macroscopic deformation tensor, $\langle k \rangle^i = \langle \overline{\mathbf{u}' \cdot \mathbf{u}'} \rangle^i / 2$ is the intrinsic turbulent kinetic energy, and μ_{t_ϕ} is the turbulent viscosity, which is modeled in de Lemos and Pedras (2001) similarly to the case of clear flow, in the form,

$$\mu_{t_\phi} = \rho c_\mu \frac{\langle k \rangle^{i2}}{\langle \varepsilon \rangle^i} \quad \ldots(5)$$

The intrinsic turbulent kinetic energy per unit mass and its dissipation rate are governed by the following equations,

$$\rho \left[\frac{\partial}{\partial t} (\phi \langle k \rangle^i) + \nabla \cdot (\bar{\mathbf{u}}_D \langle k \rangle^i) \right]$$
$$= \nabla \cdot \left[\left(\mu + \frac{\mu_{t_\phi}}{\sigma_k} \right) \nabla (\phi \langle k \rangle^i) \right] - \rho \langle \overline{\mathbf{u}' \mathbf{u}'} \rangle^i : \nabla \bar{\mathbf{u}}_D + c_k \rho \frac{\phi \langle k \rangle^i |\bar{\mathbf{u}}_D|}{\sqrt{K}} - \rho \phi \langle \varepsilon \rangle^i \quad \ldots(6)$$

$$\rho \left[\frac{\partial}{\partial t} (\phi \langle \varepsilon \rangle^i) + \nabla \cdot (\bar{\mathbf{u}}_D \langle \varepsilon \rangle^i) \right]$$
$$= \nabla \cdot \left[\left(\mu + \frac{\mu_{t_\phi}}{\sigma_\varepsilon} \right) \nabla (\phi \langle \varepsilon \rangle^i) \right] + c_1 (-\rho \langle \overline{\mathbf{u}' \mathbf{u}'} \rangle^i : \nabla \bar{\mathbf{u}}_D) \frac{\langle \varepsilon \rangle^i}{\langle k \rangle^i}$$
$$+ c_2 c_k \rho \frac{\phi \langle \varepsilon \rangle^i |\bar{\mathbf{u}}_D|}{\sqrt{K}} - c_2 \rho \phi \frac{\langle \varepsilon \rangle^{i2}}{\langle k \rangle^i} \quad \ldots(7)$$

where, c_k, c_1, c_2 and c_μ are non-dimensional constants. The second terms on the right-hand-side of Eqs. (6) and (7) represent the generation rate of $\langle k \rangle^i$ and $\langle \varepsilon \rangle^i$, respectively, due to the mean gradient of $\bar{\mathbf{u}}_D$. The third terms in both equations are related to the generation rates due to the action of the porous matrix (see Pedras and de Lemos, 2001).

Similarly, macroscopic energy equations are obtained for both the fluid and solid phases by applying time and volume average operators to the local instantaneous equations. As in the flow case, volume integration is performed over a Representative Elementary Volume (REV), resulting in:

$$(\rho c_p)_f \frac{\partial \phi \langle \overline{T_f} \rangle^i}{\partial t} + (\rho c_p)_f \left[\nabla \cdot \left\{ \phi \left(\langle \bar{\mathbf{u}} \rangle^i \langle \overline{T_f} \rangle^i + \langle {}^i\bar{\mathbf{u}} \, {}^i\overline{T_f} \rangle^i + \overline{\langle \mathbf{u}' T_f' \rangle^i} \right) \right\} \right]$$

$$= \nabla \cdot \left[k_f \nabla \left(\phi \langle \overline{T_f} \rangle^i \right) + \frac{1}{\Delta V} \int_{A_i} \mathbf{n}_i \, k_f \overline{T_f} \, dA \right] + \frac{1}{\Delta V} \int_{A_i} \mathbf{n}_i \cdot k_f \nabla \overline{T_f} \, dA \quad ...(8)$$

$$(\rho c_p)_s \left\{ \frac{\partial (1 - \phi) \langle \overline{T_s} \rangle^i}{\partial t} \right\}$$

$$= \nabla \cdot \left\{ k_s \nabla \left[(1 - \phi) \langle \overline{T_s} \rangle^i \right] - \frac{1}{\Delta V} \int_{A_i} \mathbf{n}_i \, k_s \overline{T_s} \, dA \right\} - \frac{1}{\Delta V} \int_{A_i} \mathbf{n}_i \cdot k_s \nabla \overline{T_s} \, dA \quad ...(9)$$

where $\langle \overline{T_s} \rangle^i$ and $\langle \overline{T_f} \rangle^i$ denote the intrinsic average temperature of solid and fluid phases, respectively, A_i is the interfacial area within the REV and $\mathbf{n}_i$ is the unit vector normal to the fluid-solid interface, pointing from the fluid towards the solid phase. Equations (8) and (9) are the macroscopic energy equations for the fluid and the porous matrix (solid), respectively.

Further, using the *double decomposition concept*, Rocamora and de Lemos (2000) have shown that the fourth term on the left hand side of Eq. (8) can be expressed as:

$$\overline{\langle \mathbf{u}' T_f' \rangle^i} = \overline{\langle (\langle \mathbf{u}' \rangle^i + {}^i\mathbf{u}')(\langle T_f' \rangle^i + {}^i T') \rangle^i} = \overline{\langle \mathbf{u}' \rangle^i \langle T_f' \rangle^i} + \overline{\langle {}^i\mathbf{u}' \, {}^i T_f' \rangle^i} \quad ...(10)$$

Therefore, in view of Eq. (10), Eq. (8) can be rewritten as:

$$(\rho c_p)_f \left[\frac{\partial \phi \langle \overline{T_f} \rangle^i}{\partial t} + \nabla \cdot \left\{ \phi \left(\langle \bar{\mathbf{u}} \rangle^i \langle \overline{T_f} \rangle^i + \langle {}^i\bar{\mathbf{u}} \, {}^i\overline{T_f} \rangle^i + \overline{\langle \mathbf{u}' \rangle^i \langle T_f' \rangle^i} + \overline{\langle {}^i\mathbf{u}' \, {}^i T_f' \rangle^i} \right) \right\} \right]$$

$$= \nabla \cdot \left[k_f \nabla \left(\phi \langle \overline{T_f} \rangle^i \right) + \frac{1}{\Delta V} \int_{A_i} \mathbf{n}_i \, k_f \overline{T_f} \, dA \right] + \frac{1}{\Delta V} \int_{A_i} \mathbf{n}_i \cdot k_f \nabla \overline{T_f} \, dA \quad ...(11)$$

The two-energy equation model considering a heat transfer coefficient between the fluid and solid phases is then based on the following equations,

$$(\rho c_p)_f \left[\frac{\partial \phi \langle \overline{T_f} \rangle^i}{\partial t} + \nabla \cdot \left\{ \phi \left(\underbrace{\langle \overline{\mathbf{u}} \rangle^i \langle \overline{T_f} \rangle^i}_{\text{I}} + \underbrace{\langle {}^i\overline{\mathbf{u}}\, {}^i\overline{T_f} \rangle^i}_{\text{II}} + \underbrace{\overline{\langle \mathbf{u}' \rangle^i \langle T_f' \rangle^i}}_{\text{III}} + \underbrace{\langle \overline{{}^i\mathbf{u}'\, {}^iT_f'} \rangle^i}_{\text{IV}} \right) \right\} \right]$$

$$= \nabla \cdot \left[k_f \nabla \left(\phi \langle \overline{T_f} \rangle^i \right) + \frac{1}{\Delta V} \int_{A_i} \mathbf{n}_i \, k_f \overline{T_f} \, dA \right] + h_i a_i \left(\langle \overline{T_s} \rangle^i - \langle \overline{T_f} \rangle^i \right) \quad \ldots(12)$$

$$(\rho c_p)_s \left\{ \frac{\partial (1-\phi) \langle \overline{T_s} \rangle^i}{\partial t} \right\}$$

$$= \nabla \cdot \left\{ k_s \nabla \left[(1-\phi) \langle \overline{T_s} \rangle^i \right] - \frac{1}{\Delta V} \int_{A_i} \mathbf{n}_i k_s \overline{T_s} \, dA \right\} + h_i a_i \left(\langle \overline{T_f} \rangle^i - \langle \overline{T_s} \rangle^i \right) \quad \ldots(13)$$

where, h_i and $a_i = A_i / \Delta V$ are the interfacial convective heat transfer coefficient and surface area per unit volume, respectively. The terms describing the convective transport in Eq. (12) have the following physical significance (see Rocamora and de Lemos (2000) for details): (I) macroscopic convective transport, (II) thermal dispersion associated with deviations of time-averaged local velocity and temperature (Note that this term also appears when analyzing laminar convection in porous media), (III) turbulent heat flux due to the fluctuating components of macroscopic velocity and temperature and (IV) turbulent thermal dispersion in a porous medium due to both time fluctuations and spatial deviations of both microscopic velocity and temperature.

2.2. Macroscopic Two-Energy Equation Modeling

In order to apply Eqs. (12) and (13) to obtain the temperature fields for turbulent flow in porous media, unknown terms have to be modeled as a function of the intrinsically averaged temperatures of solid and fluid phases, $\langle \overline{T_s} \rangle^i$ and $\langle \overline{T_f} \rangle^i$, respectively. To accomplish this, a gradient-type diffusion model is used for all unknown terms, i.e., thermal dispersion due to spatial deviations, turbulent heat flux due to temporal fluctuations and turbulent thermal dispersion due to both temporal fluctuations and spatial deviations. Also needed is a model for local conduction.

Using these gradient type diffusion models, we can write:

Thermal dispersion:
$$-(\rho c_p)_f \left(\phi \langle {}^i\overline{\mathbf{u}}\, {}^i\overline{T_f} \rangle^i \right) = \mathbf{K}_{\text{disp}} \cdot \nabla \langle \overline{T_f} \rangle^i \quad \ldots(14)$$

Turbulent heat flux:
$$-(\rho c_p)_f \left(\phi \overline{\langle \mathbf{u}' \rangle^i \langle T_f' \rangle^i} \right) = \mathbf{K}_t \cdot \nabla \langle \overline{T_f} \rangle^i \quad \ldots(15)$$

Turbulent thermal dispersion:
$$-(\rho c_p)_f \left(\phi \langle \overline{{}^i\mathbf{u}'\, {}^iT_f'} \rangle^i \right) = \mathbf{K}_{\text{disp},t} \cdot \nabla \langle \overline{T_f} \rangle^i \quad \ldots(16)$$

Local conduction:
$$\begin{cases} \dfrac{1}{\Delta V} \displaystyle\int_{A_i} \mathbf{n}_i \, k_f \overline{T_f} \, dA = \mathbf{K}_{f,s} \cdot \nabla \langle \overline{T}_s \rangle^i \\ \dfrac{1}{\Delta V} \displaystyle\int_{A_i} \mathbf{n}_i \, k_s \overline{T_s} \, dA = \mathbf{K}_{s,f} \cdot \nabla \langle \overline{T}_f \rangle^i \end{cases} \quad \ldots(17)$$

For the above expressions, Eqs. (12) and Eq. (13) can be written as:

$$\{(\rho c_p)_f \, \phi\} \frac{\partial \langle \overline{T} \rangle^i}{\partial t} + (\rho c_p)_f \, \nabla \cdot \left(\mathbf{u}_D \langle \overline{T}_f \rangle^i \right)$$
$$= \nabla \cdot \left\{ \mathbf{K}_{\text{eff},f} \cdot \nabla \langle \overline{T}_f \rangle^i \right\} + h_i a_i \left(\langle \overline{T}_s \rangle^i - \langle \overline{T}_f \rangle^i \right) \quad \ldots(18)$$

$$\{(1-\phi)(\rho c_p)_s\} \frac{\partial \langle \overline{T} \rangle^i}{\partial t}$$
$$= \nabla \cdot \left\{ \mathbf{K}_{\text{eff},s} \cdot \nabla \langle \overline{T}_s \rangle^i \right\} + h_i a_i \left(\langle \overline{T}_f \rangle^i - \langle \overline{T}_s \rangle^i \right) \quad \ldots(19)$$

where $\mathbf{K}_{\text{eff},f}$ and $\mathbf{K}_{\text{eff},s}$ are the effective conductivity tensors for the fluid and solid phases, respectively, given by:

$$\mathbf{K}_{\text{eff},f} = [\phi k_f]\, \mathbf{I} + \mathbf{K}_{f,s} + \mathbf{K}_{\text{disp}} + \mathbf{K}_{\text{disp},t} + \mathbf{K}_t \quad \ldots(20)$$
$$\mathbf{K}_{\text{eff},s} = [(1-\phi) k_s]\, \mathbf{I} + \mathbf{K}_{s,f} \quad \ldots(21)$$

and $\mathbf{I}$ is the unit tensor.

Further, in order to be able to apply Eq. (18), it is necessary to determine the components of the conductivity tensor in Eq. (20), i.e., $\mathbf{K}_{f,s}$, $\mathbf{K}_{\text{disp}}$, $\mathbf{K}_t$, and $\mathbf{K}_{\text{disp},t}$. Following Kuwahara and Nakayama (1996) and Quintard et al. (1997), this can be accomplished for the thermal dispersion and local conduction tensors, $\mathbf{K}_{\text{disp}}$ and $\mathbf{K}_{f,s}$, by making use of a unit cell subjected to periodic boundary conditions for the flow together with an imposed linear temperature gradient on the porous medium. The dispersion and conduction tensors are then obtained directly from the distributed results within the unit cell by making use of Eqs. (14) and (17). In addition, the following correlations by Nakayama and Kuwahara (1999) for the thermal dispersion tensor, which are valid for $\text{Pe}_D \geq 10$, can be used:

$$\frac{(K_{\text{disp}})_{xx}}{k_f} = 2.1 \frac{\text{Pe}_D}{(1-\phi)^{0.1}}, \text{ for longitudinal dispersion} \quad \ldots(22)$$

$$\frac{(K_{\text{disp}})_{yy}}{k_f} = 0.052\,(1-\phi)^{0.5}\, \text{Pe}_D, \text{ for transverse dispersion} \quad \ldots(23)$$

where $(K_{\text{disp}})_{xx}$ and $(K_{\text{disp}})_{yy}$ are the longitudinal and transverse components of $\mathbf{K}_{\text{disp}}$, respectively.

The turbulent heat flux and turbulent thermal dispersion components of $\mathbf{K}_{\text{eff},f}$, namely $\mathbf{K}_t$ and $\mathbf{K}_{\text{disp},t}$, respectively, are not determined from a distributed calculation. Instead, they are modeled through the classical eddy diffusivity concept, similar to Nakayama and Kuwahara (1999). It should be noticed that these two terms arise only if the flow is turbulent within the void space, whereas the thermal dispersion term, $\mathbf{K}_{\text{disp}}$, exists for both laminar and turbulent flow regimes. Starting out from the time-averaged local energy equation coupled with the eddy diffusivity concept, ν_t, one can write:

$$-(\rho c_p)_f \overline{\mathbf{u}' T_f'} = (\rho c_p)_f \frac{\nu_t}{\sigma_t} \nabla \overline{T}_f \qquad \ldots(24)$$

where σ_t is the turbulent Prandtl number, which is taken here as a constant.

Applying the volume average to the resulting equation, one obtains the macroscopic version of the turbulent heat flux, given by:

$$-(\rho c_p)_f \phi \langle \overline{\mathbf{u}' T_f'} \rangle^i = (\rho c_p)_f \frac{\nu_{t_\phi}}{\sigma_t} \nabla \langle \overline{T}_f \rangle^i \qquad \ldots(25)$$

where we have adopted the symbol ν_{t_ϕ} to express the macroscopic eddy diffusivity. Now, adding up equations Eqs. (15) and (16) in light of Eq. (10) one has

$$-(\rho c_p)_f \left(\phi \left[\overline{\langle \mathbf{u}' \rangle^i \langle T_f' \rangle^{i'}} + \overline{\langle {}^i\mathbf{u}' \, {}^i T_f' \rangle^i} \right] \right)$$
$$= -(\rho c_p)_f \langle \overline{\mathbf{u}' T_f'} \rangle^i = (\mathbf{K}_t + \mathbf{K}_{\text{disp},t}) \cdot \nabla \langle \overline{T}_f \rangle^i \qquad \ldots(26)$$

According to Eqs. (25) and (26), the overall turbulent heat transport is the sum of the turbulent heat flux and the turbulent thermal dispersion mechanisms, as proposed by Rocamora and de Lemos (2000). As suggested by Eq. (25), both mechanisms are modeled together, giving for $\mathbf{K}_t$ and $\mathbf{K}_{\text{disp},t}$ and the expression:

$$\mathbf{K}_t + \mathbf{K}_{\text{disp},t} = \phi \, (\rho c_p)_f \frac{\nu_{t_\phi}}{\sigma_t} \mathbf{I} \qquad \ldots(27)$$

Details of interfacial convective heat transfer coefficient are presented in the next section.

2.3. Interfacial Heat Transfer Coefficient

In Eqs. (12) and (13) the heat transferred between the two phases was modeled by means of a film coefficient h_i such that:

$$h_i a_i \left(\langle \overline{T}_s \rangle^i - \langle \overline{T}_f \rangle^i \right) = \frac{1}{\Delta V} \int_{A_i} \mathbf{n}_i \cdot k_f \nabla \overline{T}_f \, dA = \frac{1}{\Delta V} \int_{A_i} \mathbf{n}_i \cdot k_s \nabla \overline{T}_s \, dA$$

$$\ldots(28)$$

where a_i, as mentioned, is the interfacial area per unit volume. In foam-like or cellular media, the high values of a_i make them attractive for transferring thermal energy via conduction through the solid followed by convection to a fluid stream.

For obtaining macroscopic transport properties, highly permeable media can be modeled as an infinite array of rods, which, in turn, can be analogous to flow across a bundle of tubes. Accordingly, two tube arrangements are generally found in the literature, *i.e.*, the tube rows in a bundle are either inline, with rod centers forming a *square* or a rectangle, or else, they are staggered, where a *triangular* shape is obtained when connecting the tube centerlines. In this work, the two forms of arrays, namely **square** (*inline*) and **triangular** (*staggered*) layouts are used in order to model flow and heat transfer in highly porous media (see Fig. 9.1).

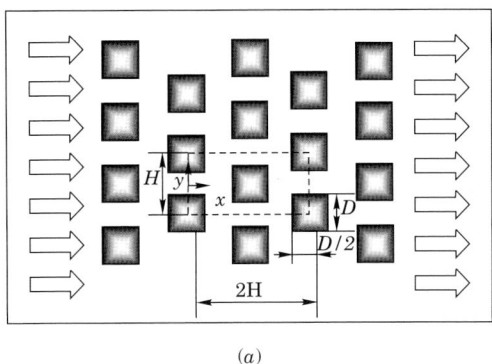

(a)

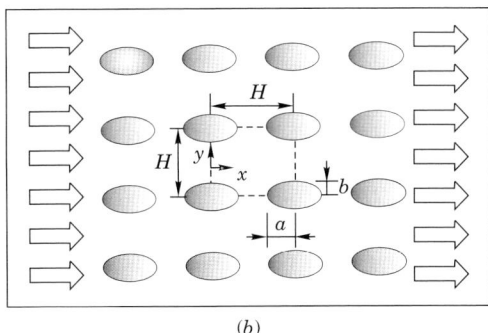

(b)

Fig. 9.1. Porous media models and coordinate systems: (a) triangular (*staggered*) array of square rods; (b) square (*inline*) array of elliptic rods

For the staggered configuration of tube banks, Zhukauskas (1972) has proposed a correlation of the form:

$$\frac{h_i D}{k_f} = 0.022 \, \text{Re}_D^{0.84} \, \text{Pr}^{0.36}, \quad \text{for } 2 \times 10^5 < \text{Re}_D < 2 \times 10^6 \quad \ldots(29)$$

where the values 0.022 and 0.84 are for tubes in cross flow.

Wakao *et al.* (1979) obtained a heuristic correlation for closely packed bed of particle diameter D and compared their results with experimental data. This correlation for the interfacial heat transfer coefficient is given by,

$$\frac{h_i D}{k_f} = 2 + 1.1 \, \text{Re}_D^{0.6} \, \text{Pr}^{1/3} \qquad ...(30)$$

For numerically determining h_i, Kuwahara et al. (2001) modeled a porous medium by considering it as an infinite number of solid square rods of size D, arranged in a regular triangular pattern (see Fig. 9.1). They numerically solved the governing equations in the void region, exploiting to advantage the fact that for an infinite and geometrically ordered medium a repetitive cell can be identified. Periodic boundary conditions were then applied for obtaining the temperature distribution under fully developed flow conditions. A numerical correlation for the interfacial convective heat transfer coefficient was proposed by Kuwahara et al. (2001) for laminar flow as:

$$\frac{h_i D}{k_f} = \left(1 + \frac{4(1-\phi)}{\phi}\right) + \frac{1}{2}(1-\phi)^{1/2} \, \text{Re}_D \, \text{Pr}^{1/3} \qquad ...(31)$$

Equation (31) is based on porosity dependency and is valid for packed beds of particle diameter, D. Saito and de Lemos (2005) obtained the interfacial heat transfer coefficient for laminar flows though an infinite square rod using the same methodology as Kuwahara et al. (2001).

3. DETERMINATION OF INTERFACIAL FILM COEFFICIENT h_i

3.1. Physical Model

Measuring flow and heat transfer characteristics within the void space in foam-like media is a challenging task. However, macroscopic behavior of permeable materials can be obtained by integrating distributed parameters calculated at pore scale. In order to follow such methodology, scientists and engineers have made use of physical models that consider a well-ordered porous medium, which is composed by regularly arranged obstacles instead of randomly distributed solid particles. Assuming further that such modeled medium is of a large size, a repetitive or unit cell can be identified, over which the balance equations are numerically solved.

Following this path, Kuwahara et al. (2001) and Nakayama et al. (2001) modeled a porous medium in terms of square obstacles displaced in a regular staggered pattern. They numerically solved the set of local governing equations in a unit or repetitive cell of that arrangement. By volume averaging the distributed parameters, they got useful information used in calculating the interfacial heat transfer coefficient h_i. Such coefficient, as seen above, is necessary to close the mathematical model when two energy equations, one for the solid and one for the saturating fluid, are solved.

Motivated by the foregoing, this work also applies the methodology of computing first distributed flow parameters in a repetitive cell followed by integration of local values. The periodic cell of volume ΔV used in this work is schematically shown in Fig. 9.1. It has dimensions $2H \times H$ for square rods [Fig. 9.1 (a)] and $H \times H$ for elliptic obstacles [Fig. 9.1 (b)]. Computations within those cells were carried out using a non-uniform grid as presented in Fig. 9.2. The Reynolds number $\text{Re}_D = \rho \bar{u}_D D/\mu$ was varied from 10^4 to 10^7.

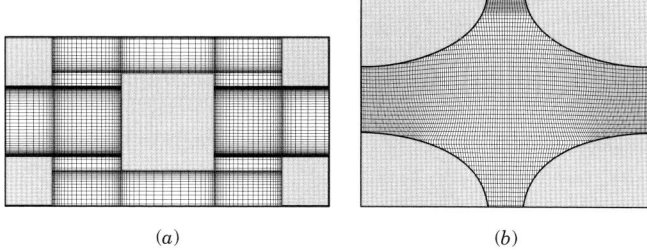

(a) (b)

Fig. 9.2. Periodic cells and computational grids: (a) Triangular (*staggered*) array of square rods, (b) Square (*aligned*) array of elliptic rods

The numerical method utilized to discretize the flow and energy equations in the unit cell was the Finite Control Volume approach. The SIMPLE method of Patankar (Patankar, 1980) was used for handling the pressure-velocity coupling. Convergence was monitored in terms of the normalized residue for all variables. The maximum residue allowed before convergence was 10^{-9}, being the variables normalized by appropriate reference values.

For fully developed flow in the cells of Fig. 9.1, the velocity at exit ($x/H = 2$ for square rods and $x/H = 1$ for elliptic rods) must be identical to that at the inlet ($x/H = 0$). Temperature profiles, however, are only identical at both cell exit and inlet if presented in terms of an appropriate non-dimensional variable. The situation is analogous to the case of forced convection in a channel with isothermal walls. Due to the periodicity of the flow, a single structural unit, as indicated in the Fig. 9.1, may be taken as a calculation domain. Turbulent flow is modeled by means of Eqs. (5), (6) and (7). Boundary conditions are given by the following:

(a) On the solid walls (laminar or low Re model):

$$\bar{u} = 0, \; k = 0, \; \varepsilon = \nu \frac{\partial^2 k}{\partial y^2}, \; \bar{T} = \bar{T}_w \qquad \ldots(32)$$

(b) On the solid walls (high Re model):

$$\frac{\bar{u}}{u_\tau} = \frac{1}{\kappa} \ln(y^+ E), \; k = k_w = \frac{u_\tau^2}{c_\mu^{1/2}}, \; \varepsilon = \frac{c_\mu^{3/4} k_w^{3/2}}{\kappa \, y_w},$$

$$q_w = \frac{(\rho c_p)_f \, c_\mu^{1/4} \, k_w^{1/2} \, (\bar{T} - T_w)}{\left(\dfrac{\sigma_t}{\kappa} \ln(y_w^+) + c_Q(\text{Pr})\right)} \qquad \ldots(33)$$

where, $u_\tau = \left(\dfrac{\tau_w}{\rho}\right)^{1/2}$, $y_w^+ = \dfrac{y_w u_\tau}{\nu}$, $c_Q = 12.5\,\text{Pr}^{2/3}$

$$+ 2.12 \ln(\text{Pr}) - 5.3 \text{ for Pr} > 0.5$$

In Eq. (33), Pr and σ_t are the Prandtl and turbulent Prandtl numbers, respectively, q_w is wall heat flux, u_τ is wall-friction velocity, y_w is the normal coordinate of the first grid point close to the wall and κ is the von Kármán's constant. Further, in Eq. (33) E is also a constant, which can accommodate different types of surface. For smooth surfaces, a standard numerical value taken for E is 9.0. The other boundary conditions are:

(c) On the symmetry planes:

$$\frac{\partial \bar{u}}{\partial y} = \frac{\partial k}{\partial y} = \frac{\partial \varepsilon}{\partial y} = 0 \qquad \ldots(34)$$

(d) On the periodic boundaries:

$$\bar{u}\big|_{\text{inlet}} = \bar{u}\big|_{\text{outlet}},\ \bar{v}\big|_{\text{inlet}} = \bar{v}\big|_{\text{outlet}},\ k\big|_{\text{outlet}},\ \varepsilon\big|_{\text{inlet}} = \varepsilon\big|_{\text{outlet}} \qquad \ldots(35)$$

$$\theta\big|_{\text{inlet}} = \theta\big|_{\text{outlet}} \Leftrightarrow \left.\frac{\bar{T} - \bar{T}_w}{\bar{T}_B(x) - \bar{T}_w}\right|_{\text{inlet}} = \left.\frac{\bar{T} - \bar{T}_w}{\bar{T}_B(x) - \bar{T}_w}\right|_{\text{outlet}} \qquad \ldots(36)$$

The bulk mean temperature of the fluid is given by:

$$\bar{T}_B(x) = \frac{\int \bar{u}\bar{T}\,dy}{\int \bar{u}\,dy} \qquad \ldots(37)$$

Computations are based on the Darcy velocity, the length of structural unit H and the temperature difference $(\bar{T}_B(x) - \bar{T}_w)$ as reference scales.

3.2. Periodic Flow

Results for velocity and temperature fields were obtained for different Reynolds numbers. In order to assure that the flow was hydrodynamically and thermally developed in the periodic cell of Fig. 9.1, the governing equations were solved repetitively in the cell, taking the outlet profiles for $\bar{u}$ and θ at the exit and plugging them back at the inlet. In the first run, uniform velocity and temperature profiles were set at the cell entrance for Pr = 1 giving $\theta = 1$ at $x/H = 0$. Then, after convergence of the flow and temperature fields, $\bar{u}$ and θ at $x/H = 2$ were used as inlet profiles for a second run, corresponding to solving again the flow for a similar cell beginning in $x/H = 2$. Similarly, a third run was carried out and again outlet results, this time corresponding to an axial position $x/H = 4$, were recorded. This procedure was repeated several times until $\bar{u}$ and θ did not differ substantially at both inlet and outlet positions, as obtained in Saito and de Lemos (2006).

For the Low Re model, the first node adjacent to the wall requires that the nondimensional wall distance be such that $vy^+ = u_\tau y \rho/\mu \leq 1$. To accomplish this requirement, the grid needs a great number of points close to the wall leading to computational meshes of large sizes.

Table 9.1. Interface (A_i) and flow (A_c) areas for distinct rod arrangements

Geometry	A_i	A_c
Square rods–triangular array	$8D$	$H - D$
Elliptic rods–square array	$2\pi\sqrt{0.5(a^2 + b^2)}$	$H - 2b$

3.3. Film Coefficient h_i

The determination of h_i is here obtained by calculating, for the unit cell of Fig. 9.1, an expression given as:

$$h_i = \frac{Q_{\text{total}}}{A_i \, \Delta T_{\text{ml}}} \qquad \ldots(38)$$

where the overall heat transferred in the cell, Q_{total}, is given by,

$$Q_{\text{total}} = A_c \, \rho \bar{u}_B \, c_p \left(\bar{T}_B \big|_{\text{outlet}} - \bar{T}_B \big|_{\text{inlet}} \right) \qquad \ldots(39)$$

and A_i, A_c are presented in Table 9.1. The bulk mean velocity of the fluid is given by,

$$\bar{u}_B(x) = \frac{\int \bar{u}\, dy}{\int dy} \qquad \ldots(40)$$

and the logarithm mean temperature difference, ΔT_{ml}, is defined as,

$$\Delta T_{\text{ml}} = \frac{\left(\bar{T}_w - \bar{T}_B \big|_{\text{outlet}} \right) - \left(\bar{T}_w - \bar{T}_B \big|_{\text{inlet}} \right)}{\ln \left[\left(\bar{T}_w - \bar{T}_B \big|_{\text{outlet}} \right) \left(\bar{T}_w - \bar{T}_B \big|_{\text{inlet}} \right) \right]} \qquad \ldots(41)$$

Equation (39) represents an overall heat balance on the entire cell and Eq. (38) associates the heat transferred to the fluid with a suitable temperature difference. Once fully developed flow and non-dimensional temperature fields were achieved, bulk temperatures were calculated according to Eq. (37), at both inlet and outlet positions. They were then used to calculate $h_i D/k_f$ using Eqs. (38)–(41).

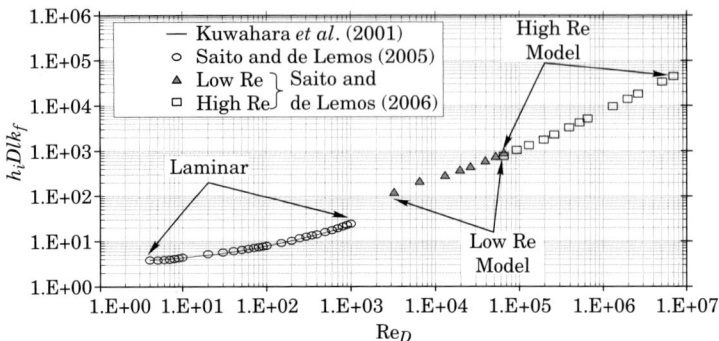

Fig. 9.3. Effect of on Re_D on h_i, $Pr = 1$ and $\phi = 0.65$ [Data: Kuwahara et al. (2001); Saito and de Lemos (2005); Saito and de Lemos (2006)]

4. RESULTS AND DISCUSSION

4.1. Correlations for Laminar and Turbulent Flows

Results for h_i covering both laminar (Saito and de Lemos (2005)) and turbulent flow regimes are plotted in Fig. 9.3 along with correlation (31) by Kuwahara et al. (2001). Numerical results by Saito and de Lemos (2006) using both Low and High Re models are also plotted showing a good overlap of values calculated with both models around $Re_D = 6 \times 10^4$.

Figure 9.4 shows numerical results for the interfacial convective heat transfer coefficient for various porosities and plotted for Re_D up to 10^7. In order to obtain a correlation for h_i in the turbulent regime, all curves were first collapsed after plotting them in terms of Re_D/ϕ, as showed in Fig. 9.5. Next, the least square technique was applied in order to determine the best correlation, which lead to a minimum overall error. Thus, the following expression was proposed in Saito and de Lemos (2006):

$$\frac{h_i D}{k_f} = 0.08 \left(\frac{Re_D}{\phi}\right)^{0.8} Pr^{1/3}; \text{ for } 1.0 \times 10^4 < \frac{Re_D}{\phi} < 2.0 \times 10^7,$$

$$\text{valid for } 0.2 < \phi < 0.9 \quad ...(42)$$

Equation (42), which gives the heat transfer coefficient for turbulent flow, is compared with numerical results obtained with Low and High Re models. Such a comparison is presented in Fig. 9.6, which also shows computations using correlations given by Zhukauskas (1972) and Wakao et al. (1979), Eq. (29) and Eq. (30), respectively. Table 9.2 summarizes correlations referred to in this chapter, through which the heat transfer coefficient h_i in porous media can be determined.

The agreement between the correlations in the literature and the numerical simulations here reported stimulates further investigation on this subject, contributing towards the building of a more general expression for the interfacial heat transfer coefficient for porous media.

Modelling and Simulation of Interfacial Heat Transfer 285

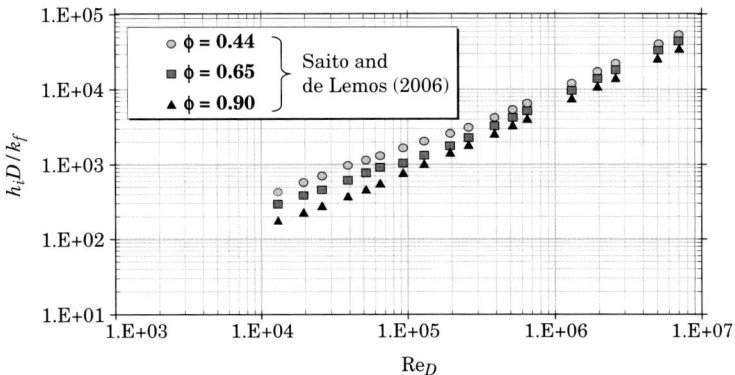

Fig. 9.4. Effect of porosity on for, square rods (Data: Saito and de Lemos, 2006)

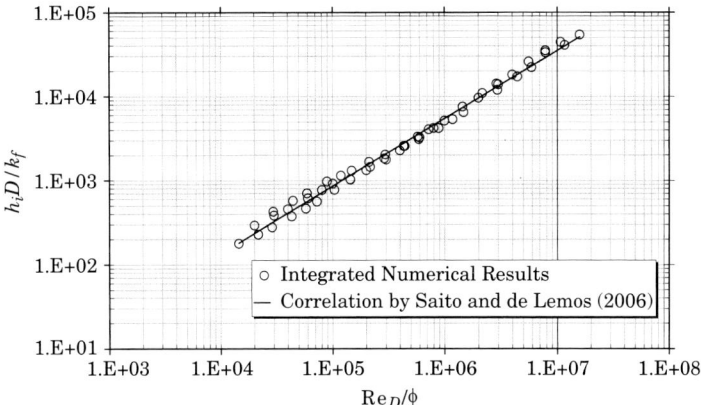

Fig. 9.5. Scaled numerical results and comparison with correlation (42) by Saito and de Lemos (2006)

Table 9.2. Correlations for interfacial heat transfer coefficient

Correlation	Validity	Equation	Reference
$\dfrac{h_i D}{k_f} = 0.022\,\mathrm{Re}_D^{0.84}\,\mathrm{Pr}^{0.36}$	$2 \times 10^5 < \mathrm{Re}_D < 2 \times 10^6$	(29)	Zhukauskas (1972)
$\dfrac{h_i D}{k_f} = 2 + 1.1\,\mathrm{Re}_D^{0.6}\,\mathrm{Pr}^{1/3}$	–	(30)	Wakao et al. (1979)
$\dfrac{h_i D}{k_f} = \left(1 + \dfrac{4(1-\phi)}{\phi}\right)$ $+ \dfrac{1}{2}(1-\phi)^{1/2}\,\mathrm{Re}_D\,\mathrm{Pr}^{1/3}$	Laminar flow, $0.2 < \phi < 0.9$	(31)	Kuwahara et al. (2001)
$\dfrac{h_i D}{k_f} = 0.08\left(\dfrac{\mathrm{Re}_D}{\phi}\right)^{0.8}\mathrm{Pr}^{1/3}$	$1.0 \times 10^4 < \dfrac{\mathrm{Re}_D}{\phi} < 2.0 \times 10^7$ $0.2 < \phi < 0.9$	(42)	Saito and de Lemos (2006)

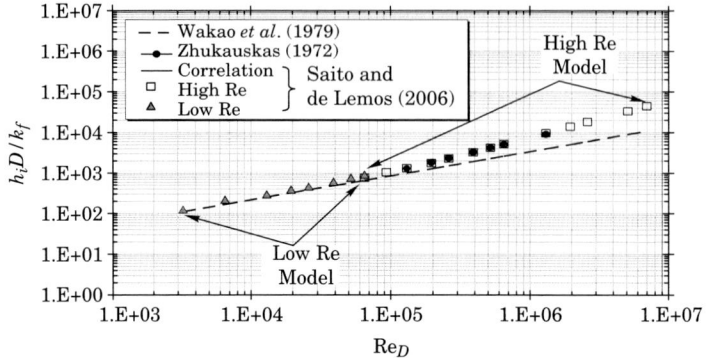

Fig. 9.6. Comparison of integrated numerical results for $\phi = 0.65$ and correlations by Saito and de Lemos (2006); Zhukauskas (1972); Wakao et al. (1979)

4.2. Macroscopic Results

A Porous Medium and a Fluid Layer

Consider a flow inside a two-dimensional channel with a height H and a length L. The x-axis is aligned along the walls of the channel while the y-axis is in the traverse direction, as shown in Fig. 9.7. A parallel plate channel partially filled with a porous and solid layer is considered, where the fluid phase is taken to be Newtonian with constant average properties. The no slip boundary condition is imposed at the walls and the conjugate heat transfer is considered with constant temperature at the boundaries.

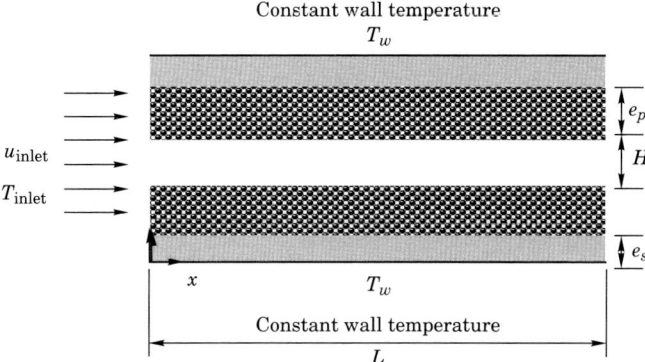

Fig. 9.7. Coordinate system of parallel plate channel partially filled with a porous layer

A comprehensive analysis of fluid flow and heat transfer for the interface region between a fluid layer and an adjacent porous layer can be found in literature. Therefore, the current study concentrates on analyzing the effects of a two-energy equation model on solid and fluid temperatures. The presentation of the results in the current study is given in terms of the velocity profile and in terms of temperature profile, as shown in Figs. 9.8 to 9.10.

Figure 9.8 shows the velocity profiles across the channel for laminar and turbulent flow. The figure clearly indicates that for laminar flows most of the fluid flows though the center of the channel, whereas for turbulent regime the higher pressure drop along the channel forces the fluid to flow thorough the porous material.

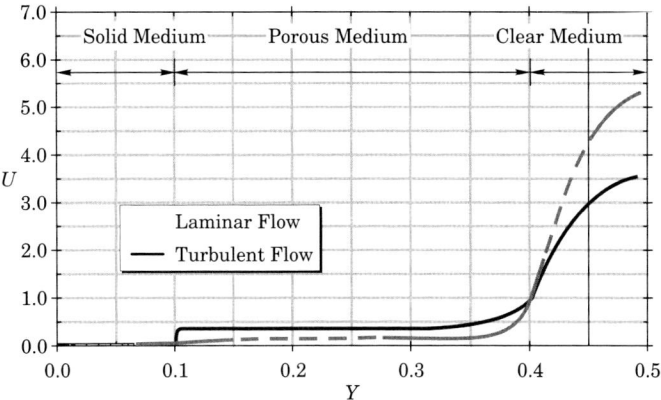

Fig. 9.8. Velocity profile for the interface between a porous medium and a fluid layer for laminar and turbulent flow

Figures 9.9 and 9.10 present the non-dimensional temperature profile across the channel, which contains a solid material, a porous medium and a fluid layer. Results are presented for laminar and turbulent flow, respectively. The conjugated problem is considered for treating the solid wall. The difference between the fluid and solid temperature profiles can be observed in Fig. 9.9. At the solid-porous interface, the fluid and solid temperatures are equal. As the porous-channel interface is approached, temperature differences increase, indicating the importance of using two-

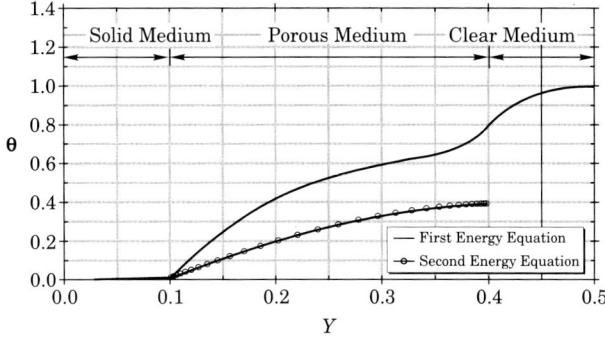

Fig. 9.9. Non-dimensional temperature profile between a porous medium, solid medium and a fluid layer for laminar flow

energy equation models, at least for the case shown in the figure. For turbulent flow (Fig. 9.10), enhancement of heat transfer between the liquid and the porous matrix decreases the temperature difference between phases.

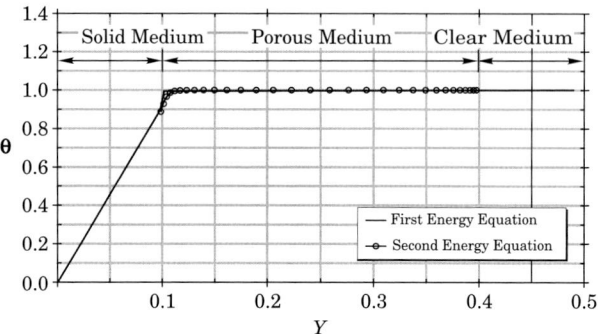

Fig. 9.10. Non-dimensional temperature profile between a porous medium, solid medium and a fluid layer for turbulent flow

5. CONCLUSIONS

A computational procedure for determining the convective coefficient of heat exchange between the porous substrate and the working fluid for a porous medium was detailed. As a result, macroscopically uniform laminar and turbulent flows through a periodic cell formed by square and elliptic rods were computed. Quantitative agreement was obtained when comparing laminar results herein with simulations by Kuwahara et al. (2001). For turbulent flows, Low and High Reynolds turbulence models were employed in order to obtain the interfacial heat transfer coefficient. Correlations for determining the heat transfer coefficient were compared. Macroscopic results for flow in a composite channel were presented.

ACKNOWLEDGEMENTS

The authors are thankful to CNPq and FAPESP, Brazil, for their invaluable financial support during the course of this and other research programs conducted at ITA.

REFERENCES

1. Bear, J. 1972. Dynamics of Fluids in Porous Media. American Elsevier Pub. Co., New York.
2. Braga, E.J. and de Lemos, M.J.S. 2006a. Turbulent heat transfer in an enclosure with a horizontal porous plate in the middle, *J. Heat Transf.* **128(11)**: 1122–9.
3. Braga, E.J. and de Lemos, M.J.S. 2006b. Simulation of turbulent natural convection in a porous cylindrical annulus using a macroscopic two-equation model, *Int. J. Heat Mass Transf.* **49(23–24)**: 4340–51.
4. Braga, E.J. and de Lemos, M.J.S. 2005a. Heat transfer in enclosures having a fixed amount of solid material simulated with heterogeneous and homogeneous models, *Int. J. Heat Mass Transf.* **48(23–24)**: 4748–65.

5. Braga, E.J. and de Lemos, M.J.S. 2005b. Laminar natural convection in cavities filled with circular and square rods, *Int. Comm. Heat Mass Transf.* **32(10)**: 1289–97.
6. Braga, E.J. and de Lemos, M.J.S. 2004. Turbulent natural convection in a porous square cavity computed with a macroscopic k-ε model, *Int. J. Heat Mass Transf.* **47(26)**: 5639–50.
7. de Lemos, M.J.S. 2006. Turbulence in Porous Media: Modeling and Applications. Elsevier, New York.
8. de Lemos, M.J.S. and Silva, R.A. 2006. Turbulent flow over a layer of a highly permeable medium simulated with a diffusion-jump model for the interface, *Int. J. Heat Mass Transf.* **49(3–4)**: 546–56.
9. de Lemos, M.J.S. 2005a. Fundamentals of the double–decomposition concept for turbulent transport in permeable media, *Materialwissenschaft und Werkstofftechnik* **36(10)**: 586–93.
10. de Lemos, M.J.S. 2005b. Turbulent kinetic energy distribution across the interface between a porous medium and a clear region, *Int. Comm. Heat Mass Transf.* **32(1–2)**: 107–15.
11. de Lemos, M.J.S. and Tofaneli, L.A. 2004. Modeling of double-diffusive turbulent natural convection in porous media, *Int. J. Heat Mass Transf.* **47(19–20)**: 4221–31.
12. de Lemos, M.J.S. and Braga, E.J. 2003. Modeling of turbulent natural convection in saturated rigid porous media, *Int. Comm. Heat Mass Transf.* **30(5)**: 615–24.
13. de Lemos, M.J.S. and Mesquita, M.S. 2003. Turbulent mass transport in saturated rigid porous media, *Int. Comm. Heat Mass Transf.* **30(1)**: 105–13.
14. de Lemos, M.J.S. and Rocamora, F.D. 2002. Turbulent transport modeling for heated flow in rigid porous media. *Proceedings of the 12th International Heat Transfer Conference*, Grenoble, France, August 18–23, pp. 791–5.
15. de Lemos, M.J.S. and Pedras, M.H.J. 2001. Recent mathematical models for turbulent flow for saturated rigid porous media, *J. Fluids Eng.* **123(4)**: 935–40.
16. Ergun, S. 1952. Fluid flow through packed columns, *Chem. Eng. Pro.* **48**: 89–94.
17. Forchheimer, P. 1901. Wasserbewegung durch Boden, *Z. Ver. Deutsch. Ing.* **45**: 1782–8.
18. Gray, W.G. and Lee, P.C.Y. 1977. On the theorems for local volume averaging of multiphase system, *Int. J. Multiphase Flow* **3**: 333–40.
19. Hsu, C.T. 1999. A closure model for transient heat conduction in porous media, *J. Heat Transf.* **121**: 733–9.
20. Hsu, C.T. and Cheng, P. 1990. Thermal dispersion in a porous medium, *Int. J. Heat Mass Transfer* **33**: 1587–97.
21. Ingham, D.B. and Pop, I. 1998. Transport Phenomena in Porous Media. Elsevier, Amsterdam, pp. 103–29.
22. Kaviany, M. 1995. Principles of Heat Transfer in Porous Media. 2nd edition, Springer, New York.
23. Kuwahara, F., Shirota, M. and Nakayama, A. 2001. A numerical study of interfacial convective heat transfer coefficient in two-energy equation model for convection in porous media, *Int. J. Heat Mass Transf.* **44**: 1153–9.
24. Kuwahara, F., Nakayama, A. and Koyama, H. 1996. A numerical study of thermal dispersion in porous media, *J. Heat Transf.* **118**: 756–61.

25. Nakayama, A., Kuwahara, F., Sugiyama, M. and Xu, G. 2001. A two-energy equation model for conduction and convection in porous media, *Int. J. Heat Mass Transf.* **44**: 4375–9.
26. Nakayama, A. and Kuwahara, F. 1999. A macroscopic turbulence model for flow in a porous medium, *J. Fluids Eng.* **121**: 427–33.
27. Nield, D.A. and Bejan, A. 1992. Convection in Porous Media. Springer, New York.
28. Ochoa-Tapia, J.A. and Whitaker, S. 1997. Heat transfer at the boundary between a porous medium and a homogeneous fluid, *Int. J. Heat Mass Transf.* **40**: 2691–2707.
29. Patankar, S.V. 1980. Numerical Heat Transfer and Fluid Flow. Hemisphere, Washington, DC.
30. Pedras, M.H.J. and de Lemos, M.J.S. 2003. Computation of turbulent flow in porous media using a Low Reynolds model and an infinite array of transversally-displaced elliptic rods, *Num. Heat Transf. Part A–Applications* **43(6)**: 585–602.
31. Pedras, M.H.J. and de Lemos, M.J.S. 2001a. Macroscopic turbulence modeling for incompressible flow through undeformable porous media, *Int. J. Heat Mass Transf.* **44(6)**: 1081–93.
32. Pedras, M.H.J. and de Lemos, M.J.S. 2001b. Simulation of turbulent flow in porous media using a spatially periodic array and a low-Re two-equation closure, *Num. Heat Transfer-Part A Applications* **39(1)**: 35–59.
33. Pedras, M.H.J. and de Lemos, M.J.S. 2001c. On the mathematical description and simulation of turbulent flow in a porous medium formed by an array of elliptic rods, *J. Fluids Eng.* **123(4)**: 941–7.
34. Pedras, M.H.J. and de Lemos, M.J.S. 2000. On the definition of turbulent kinetic energy for flow in porous media, *Int. Comm. Heat Mass Transf.* **27(2)**: 211–20.
35. Quintard, M. 1998. Modeling local non-equilibrium heat transfer in porous media. *Proceedings of the 11th Int. Heat Transfer Conf.*, Kyongyu, Korea, **1**: 279–85.
36. Quintard, M., Kaviany, M. and Whitaker, S. 1997. Two-medium treatment of heat transfer in porous media: numerical results for effective properties, *Adv. Water Res.* **20**: 77–94.
37. Rocamora, F.D. and de Lemos, M.J.S. 2000. Analysis of convective heat transfer of turbulent flow in saturated porous media, *Int. Comm. Heat Mass Transf.* **27(6)**: 825–34.
38. Saito, M.B. and de Lemos, M.J.S. 2006. A correlation for interfacial heat transfer coefficient for turbulent flow over an array of square rods, *J. Heat Transf.* **128**: 444–52.
39. Saito, M.B. and de Lemos, M.J.S. 2005. Interfacial heat transfer coefficient for non-equilibrium convective transport in porous media, *Int. Comm. Heat Mass Transf.* **32(5)**: 667–77.
40. Schumann, T.E.W. 1929. Heat transfer: liquid flowing through a porous prism, *J. Franklin. Inst.* **208**: 405–16.
41. Silva, R.A. and de Lemos, M.J.S. 2003a. Numerical analysis of the stress jump interface condition for laminar flow over a porous layer, *Num. Heat Transf. Part A-Applications*, **43(6)**: 603–17.
42. Silva, R.A. and de Lemos, M.J.S. 2003b. Turbulent flow in a channel occupied by a porous layer considering the stress jump at the interface, *Int. J. Heat Mass Transf.* **46(26)**: 5113–21.
43. Slattery, J.C. 1967. Flow of viscoelastic fluids through porous media, *AIChE J.* **13**: 1066–71.

44. Wakao, N., Kaguei, S. and Funazkri, T. 1979. Effect of fluid dispersion coefficients on particle-to-fluid heat transfer coefficients in packed bed, *Chem. Eng. Sci.* **34**: 325–36.
45. Whitaker, S. 1967. Diffusion and dispersion in porous media, *J. Amer. Inst. Chem. Eng.* **13(3)**: 420–7.
46. Whitaker, S. 1966. Equations of motion in porous media, *Chem. Eng. Sci.* **21**: 291–300.
47. Zhukauskas, A. 1972. Heat transfer from tubes in cross flow, *Advances in Heat Transfer.*, 8. Academic Press, New York.

10

Supercritical Fluids and its Applications

M. Vázquez da Silva[*]

ABSTRACT

A supercritical fluid (SCF) is any substance at a temperature and pressure above its thermodynamic critical point. It can diffuse through solids like a gas, and dissolve materials like a liquid. Additionally, close to the critical point, small changes in pressure or temperature result in great changes in density, allowing many properties to be manipulated. Supercritical fluids are suitable as a substitute for organic solvents in a variety of industrial and laboratory processes, like supercritical fluid extraction, dry cleaning, supercritical fluid chromatography, chemical reactions, nano and micro particle formation, and many others. Carbon dioxide and water are the most commonly used supercritical fluids.

NOTATION

GAS	gas anti-solvent
GC	gas chromatography
HPLC	high performance liquid chromatography
P_c	critical pressure
P_r	reduced pressure
PGSS	particles from gas-saturated solutions
RESS	rapid expansion of supercritical solutions
SC	supercritical
SCE	supercritical extraction
SCF	supercritical fluid
SCFC	supercritical fluid chromatography
T_c	critical temperature
T_r	reduced temperature

---

[*] Departamento de Ciências, Instituto Superior de Ciências da Saúde–Norte, CESPU. Rua Central de Gandra, nº 1317, 4585-116 Gandra PRD, PORTUGAL.
Centro de Estudos de Fenómenos de Transporte, Faculdade de Engenharia da Universidade do Porto. Rua Dr. Roberto Frias, s/n, 4200-465 Porto, PORTUGAL.
[*] *Corresponding author* : E-mail : mvazquez@fe.up.pt

GREEK LETTERS

ρ_c critical density
ρ_r reduced density

1. INTRODUCTION

A supercritical fluid is any substance whose temperature and pressure are greater than its critical values. The physic-chemical properties of these fluids assume intermediate values to the ones of liquids and gases. As a beginning, it is very important to understand what the supercritical state is and the phenomena associated to that state. Let us to analyze the behaviour of a pure substance for different conditions of pressure and temperature. The physical state of a pure substance depends on the conditions that it is subjected, being possible to make the substance to change from a physical state to another one through the manipulation of the temperature and pressure conditions, like it is shown in Fig. 10.1 (a).

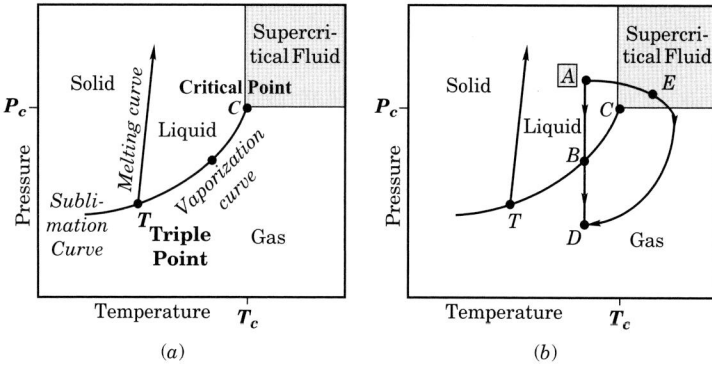

Fig. 10.1. Phase diagram for a pure substance

Considered, for example, the liquid represented by point A of Fig. 10.1 (b). Lowering the pressure isothermally, the liquid will reach point B of the vaporization curve where a liquid phase and a gaseous phase coexist in equilibrium. When crossing this curve, the properties of the substance suffer an abrupt modification, varying from the typical values of a liquid to characteristic values of a gas. Continuing to lower the pressure, the total liquid will be pulverized reaching point D, in the gaseous region.

Alternatively, the liquid can be evaporate by following the path A-E-D, where are considered the same initial and final points, but without crossing the vaporization curve. This route allows, by simultaneously modifying the pressure and temperature, to pulverize a liquid with a progressively modification of its properties, avoiding the discontinuity verified when the vaporization curve is crossed. This situation is only possible because the vaporization curve does not continue indefinitely, but finishes in point C, from which no more liquid—vapour equilibrium exists, starting to exist only one fluid phase. This point is nominated by critical point of the substance, and corresponds to the higher values of pressure and temperature where liquid-vapour equilibrium can coexist. The region above of the critical point is called supercritical or fluid region, being shaded in Fig. 10.1.

A supercritical fluid is, therefore, any substance whose temperature and pressure are above their critical values. The physical and chemical properties of these fluids assume intermediate values of those of liquids and gases.

The properties of the SCF can also be understood by examining Fig. 10.2 (De La Ossa and Serrano, 1990), which relates the reduced pressure (P_r) of a pure substance with the reduced density (ρ_r) and the reduced temperature (T_r). A reduced property is defined as the ratio between the value of the property and its value in the critical point. According to the figure, over an isotherm the density of a pure substance can vary with pressure in two different ways. Begin from a temperature lower than the critical temperature, for example $T_r = 0.90$, and a low value of pressure (point A, for example), it appears that the density of the substance increases significantly with pressure, following a typical variation of a gas reaching to point B. At this point, there is a phase transition, at pressure and

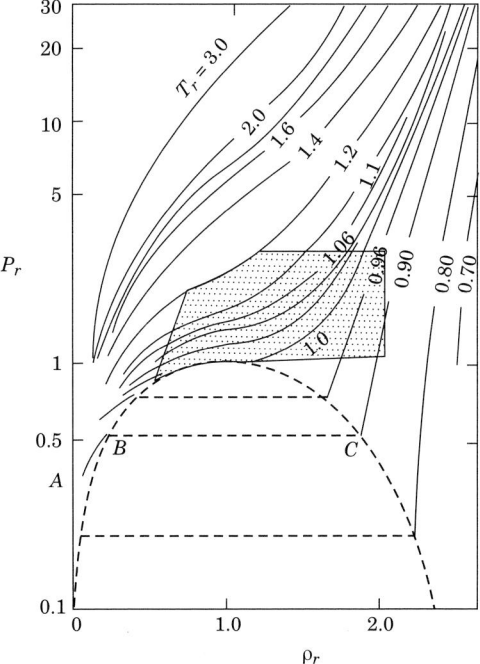

Fig. 10.2. Reduced pressure (P_r) of a pure substance as a function of the reduced density (ρ_r) and the reduced temperature (T_r)

temperature constants, in which the density of the substance changes from a low value, characteristic of the gases, to higher values, typical of the liquids (point C). From this point forward it is necessary larger increments of pressure to achieve a significant variation in the density of the substance, which is the typical behavior of little compressible fluids, like the liquids.

If one made a similar analysis for a temperature above the critical temperature, $T_r = 1.06$, for example, it is verified that an increase of pressure produces an increase in density from low values to higher values, but in this situation there is no phase transition. The ability to change the dissolution power of the SCF in a continuous way, without any transitional phase, just through the adjustment of the pressure and temperature values make these kind of fluids very attractive not only for the research community but also from the industrial point of view. It is particularly important to verify that in the vicinity of the critical point, namely for reduced temperatures in the range of 0.90 to 1.2 and reduced pressures between 1 and 4, small changes in these parameters causes enormous changes in the reduced density and, for consequence, in all the substance properties and behavior.

2. ADVANTAGES OF SUPERCRITICAL FLUIDS COMPARATIVELY TO THE TRADITIONAL (ORGANIC) SOLVENTS

The supercritical fluids have unique physical and chemical properties that make them special solvents. These properties, that affect directly the mass transfer, are the viscosity and the diffusion coefficient. In Table 10.1 are presented the typical values for the range of variation of density, viscosity and diffusion coefficient of liquids, gases and SCF.

Table 10.1. Comparison of the SCF properties with those of liquids and gases

	Liquid	SCF	Gas
Density (kg/m^3)	600–1600	200–900	0.6–2
Viscosity (Pa.s) × 10^5	20–300	1–9	1–3
Diffusion coefficient (m^2/s) × 10^9	0.2–2	20–70	10000–40000

The SCF have a density comparable to those of liquids, equivalent values of viscosity relatively to the gases, and values of diffusion coefficients intermediate between those of liquids and gases. Thus, the SCF have a dissolution power similar to those of liquids but with properties of mass transfer much more favourable, and this is one of capital advantages of the use of these fluids as solvents. In particular, the SCF are recommended for the extraction of solutes from solid matrixes because, in addition to the properties already listed, the SCF have a zero surface tension.

Some of the substances that have been frequently used as SCF are listed in Table 10.2, together with their critical properties. As shown, these solvents include a wide range of temperatures and pressures of operation.

Table 10.2. Critical properties of the most commonly used SCF (NIST, 2009)

Supercritical solvent	T_c (K)	P_c (MPa)	ρ_c (mol/L)
Inorganic			
Carbon dioxide (CO_2)	304.18	7.380	10.6
Nitrous oxide (N_2O)	309.56	7.238	10.3
Ammonia (NH_3)	405.4	11.300	not available
Water (H_2O)	647 ± 2	22.064	17.9
Hydrocarbon			
Methane (CH_4)	190.6 ± 0.3	4.61 ± 0.03	10.1 ± 0.2
Ethylene (C_2H_4)	282.5 ± 0.5	5.06 ± 0.05	7.63 ± 0.004
Ethane (C_2H_6)	305.3 ± 0.3	4.9 ± 0.1	6.9 ± 0.4
Propene (C_3H_6)	365.2 ± 0.8	4.60 ± 0.03	5.42 ± 0.03
Propane (C_3H_8)	369.9 ± 0.2	4.25 ± 0.01	5.1 ± 0.4
n-Pentane (C_5H_{12})	469.8 ± 0.5	3.36 ± 0.06	3.22 ± 0.07
n-Hexane (C_6H_{14})	507.6 ± 0.5	3.02 ± 0.04	2.71 ± 0.02
Benzene (C_6H_6)	562.0 ± 0.8	4.89 ± 0.04	3.9 ± 0.2
Toluene (C_7H_8)	593 ± 2	4.1 ± 0.1	3.17 ± 0.010
Oxygenates			
Ethoxy ethane ($C_4H_{10}O$)	467 ± 2	3.6 ± 0.1	3.5 ± 0.4
Acetone (CH_3COCH_3)	508 ± 2	4.8 ± 0.4	4.63
Methanol (CH_4O)	513 ± 1	8.1 ± 0.1	8.51 ± 0.07
Ethanol (C_2H_6O)	514 ± 7	6.3 ± 0.4	6.0 ± 0.2
Other compounds			
Chlorotrifluoromethane ($CClF_3$)	301.8 ± 0.3	3.885	5.733
Trichlorofluoromethane (CCl_3F)	471.1	4.466	4.151
Pyridine (C_5H_5N)	619 ± 2	5.660	not available

From all these solvents, carbon dioxide is the most popular, due not only to a very particular set of characteristics of physical and chemical nature, but also economic, namely: CO_2 solubilises the low molecular weight hydrocarbons and oxygenated compounds, however, it mutual solubility in water is small, so it can be used to extract organic products from aqueous solutions; the transport properties of CO_2 are interesting, since it has low viscosity and high diffusion coefficients; the critical temperature and pressure of CO_2 are moderate (304.18 K, 7.380 MPa) which are easily reached from the technologic point of view; the enthalpy of vaporization of the CO_2 is low, especially near the critical point, which means relatively low energy requirements; CO_2 is not toxic or flammable or corrosive, and is readily available on the market at low prices; finally, the use of CO_2 does not present problems of environmental contamination and can be considered to be an environmentally friendly solvent.

Not so popular, but also with many applications in the supercritical technology is the water. The critical pressure and critical temperature are of 22 MPa and 647 K, respectively, as a consequence of the high polarity of this solvent. The behaviour of water at supercritical conditions changes from one that supports only ionic species (at ambient conditions) to one that dissolves paraffin's, aromatics, gases and salts. This facility makes supercritical water very attractive to use in reaction and separation processes, or to treat toxic wastewater. Besides this, the dielectric constant of water changes from 78, at room conditions, to approximately 6 at critical conditions, which enables the control of reactions that depends on the dielectric constant of the medium.

3. APPLICATIONS OF THE SUPERCRITICAL FLUIDS

3.1. Historical Perspective

The supercritical extraction (SCE) was for many years the main area of application of the SCF, and can be defined as a unit operation that uses as separation agent a substance under conditions of temperature and pressure above its critical point.

The interest in supercritical extraction was born in the late nineteenth century, when Hannay and Hogart held a series of experiments on the solubility of some salts in SCF (Hannay and Hogart, 1879, 1880a, 1880b). These researchers measured the solubility of potassium iodide (KI), potassium bromide (KBr), calcium chloride ($CaCl_2$) and cobalt chloride ($CoCl_3$) in supercritical ethanol, and obtained values significantly higher than the expected ones, taking into account the respective vapour pressures. This fact suggests that, under supercritical conditions, ethanol significantly increased its capacity for solubilisation of these salts. Similar behaviour was found later with other supercritical fluids—methane, ethylene and carbon dioxide-for the solubilisation of different hydrocarbons, both solid and liquid (Andrews, 1887; Villard, 1896).

At the beginning of the twentieth century, namely in the 30's, a great effort was made to determine the liquid–vapour phase diagrams at high pressures for different hydrocarbons and their mixtures, which would enable the development the first industrial application of SCE: the removal of asphalt from crude oil, presented by Messmore in 1943 (Messmore, 1943).

In the 50's, it was developed, in the Kerr-McGee Refining Corporation, the ROSE process (Residuum Oil Supercritical Extraction) which is based on the use of a light, readily-available hydrocarbon solvent to extract deasphalted oil from a feedstock rich in asphaltenes. The solvent is separated from the deasphalted oil in the downstream deasphalted oil separator, then recovered and recycled; the solvent selection is based on the desired deasphalted oil purity and yield for a given feedstock. At the same time, Russian scientists (Zhuze, 1959; Zhuze and Sufrovna, 1958; Zhuze and Yushkevich, 1957a, 1957b) described similar cases to that presented by Messmore, using light hydrocarbons in the supercritical state to remove asphalt from crude oil, to the fractionation of crude oil, to the extraction of mineral waxes, and to the extraction of lanolin (Zhuze et al., 1958, 1959). In 1954, the phase behaviour of ternary systems and the solubilities of over 260 compounds in liquid carbon dioxide were studied by Francis and published in a monumental work (Francis, 1954) being proved that is possible to use liquid CO_2 near to the critical point, as a solvent for organic compounds.

In the 60's and 70's research on the SCE processes reached its peak and has been patented numerous applications of SCE for the processing of materials such as hops, coffee, tea, tobacco and spices, among others (Roselius et al., 1972a, 1972b, 1973a, 1973b; Vitzthum and Hubert, 1972, 1973, 1976; Vitzthum et al., 1975, 1976; Zosel, 1971, 1972, 1974, 1975, 1976). In the Max Plank Institute in Germany, Zosel described more than 80 different types of supercritical separations, using a wide range of solvents (Zosel, 1978).

In the sequence of all these works, several other industrial processes have been implemented, mainly in the food, pharmaceutical and cosmetic areas, using the SCE to perform the desired separation. In fact, at that time, the supercritical fluids were used almost exclusively as solvents in extraction processes. Thus, it may be interesting to look to Fig. 10.3 where the different steps that comprise a typical SCE with carbon dioxide are schematically represented.

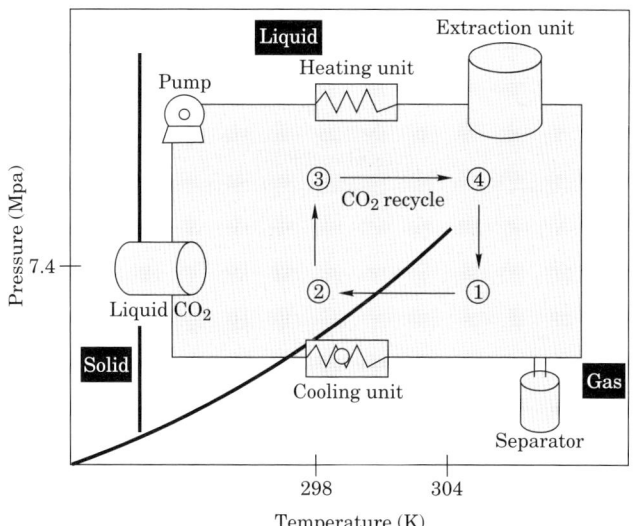

Fig. 10.3. Scheme of SCE with CO_2

Firstly, the gaseous CO_2 (point 1) is cooled and liquefied (point 2) and then compressed to a pressure above its critical value, i.e., above 7.4 MPa (point 3). Subsequently, it is heated to the operation temperature (point 4), which must be higher than the critical temperature of CO_2 (i.e.,

304.2 K). At this point, the carbon dioxide is in the supercritical state and can be fed to an extraction unit where it contacts with the material that will be processed, extracting the desired components (or solutes). The CO_2 stream that leaves the extractor is expanded to a pressure below its critical pressure, precipitating in the separation unit the dissolved components. From this unit also comes out a gaseous stream of pure CO_2 that is liquefied and recycled to the extraction unit. In Table 10.3 it can be found a lot of examples where supercritical carbon dioxide is used in extraction and fractionation processes.

Table 10.3. Examples of application of supercritical CO_2 in extraction and fractionation processes

Supercritical extraction and fractionation	
Decaffeination of coffee and tea	De Azevedo et al., 2008a, 2008b; Kim et al., 2007, 2008; Peker et al., 1992; Rajaei et al., 2005
Fractionation of crude oil	Chaffin et al.,1995; Yang and Wang, 1999; Zhao et al., 2004, 2005
Extraction of essential oils from plants, extraction of flavors, aromas and fragrances from natural products	Caredda et al., 2002; Ibañez et al., 1999; Illeís et al., 2002; Kaufmann and Christen., 2002; King, 1990; Martínez et al., 2008; Marzouki et al., 2008; Pourmortazavi et al., 2003; Ramírez et al., 2006; Reverchon et al., 1994; Scalia et al., 1999
Elimination of residual solvents from wastes and decontamination of soils	Akgerman, 1993; Anitescu and Tavlarides, 2006; Castelo-Grande et al., 2005; Sunarso and Ismadji, 2009
Nutraceuticals extraction	Kaiser et al., 2001; Perrut, 2003; Yang et al., 2002
Purification of drugs	Macnaughton et al., 1996; Maftouh et al., 2005; Petersson et al., 1994; White and Burnett, 2005; White, 2005
Fractionation of polymers	Clifford et al., 1997; Kumar et al., 1986; Yilgor and McGrath, 1984
Production of denicotined tobacco	Brunnemann et al., 1996
Cleaning of semi-conductors and ceramics	Bordet et al., 2002
Dry cleaning of fabrics	Oakes et al., 2001; Van Roosmalen et al., 2003
Extraction of natural products for cosmetics	Garrigoís et al., 1998; Lee et al., 2008; Mendes et al., 2002; Salvador et al., 2001; Scalia et al., 1995; Scalia, 2000

3.2. Recent Applications of the Supercritical Fluids

The energetic crisis of the 70s obeyed to the development of alternative technologies less intensive from the energetic point of view. This situation further increased the interest in the SCF, including the extraction of new products and the spread of their application to new areas of processing, such as nano and micro particles formation, SC chromatography, chemical reaction, impregnation, and others (Brunner, 2004; Caudell, 1999; Kiran et al., 2008; Martinez, 2007; McHugh and Krukonis, 1994; Mukhopadhyay,

2000; Sun, 2007). The discussion of these techniques will be done briefly from this point forward.

3.2.1. *Nano and micro particles formation*

The formation of small particles of a substance, in the range of micro and nano meters and with a controlled distribution of sizes is particularly important for the pharmaceutical industry (Cohen and Bernstein, 1996; Sun, 2007). According to several studies, the use of nano-particles in the production of drugs increases the efficiency and specificity of the drug therapy, can reduce the dosage, reduces the effects of toxicity and allows the use of new forms of administration of the drugs. The traditional process-milling-produces a wide range of particles sizes and may also cause the degradation of the substance, namely from the thermal point of view.

The use of SCF has developed several methods of producing particles of controlled size and morphology. The most studied is the rapid expansion of supercritical solutions (RESS), where a product is dissolved in a SCF, the solution is rapidly expanded through a sprayer with the deposition of micro-particles of the product (Peiriço *et al.*, 1999, 2000). An alternative to this process is the co-precipitation of an active ingredient with a polymer to produce particles coated by the polymer. When it dissolves slowly in water, the particles obtained can be administered to the blood stream to achieve a gradual release of drug. In Fig. 10.4 it can be seen an example of a material before and after the treatment of a SCF, being obvious the homogenous particle size distribution that can be obtained with the use of SCF.

Fig. 10.4. Photograph of a material before and after the treatment with SCF (BPD, 2002)

Another process is the crystallization of particles with a gas anti-solvent (GAS) (Krukonis *et al.*, 1994, 1995). This alternative is used to produce particles of small size when they are not soluble in the supercritical fluid, usually CO_2. For example, enzymes can be dissolved in ethanol through which passes a stream of pressurized CO_2 at controlled temperature. The bulk becomes a supercritical fluid in which the enzyme will precipitate as micro-particles, and then used in the formulation of drugs.

A third procedure called PGSS (Particles from Gas-Saturated Solutions), is the treatment of solutions, suspensions or melted materials in a similar way. A gas under pressure passes through the solution, suspension or molten material and dissolves them. The material is rapidly expanded through a sprayer, giving rise to small particles or droplets.

The dry particles (or in aqueous solution) obtained by these processes can be used in several areas, for example, for medical applications, for military purposes or for the production of paints (Kimura *et al.*, 2003; Knez and Weidner, 2003; Reverchon *et al.*, 2000; Sheu *et al.*, 2008; Sun *et al.*, 2009; Tsutsumi *et al.*, 2003; Wu *et al.*, 2006, 2007).

3.2.2. *Supercritical fluid chromatography*

Supercritical fluid chromatography (SCFC) was firstly presented by Klesper *et al.* (1962) when they described the separation of thermo-labile porphyrin derivatives using supercritical chlorofluoromethanes at pressures up to 140 bar and temperatures from 150 to 170°C.

SCFC can be used on an analytical scale, where it combines many of the advantages of HPLC and GC. SCFC separations can be done faster than HPLC separations because the diffusion of solutes in supercritical fluids is about ten times greater than that in liquids. This results in a decrease in resistance to mass transfer in the column and allows faster and higher resolution separations. SCFC can be very selective, since it is very sensitive to changes in pressure and temperature. SCFC can also be use as a purification technique since the collected product is free from solvent, as carbon dioxide vaporizes when the sample is expanded. Besides this, SCFC is an environmentally conscious technology, because it uses most of the times CO_2 as SCF with the inherent beneficial features already listed above, instead of large volumes organic chemicals such as dichloromethane or other chlorinated solvents. The main disadvantage of carbon dioxide is its inability to elute very polar or ionic compounds, which can be overcome by adding a small portion of a second fluid called a co-solvent. When compared with GC, capillary SFC can provide higher resolution chromatography at much lower temperatures, which can be vital for the analysis of thermo labile compounds.

From the operational point of view the SCFC can be described as an adaptation of either HPLC or GC, being the main different the fact that the mobile phase is not a liquid or a gas but a SCF. In SCFC the mobile phase is initially pumped as a liquid and is brought into the supercritical region by heating it above its critical temperature before reaching the analytical column. It passes through an injection valve where the sample is introduced and contacts with the supercritical stream and then into the analytical column. It is maintained supercritical as it passes through the column and into the detector by a pressure restrictor placed either after the detector or at the end of the column. The restrictor is a vital component as it keeps the mobile phase supercritical throughout the separation (Karey O'Leary, 2009). Many of the components used in SCFC are common to either HPLC or GC, namely, pumps, columns and detectors. Schematically a SCFC apparatus can be seen as it is represented in Fig. 10.5. In the literature it can be found thousands of works where the SCFC is used in many different areas like pharmaceutical, cosmetic, food, analytical chemistry, etc.

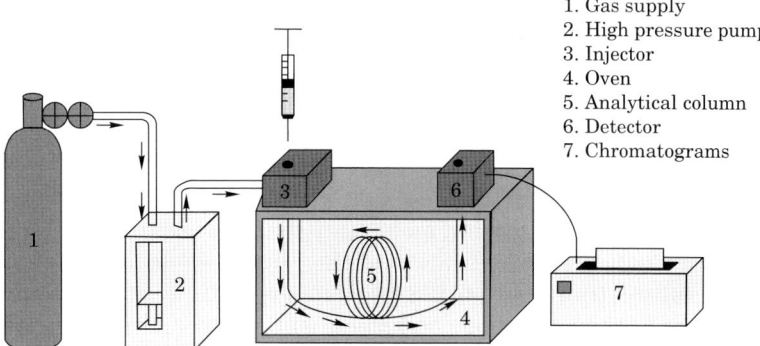

Fig. 10.5. SCFC schematic representation (adapted from Karey O'Leary, 2009)

3.2.3. *Supercritical chemical reaction*

The performance of chemical reactions in supercritical medium (Eldik and Hubbart, 1996; Eldik and Klärner, 2002; Sun, 2007) may be advantageous for several reasons: it becomes possible to shift the balance of the reaction towards the formation of the reaction products through its dissolution in another phase; when the reactions are diffusionally limited, the high diffusion coefficients characteristic of supercritical media makes the reactions to have higher velocities; it is possible to control the reaction evolution through the manipulation of the operational temperature and pressure; the use of near-critical or supercritical water (SCW) instead of organic solvents offers environmental advantages and may lead to pollution prevention.

The main motivation to carry out reactions in supercritical conditions is the fact that it is possible to control the phases balance through the manipulation of pressure and temperature, thus one can consider the presence of reagents and products in a single phase (homogenization) or in two stages (separation) as it is more convenient. The fact that a reaction occurs in a single phase (the supercritical phase) instead of a heterogeneous media, can be reflected in a more efficient and faster reaction. In other cases, the reaction control is made in order to isolate the products of the reaction, for example, in the catalytic oxidation of an hydrocarbon in presence of supercritical carbon dioxide, the operation conditions are chosen in order to promote the precipitation of the most polar products. This procedure has two main purposes, firstly it can move the chemical equilibrium toward the products formation, and secondly, it promotes the product isolation.

Nowadays, water near or above its critical point has increasing attention as a medium for organic chemistry (Savage, 1999). Near its critical point the water properties are very different from those at room conditions. The dielectric constant is much lower, and the number and persistence of hydrogen bonds are both diminished. As a result, water becomes very similar to organic solvents, with a very low polarity. The oxidation with supercritical water is a process that becomes interesting not only for the scientific community but also from the industrial point of view, being used in processes such as the elimination of organic residues, since non-polar organic compounds are soluble in supercritical water and oxygen, being rapidly and efficiently oxidized to CO_2 and H_2O (Freeman and Harris, 1995).

The use of SCF as a reaction media can also prevent from the catalysts poisoning through the solubilisation of the contaminants. Besides this, it is also an interesting alternative to the enzymatic reactions, not only because of the higher diffusion coefficients but also because of the dilation of the pores of the particles containing the molecules of enzyme, increasing also the internal diffusion. These are only some of the examples where the uses of supercritical conditions are beneficial to chemical reaction, some other examples can be easily found in the literature, for example in Savage (1999) and Wai et al. (1998).

3.2.4. *Impregnation*

The impregnation process is basically the reverse of an extraction, where a substance is dissolved in the supercritical fluid, passes through a solid matrix and is deposited or adsorbed on this matrix. This technique can be very useful in many different situations such as the impregnation of preservatives in wood (Hassan et al., 2001; Schneider et al., 2006), the impregnation with drugs for the preparation of controlled release formulations (Duarte et al., 2008; Gong et al., 2006; Loípez-Periago et al., 2009), flavors and preservatives impregnation in food, the impregnation of materials for book reinforcement polymer modification (Muth et al., 2000), etc.

Besides all the mentioned examples, the dyeing of textiles, namely, polyester fibers using non-ionic dispersed dyes is, probably, the best known example of impregnation with SCF (Clifford and Bartle, 1996; Fernandez Cid et al., 2004; Montero et al., 2000; Saus et al., 1993; Sicardi et al., 2000). The advantages of this process are not only economic but also environmental: there are no production of liquid effluents, so it is not necessary to make their treatment; the impregnation process occurs in supercritical media, so it is not necessary to use dispersants to solubilise the dyes in water; the degree of solubilisation of the dyes is easily manipulated through the control of the operating conditions, namely the pressure, so it is easy the modify the intensity and color of the dyeing; the process is faster in supercritical media because the diffusivity in those conditions is higher and the polymer fiber absorbs CO_2 which is a valuable contribution for the diffusion of the dye in the fiber, reinforced by the fact that the viscosity is lower. Initially, the research work was interested in the study of synthetic fibers, but the success of this technique expands the research to other kind of textiles, like cotton.

3.2.5. *Dry cleaning*

Dry cleaning is any cleaning process for clothing and textiles using an organic solvent rather than water. The solvent is generally known as dry cleaning fluid. The dry cleaning fluid most commonly used is tetrachloroethylene, more generally known as perchloroethylene or "perc." This solvent, however, produces toxic waste and is classified as a probable human carcinogen.

The use of supercritical carbon dioxide instead of the traditional solvents has numerous advantages: it is easily to control the solubility by changing the temperature and/or pressure; the SC CO_2 has a significant cleaning effect, but neither destroys fabrics nor dissolves dyes in fabrics; the low viscosity of SC CO_2 and its high diffusivity gives it the ability to clean in small cavities, the solvent is nonpolar so it is expected to dissolve grease and, finally, after the depressurization to atmospheric conditions there is no solvent residues in the textile.

Dry cleaning of clothes is, with no doubt, the most well known process of cleaning with this solvent, however, carbon dioxide cleaning is also used for food drying (Brown *et al.*, 2008), for fire- and water-damage restoration due to its effectiveness in removing toxic residues, soot, and associated odors of fire.

4. WHY IS NOT THE SUPERCRITICAL TECHNOLOGY GENERALIZED TO THE INDUSTRIAL PROCESSES?

As can be seen throughout the text presented, the advantages of using supercritical fluids are undeniable, both from the technologic point of view, but also from the environmental and efficiency point of view. In general, these processes occur at relatively low temperatures, especially if the SCF used is carbon dioxide (T_c = 304.18 K), being particularly interesting for the treatment of products that are sensitive to high values of temperature. Besides this, it is avoided the use of traditional organic solvents and the consequent environmental problems. Thus, one would expect that the use of SCF would be extended to thousands of industrial processes. This, however, is not a reality because this type of technology requires large capital investments, since it operates with facilities subject to very high values of pressure and, in some cases, temperature. These requirements make, in most of the cases, the current technology not to be appropriate, and it is necessary to built new facilities in order to support the specifications required for the operative conditions that will be placed. This initial capital investment has probably been the biggest obstacle to wider use of supercritical fluids in the industry.

However, the situation tends to be changed, mainly, because the global concerns about ambient problems is raising and the use of a 'green technology', such is the case of supercritical technology, has good hypotheses of progression and generalized application to almost all the industrial areas.

5. CONCLUSIONS

The supercritical fluids are a new class of solvents with unique chemical and physical properties, namely those related to solubilisation and transport. The use of supercritical fluids as an alternative to traditional solvents has increased in the recent years, which is a reflection of a greater knowledge and scientific research on the subject. An example of this situation are the large number of meetings and conferences that occurs annually strictly dedicated to the SCF problematic, such as European Meeting on Supercritical Fluids, International Symposium on Supercritical Fluids, International Conference Supercritical Fluid Chromatography (SFC), Italian Conference on Supercritical Fluids and their Applications, Meeting on Supercritical Fluids, Conference on Supercritical Fluids and Their Applications, Brazilian Meeting on Supercritical Fluids, Ibero-American Meeting on Supercritical Fluids and many others.

Nowadays, there are already many industries that make use of the available information and apply supercritical fluids to its processes. SCF can be used in areas as diverse as the formation of nanoparticles, extraction, chromatography, chemical reaction, impregnation, and so on. As it was stated above the effort made by the research community and the environmental concerns are two main points that can be decisive to impulse the creation of new unit plants working under supercritical conditions.

REFERENCES

1. Akgerman, A. 1993. Supercritical fluid extraction of contaminants from environmental matrices, *Waste Manage.* **13(5–7)**: 403–15.
2. Andrews, T. 1887. On the properties of matter in the gaseous and liquid state under various conditions of temperature and pressure, *Philos. T. R. Soc. A* **178**: 45.
3. Anitescu, G. and Tavlarides, L.L. 2006. Supercritical extraction of contaminants from soils and sediments, *J. Supercrit. Fluid.* **3(2)**: 167–80.
4. Bordet, F., Chartier, T. and Baumard, J.F. 2002. The use of co-solvents in supercritical debinding of ceramics, *J. Eur. Ceram. Soc.* **22(7)**: 1067–72.
5. BPD, http://www.bpd.co.uk/technology.htm, acceded 06/08/2002.
6. Brown, Z.K., Fryer, P.J., Norton, I.T., Bakalis, S. and Bridson, R.H. 2008. Drying of foods using supercritical carbon dioxide-Investigations with carrot, *Innove. Food Sci. Emerg. Tech.* **9(3)**: 280–9.

7. Brunnemann, K.D., Prokopczyk, B., Djordjevic, M.V. and Hoffmann, D. 1996. Formation and analysis of tobacco-specific N-nitrosamines, *Crit. Rev. Toxicol.* **26(2)**: 121–37.
8. Brunner, G.H. 2004. Supercritical Fluids as Solvents and Reaction Media. Elsevier Science, pp. 650.
9. Caredda, A., Marongiu, B., Porcedda, S. and Soro, C. 2002. Supercritical carbon dioxide extraction and characterization of *Laurus nobilis* essential oil, *J. Agr. Food Chem.* **50(6)**: 1492–6.
10. Castelo-Grande, T., Augusto, P.A. and Barbosa, D. 2005. Removal of pesticides from soil by supercritical extraction-A preliminary study, *Chem. Eng. J.* **111(2–3)**: 167–71.
11. Caudell, T. 1999. Practical Supercritical Fluid Chromatography and Extraction. CRC, pp. 456.
12. Chaffin, J.M., Davison, R.R., Glover, C.J. and Bullin, J.A. 1995. Supercritical refining of asphalt to produce asphalt recycling agents, *Am. Chem. Soc. Div. Petrol. Chem.* **40(4)**: 799–803.
13. Clifford, A.A. and Bartle, K.D. 1996. Supercritical fluid dyeing, *Textile Technol. Int.*, pp. 113–7.
14. Clifford, A.A., Bartle, K.D., Gelebart, I. and Zhu, S. 1997. Fractionation of polymers using supercritical fluid extraction, *J. Chromatogr. A* **785(1–2)**: 395–401.
15. Cohen, S. and Bernstein, H. 1996. Microparticulate systems for the delivery of proteins and vaccines in drugs and the pharmaceutical sciences series. *CRC.* pp. 552.
16. De Azevedo, A.B.A., Kieckbush, T.G., Tashima, A.K., Mohamed, R.S., Mazzafera, P. and Melo, S.A.B.V. 2008a. Extraction of green coffee oil using supercritical carbon dioxide, *J. Supercrit. Fluid.* **44(2)**: 186–92.
17. De Azevedo, A.B.A., Mazzafera, P., Mohamed, R.S., Vieira De Melo, S.A.B. and Kieckbusch, T.G. 2008b. Extraction of caffeine, chlorogenic acids and lipids from green coffee beans using supercritical carbon dioxide and co-solvents, *Braz. J. Chem. Eng.* **25(3)**: 543–52.
18. De La Ossa, E.M. and Serrano, M.A.G. 1990. Extracción con fluidos supercríticos: fundamentos, *Ingeniería Química*, **7**: 169–75.
19. Duarte, A.R.C., Simplicio, A.L., Vega-Gonzàlez, A., Subra-Paternault, P., Coimbra, P., Gil, M.H., De Sousa, H.C. and Duarte, C.M.M. 2008. Impregnation of an intraocular lens for ophthalmic drug delivery, *Curr. Drug Deliv.* **5(2)**: 102–7.
20. Eldik, R.V. and Hubbart, C.D. 1996. Chemistry under extreme and non-classic conditions. Wiley, John & sons Inc. pp. 555.
21. Eldik, R.V. and Klärner, F.G. 2002. High pressure chemistry: synthetic, mechanistic, and supercritical applications. Wiley-VCH. pp. 474.
22. Fernandez Cid, M.V., Van Der Kraan, M., Veugelers, W.J.T., Woerlee, G.F. and Witkamp, G.J. 2004. Kinetics study of a dichlorotriazine reactive dye in supercritical carbon dioxide, *J. Supercrit. Fluid.* **32(1–3)**: 147–52.
23. Francis, A.W. 1954. Ternary systems of liquid carbon dioxide, *J. Phys. Chem.* **58(12)**: 1099–1114.
24. Freeman, H.M. and Harris, E.F. 1995. Hazardous waste remediation: innovative treatment technologies. CRC Press. pp. 361.
25. Garrigoís, M.C., Reche, F., Pernías, K., Sànchez, A. and Jimeínez, A. 1998. Determination of some aromatic amines in finger-paints for children's use by supercritical fluid extraction combined with gas chromatography, *J. Chromatogr. A* **819(1–2)**: 259–66.

26. Gong, K., Darr, J.A. and Rehman, I.U. 2006. Supercritical fluid assisted impregnation of indomethacin into chitosan thermo sets for controlled release applications, *Int. J. Pharm.* **315(1–2)**: 93–98.
27. Hannay, J.B. and Hogart, J. 1879. On the solubility of solids in gases. *Proceedings of the Royal Society* **29**: 324–6.
28. Hannay, J.B. and Hogart, J. 1880a. On the solubility of solids in gases. *Proceedings of the Royal Society* **30**: 178–88.
29. Hannay, J.B. and Hogart, J. 1880b. On the solubility of solids in gases. *Proceedings of the Royal Society* **30**: 484–9.
30. Hassan, A., Levien, K.L. and Morrell, J.J. 2001. Modeling phase behavior of multicomponent mixtures of wood preservatives in supercritical carbon dioxide with cosolvents, *Fluid Phase Equilibr.* **179(1–2)**: 5–22.
31. Ibàñez, E., Oca, A., De Murga, G., Loípez-Sebastiàn, S., Tabera, J. and Reglero, G. 1999. Supercritical fluid extraction and fractionation of different pre-processed rosemary plants, *J. Agr. Food Chem.* **47(4)**: 1400–4.
32. Illeís, V., Daood, H.G., Perneczki, S., Szokonya, L. and Then, M. 2002. Extraction of coriander seed oil by CO_2 and propane at super- and subcritical conditions, *J. Supercrit. Fluid.* **17(2)**: 177–86.
33. Kaiser, C.S., Rompp, H. and Schmidt, P.C. 2001. Pharmaceutical applications of supercritical carbon dioxide, *Pharmazie* **56(12)**: 907–26.
34. Karey, O'Leary. Supercritical Fluid Chromatography (SFC), URL: http://www.cee.vt.edu/ewr/environmental/teach/smprimer/sfc/sfc.html, accessed 26/03/2009.
35. Kaufmann, B. and Christen, P. 2002. Recent extraction techniques for natural products: Microwave-assisted extraction and pressurised solvent extraction, *Phytochem. Analysis* **13(2)**: 105–13.
36. Kim, W.J., Kim, J.D. and Oh, S.G. 2007. Supercritical carbon dioxide extraction of caffeine from Korean green tea, *Sep. Sci. Technol.* **42(14)**: 3229–42.
37. Kim, W.J., Kim, J.D., Kim, J., Oh, S.G. and Lee, Y.W. 2008. Selective caffeine removal from green tea using supercritical carbon dioxide extraction, *J. Food Eng.* **89(3)**: 303–9.
38. Kimura, Y., Abe, D., Ohmori, T., Mizutani, M. and Harada, M. 2003. Synthesis of platinum nano-particles in high-temperatures and high-pressures fluids, *Colloid Surface A: Physicochem. Eng. Asp.* **231(1–3)**: 131–41.
39. King, J.W. 1990. Applications of capillary supercritical fluid chromatography-supercritical fluid extraction to natural products, *J. Chromatogr. Sci.* **28(1)**: 9–14.
40. Kiran, E., Debenedetti, P.G. and Peters, C.J. 2008. Supercritical fluids: fundamentals and applications (NATO Science Series E: (closed)). Springer, pp. 616.
41. Klesper, E., Corwin, A.H. and Turner, D.A. 1962. High pressure gas chromatography above critical temperatures, *J. Org. Chem.* **27**: 700–1.
42. Knez, Z. and Weidner, E. 2003. Particles formation and particle design using supercritical fluids, *Curr. Opin. Solid St. M.* **7(4–5)**: 353–61.
43. Kruknois, V.J., Gallagher, P.M. and Coffey, M.P. 1995. Gas anti-solvent recrystallization and application for the separation and subsequent processing of RDX and HMX. *American Patent* 5389263.
44. Kruknois, V.J., Gallagher, P.M. and Coffey, M.P. 1994. Gas anti-solvent recrystallization process. *American Patent* 5360478.
45. Kumar, S.K., Suter, U.W. and Reid, R.C. 1986. Fractionation of polymers with supercritical fluids, *Fluid Phase Equilibr.* **29(C)**: 373–82.

46. Lee, H.Y., Kim, Y.J., Kim, E.J., Song, Y.K. and Byun, S.Y. 2008. Red pigment from Lithospermum erythrorhizon by supercritical CO_2 extraction, *J. Cosmet. Sci.* **59(5)**: 431–40.
47. Lòpez-Periago, A., Argemí, A., Andanson, J.M., Fernàndez, V., García-Gonzàlez, C.A., Kazarian, S.G., Saurina, J. and Domingo, C. 2009. Impregnation of a biocompatible polymer aided by supercritical CO_2: Evaluation of drug stability and drug-matrix interactions, *J. Supercrit. Fluid.* **48(1)**: 56–63.
48. Macnaughton, S.J., Kikic, I., Foster, N.R., Alessi, P., Cortesi, A. and Colombo, I. 1996. Solubility of anti-inflammatory drugs in supercritical carbon dioxide, *J. Chem. Eng. Data* **41(5)**: 1083–6.
49. Maftouh, M., Granier-Loyaux, C., Chavana, E., Marini, J., Pradines, A., Vander Heyden, Y. and Picard, C. 2005. Screening approach for chiral separation of pharmaceuticals: Part III. Supercritical fluid chromatography for analysis and purification in drug discovery, *J. Chromatogr. A* **1088(1–2)**: 67–81.
50. Martinez, J.L. 2007. Supercritical fluid extraction of nutraceuticals and bioactive compounds. CRC, pp. 424.
51. Martínez, M.L., Mattea, M.A. and Maestri, D.M. 2008. Pressing and supercritical carbon dioxide extraction of walnut oil, *J. Food Eng.* **88(3)**: 399–404.
52. Marzouki, H., Piras, A., Marongiu, B., Rosa, A. and Dessiì, M.A. 2008. Extraction and separation of volatile and fixed oils from berries of *Laurus nobilis* L. by supercritical CO_2, Molecules **13(8)**: 1702–11.
53. McHugh, M.A. and Krukonis, V.J. 1994. Supercritical Fluid Extraction: Principles and Practice. Butterworth-Heinemann, pp. 608.
54. Mendes, M.F., Pessoa, F.L.P. and Uller, A.M.C. 2002. An economic evaluation based on an experimental study of the vitamin E concentration present in deodorizer distillate of soybean oil using supercritical CO_2, *J. Supercrit. Fluid.* **23(3)**: 257–65.
55. Messmore, H.E. 1943. Asphaltic materials, *United States Patent* 2420185.
56. Montero, G.A., Smith, C.B., Hendrix, W.A. and Butcher, D.L. 2000. Supercritical fluid technology in textile processing: An overview, *Ind. Eng. Chem. Res.* **39(12)**: 4806–12.
57. Mukhopadhyay, M. 2000. Natural extracts using supercritical carbon dioxide. CRC, pp. 360.
58. Muth, O., Hirth, Th. and Vogel, H. 2000. Polymer modification by supercritical impregnation, *J. Supercrit. Fluid.* **17(1)**: 65–72.
59. NIST Livro de Química na Web, URL: http://webbook.nist.gov/chemistry/, accessed 21/01/2009.
60. Oakes, R.S., Clifford, A.A. and Rayner, C.M. 2001. The use of supercritical fluids in synthetic organic chemistry, *Journal of the Chemical Society. Perkin Trans.* **1(9)**: 917–941.
61. Peiriço, N., Matos, H. and Gomes de Azevedo, E. 1999. Particle formation of pharmaceutical drugs using supercritical carbon dioxide in *Proceeding of 17th European Seminar on Applied Thermodynamics*, Vilamoura, Portugal.
62. Peiriço, N., Matos, H. and Gomes de Azevedo, E. 2000. A new apparatus for encapsulation of pharmaceutical drugs using supercritical carbon dioxide in *Proceedings of the 5th International Symposium on Supercritical Fluids*; Atlanta-EUA, CD-ROM.
63. Peker, H., Srinivasan, M.P., Smith, J.M. and McCoy, B.J. 1992. Caffeine extraction rates from coffee beans with supercritical carbon dioxide, *AIChE J.* **38(5)**: 761–70.
64. Perrut, M. 2003. Supercritical fluids applications in the pharmaceutical industry, *S.T.P. Pharma Sci.* **13(2)**: 83–91.

65. Petersson, P., Malmquist, J., Markides, K.E. and Sjoberg, S. 1994. Determination of enantiomeric purity of (S)-carboranylalanine using capillary column supercritical fluid chromatography, *J. Chromatogr. A* **670(1–2)**: 239–42.
66. Pourmortazavi, S.M., Sefidkon, F. and Hosseini, S.G. 2003. Supercritical carbon dioxide extraction of essential oils from Perovskia atriplicifolia Benth, *J. Agr. Food Chem.* **51(18)**: 5414–9.
67. Rajaei, A., Barzegar, M. and Yamini, Y. 2005. Supercritical fluid extraction of tea seed oil and its comparison with solvent extraction, *Eur. Food Res. Technol.* **220(3–4)**: 401–5.
68. Ramírez, P., García-Risco, M.R., Santoyo, S., Señoràns, F.J., Ibàñez,, E. and Reglero, G. 2006. Isolation of functional ingredients from rosemary by preparative-supercritical fluid chromatography (Prep-SFC), *J. Pharmaceut. Biomed.* **41(5)**: 1606–13.
69. Reverchon, E., Ambruosi, A. and Senatore, F. 1994. Isolation of peppermint oil using supercritical CO_2 extraction, *Flavour Frag. J.* **9(1)**: 19–23.
70. Reverchon, E., Della Porta, G., De Rosa, I., Subra, P. and Letourneur, D. 2000. Supercritical antisolvent micronization of some biopolymers, *J. Supercrit. Fluid.* **18(3)**: 239–45.
71. Roselius, W., Vitzthum, O. and Hubert, P. 1972a. Nicotine removal from tobacco, German Patent 2043537.
72. Roselius, W., Vitzthum, O. and Hubert, P. 1972b. Recovery of coffee oil containing aroma ingredients from roasted coffee, German Patent 2106133.
73. Roselius, W., Vitzthum, O. and Hubert, P. 1973a. Selective extraction of nicotine from tobacco, German Patent 2142205.
74. Roselius, W., Vitzthum, O. and Hubert, P. 1973b. Extracting cocoa butter, German Patent 3923847.
75. Salvador, A., Gadea, I., Chisvert, A. and Pascual-Martí, M.C. 2001. Supercritical fluid extraction and high performance liquid chromatography determination of homosalate in lipsticks, *Cromatogr.* **54(11–12)**: 795–7.
76. Saus, W., Knittel, D. and Schollmeyer, E. 1993. Dyeing of textiles in supercritical carbon dioxide, *Textile Res. J.* **63(3)**: 135–42.
77. Savage, P.E. 1999. Organic Chemical Reactions in Supercritical Water, *Chem. Reviews* **99(2)**: 603–22.
78. Scalia, S. 2000. Determination of sunscreen agents in cosmetic products by supercritical fluid extraction and high-performance liquid chromatography, *J. Chromatogr. A* **870(1–2)**: 199–205.
79. Scalia, S., Giuffreda, L. and Pallado, P. 1999. Analytical and preparative supercritical fluid extraction of Chamomile flowers and its comparison with conventional methods, *J. Pharmaceut. Biomed.* **21(3)**: 549–58.
80. Scalia, S., Renda, A., Ruberto, G., Bonina, F. and Menegatti, E. 1995. Assay of vitamin A palmitate and vitamin E acetate in cosmetic creams and lotions by supercritical fluid extraction and HPLC, *J. Pharmaceut. Biomed.* **13(3)**: 273–7.
81. Schneider, P.F., Levien, K.L. and Morrell, J.J. 2006. Effect of wood characteristics on pressure responses during supercritical carbon dioxide treatment, *Wood Fiber Sci.* **38(4)**: 660–71.
82. Sheu, S.R., Jang, M.J. and Wang, C.C. 2008. A study of manufacturing micro-nano particles of green tea by rapid expansion of supercritical solutions method, *Proceedings of the ASME International Design Engineering Technical Conferences and Computers and Information in Engineering Conference*, DETC2007 3 PART A: 687–91.

83. Sicardi, S., Manna, L. and Banchero, M. 2000. Diffusion of disperse dyes in PET films during impregnation with a supercritical fluid, *J. Supercrit. Fluid.* **17(2)**: 187–94.
84. Sun, F., Pan, Y., Wang, J., Wang, Z., Hu, C. and Dong, Q. 2009. Preparation of conducting polyaniline and polyaniline-Fluorinated montmorillonite nanocomposites in supercritical carbon dioxide, *J. Macromol. Sci. A: Pure Appl. Chem.* **46(1)**: 37–45.
85. Sun, Y.P. 2007. Supercritical fluid technology in materials science and engineering-Synthesis, properties and applications. Taylor & Francis; pp. 600.
86. Sunarso, J. and Ismadji, S. 2009. Decontamination of hazardous substances from solid matrices and liquids using supercritical fluids extraction: A review, *J. Hazard. Mater.* **161(1)**: 1–20.
87. Tsutsumi, A., Ikeda, M., Chen, W. and Iwatsuki, J. 2003. A nano-coating process by the rapid expansion of supercritical suspensions in impinging-stream reactors, *Powder Technol.* **138(2–3)**: 211–5.
88. Van Roosmalen, M.J.E., Woerlee, G.F. and Witkamp, G.J. 2003. Dry-cleaning with high-pressure carbon dioxide-The influence of process conditions and various co-solvents (alcohols) on cleaning-results, *J. Supercrit. Fluid.* **27(3)**: 337–44.
89. Villard, D. 1896. Solubility of liquids and solids in gas, *J. Phys.* **5**: 455–9.
90. Vitzthum, O. and Hubert, P. 1972. Recovering fats and oils from plant seeds, German Patent 2127596.
91. Vitzthum, O. and Hubert, P. 1973. Extraction of caffeine from coffee, German Patent 2212281.
92. Vitzthum, O. and Hubert, P. 1976. Spice extracts, Canadian Patent 989662.
93. Vitzthum, O., Hubert, P. and Sirtl, W. 1975. Decaffeination of crude coffee, German Patent 2357590.
94. Vitzthum, O., Hubert, P. and Sirtl, W. 1976. Hop extracts, Canadian Patent 987250.
95. Wai, C.M., Hunt, F., Ji, M. and Chen, X. 1998. Chemical reactions in supercritical carbon dioxide, *J. Chem. Educ.* **75(12)**: 1641–5.
96. White, C. 2005. Integration of supercritical fluid chromatography into drug discovery as a routine support tool: Part I. Fast chiral screening and purification, *J. Chromatogr. A* **1074(1–2)**: 163–73.
97. White, C. and Burnett, J. 2005. Integration of supercritical fluid chromatography into drug discovery as a routine support tool: II. Investigation and evaluation of supercritical fluid chromatography for achiral batch purification, *J. Chromatogr. A* **1074(1–2)**: 175–85.
98. Wu, H.T., Lee, M.J. and Lin, H.M. 2006. Nano-particles formation for pigment red 177 via a continuous supercritical anti-solvent process, *J. Supercrit. Fluid* **33(2)**: 173–82.
99. Wu, H.-T., Lin, H.M. and Lee, M.J. 2007. Ultra-fine particles formation of C.I. Pigment Green 36 in different phase regions via a supercritical anti-solvent process, *Dyes Pigments* **75(2)**: 328–34.
100. Yang, C., Xu, Y.R. and Yao, W.X. 2002. Extraction of pharmaceutical components from *Ginkgo biloba* leaves using supercritical carbon dioxide, *J. Agr. Food Chem.* **50(4)**: 846–9.
101. Yang, G. and Wang, R.A. 1999. The supercritical fluid extractive fractionation and the characterization of heavy oils and petroleum residua, *J. Petrol. Sci. Eng.* **22(1–3)**: 47–52.
102. Yilgör, I. and McGrath, J.E. 1984. Novel supercritical fluid techniques for polymer fractionation and purification-1. Background, *Polym. Bull.* **12(6)**: 491–7.

103. Zhao, S., Xu, Z., Xu, C. and Chung, K.H. 2004. Feedstock characteristic index and critical properties of heavy crudes and petroleum residua, *J. Petrol. Sci. Eng.* **41(1–3)**: 233–42.
104. Zhao, S., Xu, Z., Xu, C., Chung, K.H. and Wang, R. 2005. Systematic characterization of petroleum residua based on SFEF, *Fuel* **84(6)**: 635–45.
105. Zhuze, T.P. 1959. Use of Compressed Hydrocarbon Gases as Solvents, *Vestnik Akad. Nauk, S.S.S.R.* **29(11)**: 47–52.
106. Zhuze, T.P. and Sufrovna, N. 1958. Solubility of natural gas components in crude oil at elevated temperatures and pressures, *Isvest. Akad. Nauk S.S.S.R. Otdel. Tekh Nauk* **3**: 104.
107. Zhuze, T.P. and Yushkevich, G.N. 1957a. Compressed hydrocarbon gases as solvent for crude oils and residuum I, *Isvest. Akad. Nauk S.S.S.R. Otdel. Tekh. Nauk* **11**: 63.
108. Zhuze, T.P. and Yushkevich, G.N. 1957b. Compressed hydrocarbon gases as solvent for crude oils and residuum II, *Isvest. Akad. Nauk S.S.S.R. Otdel. Tekh. Nauk* **12**: 83.
109. Zhuze, T.P., Yushkevich, G.N. and Grekker, J.E. 1958. Extraction of lanolin from wool fat with the aid of compressed gas, *Maslob.-Zhir. Prom.,* **24**: 34.
110. Zhuze, T.P., Yushkevich, G.N., Grekker, J.E., Vainshtok, V.V. and Bondarevskii, G.D. 1959. Complex reprocessing of wool fat, *Maslob.-Zhir. Prom.*, **25**: 25.
111. Zosel, K. 1971. Decaffeinizing coffee, *French Patent* 2079261.
112. Zosel, K. 1972. Caffeine from crude coffee, German Patent 2221560.
113. Zosel, K. 1974. Extraction of oils from animals and plants materials, German Patent 2363418.
114. Zosel, K. 1975. Simultaneous hydrogenation and deodorization of fats and/or oils, German Patent 2441152.
115. Zosel, K. 1976. Process for the separation of mixtures of substances, United States Patent 3969196.
116. Zosel, K. 1978. Separation with supercritical gases: practical applications, *Angewandte Chemie International Edition*, **17**: 702.

11

Biodiesel Production Processes

T.M. Mata[*] and A.A. Martins

ABSTRACT

This work presents the current state of development concerning different routes for biodiesel production, focusing on their chemical and technological aspects. Their relative advantages and disadvantages are presented and discussed in detail, with a focus on their industrial implementation and exploration. The full biodiesel production chain is taken into account, with a brief description of the potential feedstocks that can be used. Not only is the conventional production process analysed, but also new developments that can improve significantly the performance of current commercial production units. In the final sections expected future development for the production processes are presented and discussed.

1. INTRODUCTION

Biofuels for transportation are not new. The first automobiles used it. For example, Rudolph Diesel used peanut oil to power up its groundbreaking combustion engine presented in the World Exhibition in Paris in 1900, though he could have used a wide range of fuels. However in that time the vegetable oil fuels were more expensive than petroleum fuels and in the first decades of the twentieth century there was a significant development of the petrochemical industry. Besides the widespread availability of cheap and easy to explore sources of oil, fuels produced were clearly superior in terms of performance and power output when compared to other fuels. On the other hand, oil processing allowed the production of many other chemical compounds, as for example aromatics, that were increasingly used in the production of other chemical compounds and products used in many human activities, such as plastics, detergents, paints, among many others. Therefore, fossil fuels became one of the foundations or modern societies.

Since 1990's and mainly today due to recent increases in petroleum prices and uncertainties about petroleum availability, there is a growing

---

[*] Faculty of Engineering, University of Porto, Portugal, Rua Dr. Roberto Frias, 4200 465 Porto, Portugal.
[*] *Corresponding author* : E-mail : tmata@fe.up.pt

interest and awareness about the importance of alternative energy from renewable sources, in particular for transportation fuels, leading to an increase in the investment in research and development in this area. Examples include bio-ethanol from the fermentation of starch or sugar rich agricultural crops, vegetable oils from plants or microalgae and animal fats, gasification of biomass, wind, hydroelectric power, solar power including photovoltaic energy, among many others (Dewulf and Van Langenhove, 2006; Gilbert and Perl, 2008).

Nowadays biodiesel is the most important alternative diesel fuel in the EU representing 82% of the total biofuel production (Bozbas, 2008) and has been reported to emit substantially lower quantities of most of the regulated pollutants compared to mineral diesel. When compared to fossil fuels, their relative degree of development varies widely, and its potential strongly depends on local conditions and sectors of activity where the energy is used, making it difficult to select the best option, and the design and operation of renewable production systems. These questions are particular acute in the transportation sector, a cornerstone of modern society and one of the drivers of economic development, currently almost dependent on fossil fuels and one of the main contributor to air pollution, in particular to greenhouse gas (GHG) emissions (Akoh et al., 2007).

In fact fossil fuels are still the dominant source of energy used either for domestic or industrial consumption and in the medium to long term this situation will pose a growing threat, economic, societal, or environmentally, to the ongoing development of human activities. There is still no consensus among specialists, but it is common belief that the oil peak is close and the existing reserves will be increasingly harder and more expensive to explore in future (Laherrere, 2005). For example, tar sands and coal are seen as valuable resources that may substitute oil, but they will create far-reaching environmental problems, such as the destruction of ecosystems and a significant increase in the carbon dioxide and sulphur oxides emissions (Toman et al., 2008).

The soaring oil prices and growing concerns for the environment lead to an increased interest in other fuel sources. Among them, special attention is given to energy sources such as biodiesel, hydrogen, and bio-ethanol, in particular when produced from renewable raw materials (Girard and Fallot, 2006). However biofuel's industry, too, have their critics, notably among those concerned about the impact on food prices (when farm commodities like maize, wheat, and sugarcane are diverted to energy), the conversion of forests and other critical habitats for biofuel's feedstocks cultivation with the associated damage on biodiversity (depending on local biodiversity), loss of soil quality or land fertility (depending on initial soil conditions), emissions from carbon stock change (including both soil and vegetation), and land competition (depending on local availability of land).

The increasing demand for biofuels and the fast-rising of energy costs will require new and independent energy sources to be developed. This means for biofuels that in addition to the feedstocks presently used, new raw materials will have to be applied to cover the growing demand. For example, one can expect that next-generation lignocellulosic bio-ethanol and algal biodiesel technologies will become commercially significant in the longer term. Together with the search for new feedstocks, new technologies and applications need to be developed with a focus on the continuous process adaptation to future requirements in order to guarantee a steady quality of biofuels (Geissler, 2008; Eijsink et al., 2008).

Due to the wide variety of biofuel alternatives and different stages of development, this article focus its attention to biodiesel, a fuel that is already being produced at industrial scale and can have a significant impact in the near future. This will allow that other transportation fuels to complement each other and prevent some of the problems associated with the widespread production and consumption of biofuels.

Due to its umbilical relationship with the petrochemical industry, responsible for almost of the fuels produced and consumed for transportation in a world basis, chemical engineering has a key role to play in the developing biodiesel production sector. There is already extensive knowledge about various industrial processes that can be applied successfully in the development, design, and effective operation of biodiesel production facilities. Therefore, as the field is evolving fast, it is expected that chemical engineering as a discipline will have a profound impact.

This article starts by a brief description of what it is biodiesel, advantages and disadvantages of its utilization in the transportation sector, and a brief description of its production chain. Then, attention will be placed in the production of biodiesel itself, as it is in fact the key part and where chemical engineering can have the most profound impact. The questions linked with the type of alcohol used and how it impacts the process, and the potential destination of glycerol will be considered. To conclude, the future evolution and interactions with the growing field of biofuels and bio based economy is discussed, as well its potential contribution to sustainable development.

2. BIODIESEL-WHAT IT IS?

Biodiesel is a mixture of fatty acid alkyl esters obtained from transesterification (ester exchange) of vegetable oils and animal fats. These lipid feedstocks are composed by 90–98% (weight) of triglycerides and small amounts of mono and diglycerides, free fatty acids (generally 1–5%), and residual amounts of phospholipids, phosphatides, carotenes, tocopherols, sulphur compounds, and traces of water. In the transesterification reaction, a homogeneous or heterogeneous, acid or basic catalyst is normally used to enhance the reaction rate, although for some processes it may be not necessary to use a catalyst.

The quality of fatty acid alkyl esters are characterized by their physical and fuel properties (standardized by the EN 14214 (2003) in EU and the

ASTM D 6751 in USA) such as density, viscosity, composition of fatty acid esters, iodine value, acid value, cloud and pour points, flash point, gross heating value, volatility, cetane number, copper corrosion, water content, carbon residue, ash, and sulphur.

Biodiesel and mineral diesel have comparable energy density, cetane number, heat of vaporization, and stoichiometric air/fuel ratio (Agarwal, 2007). However, biodiesel has a higher proportion of multi-bonded carbon compounds that pyrolyse more readily and engines can suffer coking of the combustion chamber and injector nozzles, gumming and sticking of the piston rings. The flash point is normally above 170°C and the heating values of biodiesel (39–40 MJ/kg) are about 10% lower than diesel fuels (42–46 MJ/kg). The cetane numbers of biodiesel normally range from 40 to 70 while of diesel range from 47 to 55. The iodine value ranges from 0 to 200 depending upon unsaturation (Mondal et al., 2008). The cloud and pour points of biodiesel are higher than those of diesel fuels. Also, the viscosity of biodiesel is slightly greater than that of petroleum diesel but approximately an order of magnitude less than that of the parent vegetable oil. For example, at 40°C the viscosity values of methyl esters range from 2.8 to 3.7 mm^2/s, the viscosity of vegetable oils between 25 and 45 mm^2/s, and viscosity of diesel fuel is about 2.7 mm^2/s (Balat, 2008).

Among the many important advantages of biodiesel fuel some can be pointed when comparing it with petroleum diesel for transportation (Agarwal, 2007; Bozbas, 2008; Delucchi, 2003; Canakci and Sanli, 2008). They are listed below:

- Biodiesel is non-aromatic, almost sulfurless, produced from renewable sources, presently vegetable oils or animal fats. In another hand the lubricity property of biodiesel is much better than that of low-sulfur diesel fuel. Little biodiesel additive (about 1%) is enough to significantly improve the conventional diesel fuel's lubricity.
- Full life cycle analysis indicates that, on average, biodiesel emit less CO_2 than conventional fossil fuels. The use of biomass energy has the potential to greatly reduce GHG emissions and thus biodiesel have a significant smaller contribution to global warming when compared to fossil fuels. Although fossil fuels release carbon dioxide captured by photosynthesis millions of years ago, biomass releases carbon dioxide that is largely balanced by the carbon dioxide captured in its own growth, depending on how much energy was used to grow, harvest, and process the fuel.
- Biodiesel tailpipe CO_2 emissions per mass unit of biodiesel burnt are lower than for petroleum diesel. It has reduced visible smoke, noxious fumes and odours, and emits 40–50% less particulate matter (PM), 30–70% less HC, 20–50% less CO, and less 50% in soot emissions. Although the NOx may be about 10–15% higher this problem can be overcome by retarding the injection timing.

- Raw materials can be produced in marginal land, or in such a way that increases the soil productivity due to the utilization of crops that allow the fixation of nitrogen in the soil.
- Many different types of crops can be used depending on the local soil and climatic conditions. Also, wastes that are hard to deal with, such as animal fats, can be used as well. The broad variety of raw materials makes it possible to produce biodiesel at local scale, to be used in a farm or at an industrial production, for example to supply transportation fuels at a regional or even at national scale.
- When compared to other biofuels, such as bio-ethanol from corn, biodiesel has higher energy content and requires less energy, water and materials per unit of energy produced.
- The technologies associated with the production of biodiesel and pre-processing of raw materials are simple and easy to implement based on currently available process units.
- It is safe for use in all conventional diesel engines, offers similar performance and engine durability as petroleum diesel fuel. Therefore, no new types of engines are necessary, or significant changes to existing, making the adoption of biofuel simpler and reducing the resistance to change.

Some of the advantages listed above, in particular those directly linked with the reduced environmental impacts associated with biodiesel production and the possibility of using existing engines, are the main drivers behind the currently push to biodiesel.

However, pure vegetable oils have some significant drawbacks, such as high fuel viscosity that makes impractical its utilization in low temperatures, low power input, oxidation instabilities and the formation of deposits in the engine combustion chambers leading to corrosion and inefficient operation, just to name a few (Akoh *et al.*, 2007; Agarwal, 2007). Other disadvantages are included below:

- They are not economically feasible yet, at least when competing with currently used fossil fuels.
- Feedstock costs account for a large percent of the direct biodiesel production costs, including capital cost and return.
- There is still a research effort needed for the development of biodiesel through all its life cycle stages. In the production of raw materials stage there is a need for better and more efficient crops. In the biodiesel production stage there is a need for example to develop faster and more efficient production processes. Moreover, different strategies are needed to increase the robustness of existing processes and improve their capacity for using feedstocks with significant variations in their characteristics, as is normally the case of petrochemical industry.

- The utilization of vegetable oils for biodiesel production, many of them used in human consumption, is impacting the human food chain and increasing the price of food. Thus, the development of this area has to take into account not only technical aspects, but also its impact in the economic and societal dimensions.
- Modifications in currently used engines may be necessary to better use biodiesel, namely changes in the injection engine system and the utilization of heated fuel lines. As triglycerides have larger molecules, when compared to normal diesel, its viscosity will be higher and this is will impact especially the pump and injection system, influencing the engine performance. However, as stated before, current engines can used biodiesel without modifications, yet the conditions or its usage may be more restrictive and not optimal.
- New additives may have to be developed to blend with biodiesel to avoid some of biodiesel limitations, and reduce the potential environmental impact of its utilization in currently used engines.

Some of the advantages represent real problems that will have or are already having an impact in the development of biodiesel, as for example the influence that the utilization of agricultural crops had on food prices increases. Chemical Engineering can have, and in some regards it is starting to have an impact on the biodiesel industry, in particular for the development of new and more efficient production processes.

3. BIODIESEL PRODUCTION CHAIN

In a simplified way the biodiesel production chain stages can be represented as in Fig. 11.1, including the following main steps: (1) feedstocks production, for example of vegetable oils, animal fats (*e.g.* lard, tallow, poultry fat, fish oil) and microalgae; (2) feedstocks processing, including oil extraction and pre-treatment; (3) biodiesel production; (4) biodiesel post-treatment and blending; (5) distribution and final use.

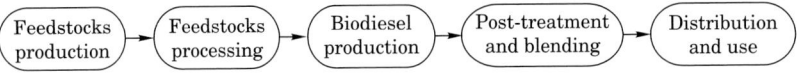

Fig. 11.1. Biodiesel production chain stages

3.1. Feedstocks Production

Liquid biofuels, of which biodiesel is a significant part, can be manufactured from a wide variety of lipid feedstocks, such as vegetable oil crops, waste fat materials from slaughter houses, waste frying oils, among many others. Currently most of the biodiesel is produced from a range of crops that are relatively specific to geographic location. In temperate regions, rapeseed, sunflower, corn, and other legumes or cereals are used as feedstock, whereas in tropical regions, palm oil, jatropha, and to a lesser degree soybeans are used. Recently new feedstocks that can be considered as independent of any specific climate and geographical are being considered, such as cellulosic biomass or microalgae.

Estimates of the 2007/2008 world production of major oilseeds, both for the food and biofuel industries, include (FAO, 2009): 220.8 million tonnes of soybeans, 48.9 million tonnes of rapeseed, 20.2 million tonnes of sunflower seeds, and 10.1 million tonnes of palm kernels. Europe has dominated the biodiesel industry to date, generating around 90 percent of global production using rapeseed oil as the main feedstock. The oilseed crops harvested worldwide are crushed in order to obtain oils and fats for human consumption or industrial use, and cakes and meals are used as feed ingredients for livestock feed.

The way crops cultivation and harvesting is performed is not significant from a chemical engineering perspective, as agronomic engineers and other agricultural specialists are the people looking for that area. Good descriptions of the requirements and cultivation practices for the most important biodiesel feedstocks can be found in literature (FAO, 2008; Kurki et al., 2006). Yet, depending on the feedstock different pre-processing procedures should be considered, to ensure an optimal operation of the biodiesel production processes and to meet final fuel quality requirements (Canakci and Sanli, 2008).

The feedstock selection can have a profound impact in the production process. Moreover, as the environmental issues are becoming evermore important, other aspects ought to be considered. Some vegetable oil crops, such as soy and palm oil, are also basic commodities used for human consumption, either directly or indirectly through to livestock feeding. Therefore, their use for biodiesel production will increase its price, leading to increased production costs. Also, the clearing of forests and the other unoccupied land reduces the biodiversity and reduces the production of food for human consumption (Bickel and Dros, 2003).

Recently much attention has been given to microalgae as a biodiesel feedstock that may greatly reduces many problems associated with vegetable oil crops. The growth and harvest of microalgae need much less land area than other biodiesel feedstocks of agricultural origin, up to 49 or 132 times less when compared to rapeseed or soya crops, for a 30% oil content in biomass weight (Chisti, 2007). From a practical point of view, microalgae are easy to cultivate, can grow at low cost with little attention, using water unsuitable for human consumption.

Combined with their ability to grow under harsher conditions, and their reduced needs for nutrients, microalgae can be cultivated in areas unsuitable for agricultural purposes (on marginal or non-arable land), independently of the seasonal weather changes. This fact greatly reduces the competition for arable soil with other crops, in particular for human consumption, and can open up new economic opportunities for arid drought or salinity-affected regions (Schenk et al., 2008).

Moreover, microalgae are becoming an alternative oil source with increased oil yields and much higher growth rates when compared to conventional biodiesel feedstocks. They reproduce themselves using photosynthesis to convert solar energy into chemical energy, completing an entire growing cycle every few days. Some studies have shown that can even be used simultaneously for pollution control, as for the example to capture CO_2 from flue gas emissions or to remove nitrogen and phosphorus from wastewater effluents, or for the production of other high value chemicals (Murakami et al., 1998; Hodaifa et al., 2008). Thus, microalgae are environmentally preferable and cost effective when compared with other feedstocks.

Waste vegetable oil sourced from restaurants or industrial food, with the advantage of having a lower price than the raw oil, is often used for biodiesel production. Animal fats such as tallow, lard, poultry fat, fish oils are by-products of the meat and fish processing industries that can also be used for biodiesel production.

3.2. Feedstocks Processing

3.2.1. Oil extraction

After the harvest the oil seeds are transported to the oil extraction unit for which two main processes can be used, either alone or in combination: mechanical press extraction and solvent extraction. In mechanical press extraction the seeds are first heated to 40–50°C and then crushed in a screw press. The mechanical extraction can be combined with solvent extraction to produce oil with a higher degree of purity but it is more expensive. In the solvent extraction process hexane is normally used as solvent to dissolve the oil, which can then be recovered and separated from the oil by distillation. The solvent can be reused in the process. The resulting seed straw and cake can be used for animal feed or burned for energy.

In the case of microalgae the oil extraction process differs from the other vegetable oils feedstocks (Grima et al., 2003). After separation from the culture medium algal biomass (5–15% dry weight) must be quickly processed lest it should get spoiled in only a few hours in a hot climate. Processing represents a major economic limitation to the production of low cost commodities (foods, feeds, fuels) and also to higher value products (b-carotene, polysaccharides). It is difficult to discuss processing, since it is highly specific and strongly depends on the desired product. It is common to do the dehydration of the biomass that also increases its shelf-life and of the final product. Several methods have been employed to dry microalgae where the most common include spray-drying, drum-drying, freeze-drying and sun-drying (Richmond, 2004). Because of the high water content of algal biomass sun-drying is not a very effective method for algal powder production and the spray-drying is not economically feasible for low value products, such as biofuels or protein.

After drying cell disruption follows of the microalgae cells for release of the metabolites of interest. Several methods can be used depending on the microalgae wall and on the nature of the product to be obtained, either based on mechanical action (*e.g.* cell homogenizers, bead mills and ultrasounds, autoclave, spray drying), or non-mechanical action (*e.g.* freezing, organic solvents and osmotic shock, acid, base, enzyme reactions). Lyophilization breaks up the cells and turns the algal material into a loose and fine powder, making other treatment unnecessary.

For biodiesel production, lipids and fatty acids have to be extracted from the microalgal biomass. So, for lipids a solvent extraction is normally done directly from the lyophilized biomass. Several solvents can be used (Richmond, 2004), for example *n*-hexane or ethanol (96%), albeit *n*-hexane is preferred. Ultrasound-assisted extraction (UAE) and microwave-assisted extraction (MAE) techniques have also been employed as complementary techniques to extract oils from vegetable sources (Cravotto *et al.*, 2008; Deshmane *et al.*, 2009).

3.2.2. Pre-treatment

The transesterification reaction is sensitive to the feedstock purity requiring usually some pre-treatment operations. The refined vegetable oils do not need a pre-treatment for biodiesel production. However, waste oils and animal fats have a lot of impurities such as free fatty acids (FFA) and water that negatively affect the reaction performance. They can reduce the reaction rate by orders of magnitude, even in small amounts (Canakci, 2007). Also, FFA cannot be converted to biodiesel, forming instead soap that limits the mass transfer between phases, reducing significantly the chemical reaction rate and the selectivity to biodiesel, and further complicating the separation of phases after the reaction completion (Aranda *et al.*, 2008).

The FFA content of waste oils and animal fats vary widely, being normally higher than 2% (w/w) (Watanabe *et al.*, 2001), typically 15%, but can be as high as 40% (Van Gerpen *et al.*, 2004; Canakci, 2007). In some cases, FFA are the by-products of the food processing and oleochemical industry, which can produce 4–8% of FFA from the total oil during its physical refining (Watanabe *et al.*, 2007). The recovery of the FFA residue is difficult and not economically feasible (Aranda *et al.*, 2008). Thus, its alternative usage as feedstock for biodiesel production looks promising.

Fatty acids vary in their carbon chain length and number of double bonds. However, there is a remarkable dominance of fatty acids with an even number of carbon atoms (Allen *et al.*, 1999). Different types of oils have different types of fatty acids with different carbon chain lengths, typically from C12 to C20, with a preponderance of C16 and C18 molecules. Palmitic (C16:0), stearic (C18:0), oleic (C18:1), linoleic (C18:2) and linolenic (C18:3) acids are the fatty acids most commonly found in vegetable oils

(Balat, 2008). The physical and chemical fuel properties of biodiesel depend on the fatty acids distribution of the triglyceride used in the production. For example, the high melting point and high viscosity of beef tallow may result from higher stearic (C18:0) and palmitic (C16:0) acid contents (Mondal et al., 2008; Balat, 2008).

Feedstocks with FFA contents greater than about 1% (w/w) must be pre-treated to either remove the FFA or convert the FFA to esters before performing the biodiesel generation reaction. Otherwise, the catalyst will react with the FFA to form soap and water. The soap formation reaction is very fast and it complete before any relevant esterification occurs (Van Gerpen et al., 2004). Besides consuming the catalyst the saponification reaction also promotes the formation of emulsions that create downstream problems in the post-treatment processes and purification of biodiesel. The following equations represent respectively the saponification reactions of FFA and esters.

$$R-COOH + NaOH \xrightarrow{heat} R-COONa + H_2O \quad ...(1)$$
$$\text{FFA} \quad \text{metallic alkoxide} \quad\quad\quad \text{salt} \quad\quad \text{water}$$

$$R-COOR' + NaOH \xrightarrow{water} R-COONa + H_2O \quad ...(2)$$
$$\text{Ester} \quad \text{metallic alkoxide} \quad\quad\quad \text{salt} \quad\quad \text{water}$$

FFA content higher than 0.5% (w/w) is reported to affect the yield of the transesterification reaction. Rice et al. (1997) reported that a reduction of the FFA from 3.6% to 0.5% increased yields from 73% to 87%. Canacki and Van Gerpen (1999) refer that if FFA levels above 5% can lower the ester conversion rate below 90%.

The esterification can be used as a pre-treatment for basic transesterification reaction to convert the FFA into methyl esters (Aranda et al., 2008; Issariyakul et al., 2007). The esterification reaction is acid catalysed and may be represented as follows:

$$R-COOH + R'-OH \xrightarrow{acid\ catalyst} R-COOR' + H_2O \quad ...(3)$$
$$\text{FFA} \quad \text{alcohol} \quad\quad\quad \text{esters} \quad\quad \text{water}$$

An excess methanol (20:1 ratio) is generally necessary to ensure full conversion of FFA. Direct acid esterification of a high free fatty acid feed requires the by-product water removal during the reaction or it will be quenched prematurely, also affecting the biodiesel production further downstream. Thus, the acid esterification system needs to have a water management strategy to minimize the amount of methanol required for the reaction. One possible approach is to remove water while the reaction occurs using for example a membrane reactor.

Another approach proposed by Van Gerpen et al. (2004) and Canakci and Van Gerpen (1999, 2001, 2003) is to perform the reaction in a two rounds, with the removal of the methanol, sulphuric acid, and water phase in between, followed by the addition of more fresh reactant to perform a second round driving it closer to completion. Zhang *et al.* (2003) suggest the addition of glycerine after the second round reaction to remove all the water from the oil stream, having the additional advantage of removing also the acid catalyst which may cause the neutralization of the alkalis catalyst during the transesterification reaction.

Typical water content of waste oils and animal fats may vary considerably depending on the origin. For example Rice *et al.* (1997) reports a range of 1–5% (w/w) of water contents in oils. The presence of water causes esterification and transesterification reaction inhibition, favours the hydrolysis of triglycerides and FFA, lowers the yield of esters, and renders the separation of ester and glycerol and water washing difficult (Canakci, 2007; Aranda *et al.*, 2008). If the water concentration is greater than 0.5%, the ester conversion rate may drop below 90% (Canakci and Van Gerpen, 1999). Also, water promotes the formation of soap in the presence of the alkalis catalysts, increasing catalyst consumption and diminishing its efficiency. The water content in the feedstock should be lower than 0.06% (w/w) (Rice *et al.*, 1997). Heating the waste cooking oil or tallow over 100°C, to about 120°C, can boil off any excess water present in the feedstock.

Additionally to water and FFA, the waste cooking oils also have other impurities such as solid particles resulting from food frying and sodium chloride that is added to the fried food. The separation of solid particles may be accomplished for example, by filtration, pressing, and centrifugation, depending on the feedstock characteristics.

The presence of chlorides may cause corrosion problems in the process equipment and piping system. One needs to balance the trade-off between increasing the process operating costs removing the chlorides and increasing the capital costs investing in process equipment materials resistant to corrosion such as stainless steel.

After pre-treatment the next step is the production of biodiesel. Due to their high viscosity Canakci and Van Gerpen (1999) suggest to store greases in agitated tanks heated at 55–60°C.

3.3. Biodiesel Production Processes

As referred above, at its core the biodiesel production process is based on the transesterification reaction between triglycerides and alcohols. As most of the biodiesel feedstocks have similar characteristics, in particular those of vegetal origin, any improvement in the way the reaction is done, in particular when related to the reaction time and final product quality will have a profound impact in the production capacity and the overall process.

The transesterification reaction can be performed in various ways. Currently most processes involve homogeneous catalysis using normally alkali as catalysts and based in stirred reactors operating in batch models. Recently some improvements were proposed for this process, in particular to be able to operate in continuous mode in order to reduce the reaction time. Other possibilities currently being considered include the utilization of:

- Reactors with improved mixing such as static mixers (Noureddini et al., 1998), continuous high-shear mixing reactors, microwave assisted reaction (Azcan and Danisman, 2008; Cravotto et al., 2008), cavitation reactors (Gogate, 2008; Gogate and Pandit, 2000a, 2000b; Gogate et al., 2001; Gogate et al., 2006; Gogate and Kabadi, 2009; Kelkar et al., 2008), and ultrasonic reactors (Quintana, 2002; Santos et al., 2009; Colucci et al., 2005; Kalva et al., 2008; Deshmane et al., 2009);
- Heterogeneous catalysts (inorganic chemical or enzymes), to avoid the need for the removal and recycling of the catalyst (Watanabe et al., 2000; Alonzo, 2007; Al-Zuhair, 2007; Akoh et al., 2007; Barakos et al., 2007; Antunes et al., 2008; Albuquerque et al., 2008; Benjapornkulaphong et al., 2009; Ranganathan et al., 2008; Ibrahim et al., 2008);
- Non-catalytic supercritical methanol or ethanol for the transesterification reaction (Demirbas, 2009; Madras et al., 2004; Chen et al., 2009). This is a simpler process allowing a high yield because transesterification of triglycerides and methyl esterification of fatty acids occurs simultaneously without any catalyst.
- Co-solvents to enhance the solubility of reactants, by diminishing the mixture polarity, and increasing the reaction rate (Boocock et al., 1996; Boocock, 2003e, 2004; Shi and Bao, 2008; Kuramochi et al., 2008; Guan et al., 2009a, 2009b).

In the next sub-sections the relative advantages and disadvantages of the different routes for biodiesel production are presented and discussed, focusing on their chemical and technological aspects.

3.3.1. *Conventional processes for biodiesel production*

3.3.1.1. Homogeneous catalyzed biodiesel production process

There are several routes to obtain biodiesel from lipid feedstocks. The most widely used is the transesterification of triglycerides with low molecular weight alcohols in the presence of a homogeneous catalyst (acid or alkalis) and operated in batch mode (Demirbas, 2008).

This process has the advantages of being easy to implement, operate and control, and the reactions occur in liquid phase under mild conditions of temperature and pressure. Since alkali-catalyzed transesterification is much faster by an order of magnitude when compared with acid-catalyzed transesterification, alkaline metal hydroxides (*e.g.* NaOH, KOH) are the most often used commercially, and to a lesser extent methoxides and carbonates (Agarwal, 2007; Demirbas, 2008; Aranda et al., 2009; Han et al., 2009).

Transesterification is a multiple reaction including three reversible steps in series, where triglycerides are converted to diglycerides, then diglycerides are converted to monoglycerides, and monoglycerides are converted to esters and glycerol. From this reaction it results three moles of fatty acid alkyl monoester (biodiesel) and a mole of glycerol as a by-product, as seen in Fig. 11.2.

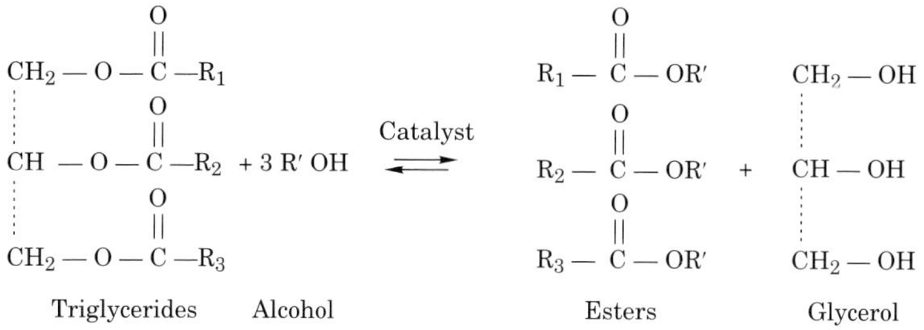

$$\begin{array}{c} CH_2-O-\overset{O}{\overset{\|}{C}}-R_1 \\ CH-O-\overset{O}{\overset{\|}{C}}-R_2 + 3\,R'OH \\ CH_2-O-\overset{O}{\overset{\|}{C}}-R_3 \end{array} \overset{Catalyst}{\rightleftharpoons} \begin{array}{c} R_1-\overset{O}{\overset{\|}{C}}-OR' \\ R_2-\overset{O}{\overset{\|}{C}}-OR' \\ R_3-\overset{O}{\overset{\|}{C}}-OR' \end{array} + \begin{array}{c} CH_2-OH \\ CH-OH \\ CH_2-OH \end{array}$$

Triglycerides   Alcohol   Esters   Glycerol

Fig. 11.2. Overall transesterification reactions of triglycerides

The relationship between the feedstock mass input and biodiesel mass output is about 1:1 ratio. Because of the transesterification reactions are reversible, an excess of a primary alcohol is required to shift the equilibrium to the product side and ensure a full conversion of the triglycerides. The most suitable alcohol amount may vary case to case. Although the theoretical stoichiometric alcohol/oil molar ratio is 3:1, molar ratios of 6:1 if alkalis catalysed (Van Gerpen, 2005), or 12:1 if acid catalysed (Han et al., 2009), or even higher depending on the process conditions may be used. Generally, the biodiesel yield increases with the excess of the alcohol, but production costs rise, in particular due to an increase in the reactor volume needed for the reactor and the separation of glycerol that becomes more difficult.

After the chemicals are mixed, two essentially immiscible phases are formed: one apolar containing triglycerides and esters, and the other polar containing glycerol and alcohol. When the alkalis catalyst is used some emulsification occurs due to saponification reaction (Ataya et al., 2008). As the process involves the formation of two immiscible phases, the reactor vessels are intensely stirred to promote mass transfer (Ataya, 2008).

Under the conditions normally used in practice, especially for temperatures between 50 to 70°C (higher temperatures are needed in ethanol instead of methanol is used), the conversion of the oil is complete in a few hours (Akoh et al., 2007). After the reaction is finished the glycerol is removed by allowing the two phases to form and settle. Then, any excess alcohol that did not react and catalyst are removed from both phases (of esters and glycerol) and recycled back to the reactor. The removal and recycling of the catalyst and unreacted excess alcohol increase the complexity and operating costs of the process (see Fig. 11.3).

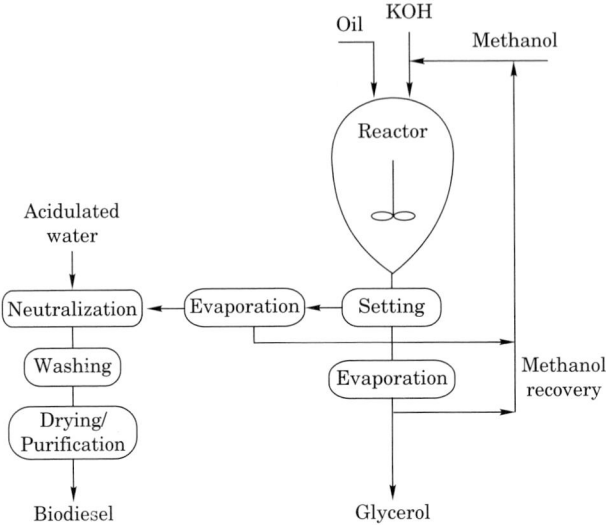

Fig. 11.3. Schematic of typical biodiesel production process with alkaline catalysis

3.3.1.2. Conventional process improvement

Different strategies are being considered to enhance the performance of the homogeneous production process, by addressing particular issues identified before.

One possibility involves the utilization of heterogeneous catalysts of sodium or potassium alkoxides, as well as their carbonates, instead of the homogeneous alkali catalysts (Arzamendi *et al.*, 2007, 2008). Alkoxides are very reactive and can reduce the total reaction time to less than 1 h. However, they are water sensitive and have to be anhydrous to avoid their hydrolysis.

On the other hand, although acid catalysts are seldom used due to their lower reaction rates, they are better suited for oils with high FFA contents, with no soap formation (Canakci and Van Gerpen, 1999; Han *et al.*, 2009). Nevertheless, they normally require higher temperatures and higher oil: alcohol molar ratios when compared to the alkali catalysed processes, and are more corrosive to the process equipment (Akoh *et al.*, 2007).

The effects of mixing are also significant to the process, as two phases are formed and the diverse reactants are presented essentially in different phases. The interfacial area between phases increases with higher mixing intensity, leading to more mass transfer between phases and naturally higher reaction rates (Ataya, 2008). The experimental results of Noureddini and Zhu (1997) confirm the previous conclusions and show that depending on the reaction stage either the mass transport or the reaction kinetics are the dominant aspects controlling the process performance.

Other possibility considered in literature involves the utilization of different co-solvents to enhance the reaction and mass transfer such as dimethyl ether (DME), diethyl ether (DEE), methyl tert-butyl ether (MTBE) and tetrahydrofuran (THF). The use of co-solvents has attracted much attention since they allow one to increase the reaction rate, under milder conditions, by diminishing the mixture polarity (Mao, 1995; Boocock et al., 1996; Boocock, 2003e, 2004; Shi and Bao, 2008; Kuramochi et al., 2008; Guan et al., 2009a, 2009b). Moreover, transesterification in supercritical methanol, employing propane and CO_2 as co-solvents was also developed (Cao et al., 2005; Han et al., 2005; Yin et al., 2008; Sawangkeaw et al., 2007).

Therefore, the selection of the appropriate co-solvent and the mixing intensity are critical factors contributing to the correct operation and performance of the reaction system. The use of co-solvents and high mixing intensity for the reaction reduces the need to use higher temperatures to enhance the solubility among reactants and the mass transfer between both phases (Ataya et al., 2008a).

3.3.2. *New and emerging processes for biodiesel production*

The conventional processes for biodiesel production have several drawbacks, in particular the large reaction times needed to obtain a complete conversion of oil, their operation mainly in batch mode, and their complex separation stages after the reaction.

The several by-products and residues (*e.g.,* straw, grain dust, seed cake, rind and greaves, glycerine, soaps, wastewater) resulting from the biodiesel production chain phases and the potential environmental impacts associated to biodiesel production, including the upstream and downstream operations, are other problems that need to be addressed (Zabaniotou et al., 2008; Ioannidou et al., 2009).

More efficient production processes are needed in order to respond to the expected increase in global demand for biodiesel, market competitiveness, and need to fulfil the goals defined at governmental and regional levels.

Many possibilities are being considered and studied in detail, some of them already implemented and used in industrial settings. With varying degrees of success, they try as much as possible to reduce the reaction time and to allow for a continuous process operation mode, essential for a production capacity that can meet the expected demand for biodiesel, while reducing operating costs, namely of the raw materials stocks and of the biodiesel cleaning operations associated to the conventional batch operation processes (Ataya et al., 2008).

As referred above, examples of processes being studied include the enzymatic catalysis, reaction in supercritical fluids, reactor designs that promote improved mixing by new methods, such as static mixers, ultrasounds or cavitations, among others. In the next sections some of these possibilities are presented and discussed in detail.

3.3.2.1. *Heterogeneous catalyzed biodiesel production process*
The present utilization of homogeneous catalyst in which the catalyst has to be removed from the final products and recycled back to the reactor, adds complexity and produces a neutralization waste, leading to increased process equipment, operating costs and environmental impacts.

The possibility of replacing homogeneous catalysts with solid hetero-geneous catalysts greatly simplify the process by facilitating the catalyst separation and by eliminating the complex purification steps involved in the recycling of the catalyst, yielding a cleaner product, greatly decreasing the cost of production, and making it easier to operate in a continuous mode.

There is already an enormous body of knowledge related to heterogeneous catalysis, due to its widespread use in a wide variety of processes (*e.g.* catalytic converter for automotive exhaust systems and packed bed reactors) mainly associated to the petrochemical and chemical industry. Thus, the use of heterogeneous catalysis is a natural step with the potential to greatly improve existing processes.

Various types of heterogeneous catalysts are being considered for biodiesel production. An extensive and updated review of the current status on the use of solid base catalysts is provided by Liu *et al.* (2007). Another review concerning acid and basic heterogeneous catalyst performances for biodiesel production, examining both scientific and patent literature, has been presented by Di Serio *et al.* (2008).

Although the extensive use of solid catalysts for the esterification reaction in the organic chemical industry, for example to produce several types of fatty acid esters, polyesters, and other esters-based compounds (Otera, 2003), its utilization in the transesterification reaction for biodiesel production is almost inexistent at industrial scale. Some examples of solid catalysts that have been studied for biodiesel production include, but are not limited to:

- Zeolites, such as the ZSM-5 (MFI), mordenite (MOR), faujasite (FAU), beta (BEA), Y, and silicalite (Brito *et al.*, 2007; Chung *et al.*, 2008).
- Layered nitrate and oxides with the advantage of having tuning properties (Cordeiro *et al.*, 2008).
- Various types of simple and complex hydrotalcites (Barakos *et al.*, 2007).
- Various inorganic oxides, such as MgO, CaO in powder or nanocrystaline forms, WO (tugsten oxide), aluminium oxides, complex oxides containing Mg, Al, Ca, alkaline metals hydroxides, and other elements

(Albuquerque *et al.*, 2008; Arzamendi *et al.*, 2008; Benjapornkulaphong *et al.*, 2009; Boz and Kara, 2008; Montero *et al.*, 2009).

- Ion-exchange resins such as Amberlyst-15, Nafion-NR50, and other kinds of anionic exchange resins (Kiss *et al.*, 2006).
- Sulphonated metal oxides, such as zirconia or titanium oxide (Almeida *et al.*, 2008).

Brito *et al.* (2007) performed studies with zeolites Y of transesterification of different feedstocks such as for example fried oils, using various reaction conditions, reactant ratios, and co-solvents.

Chung *et al.* (2008) studied the possibility of using different types of zeolites in the esterification of FFA existing in fried oils, in particular the influence of the pore structure and the acidic properties in the conversion of FFA.

Albuquerque *et al.* (2008) concluded that the catalyst activity strongly depends on the metal composition of the oxides used, and new materials with well defined structures, high surface area, and adequate basic or acid properties have yet to be developed.

Although many of the solid catalysts proposed in literature for biodiesel production have good catalytic performances, they require high temperatures and pressure to work properly. Also, it is necessary to address the questions of deactivation, reusability and regeneration of the catalysts in practical conditions to assess their real potential for using in commercial applications.

To the best of the authors' knowledge, presently there are no commercial processes employing heterogeneous catalysts operating under mild reaction conditions, in the temperature range of 20–60°C. The Esterfip-HTM is an example of a heterogeneous catalyzed process, developed by the French Institute of Petroleum (IFP) in 2005, which is operated at industrial scale using high temperature and pressure (200°C and 60 bar). This process catalyst consist of a spinel mixed oxide of two non-noble metals, Al and Zn. Recently other processes combining solid-acid and homogeneous-base catalysts have been introduced by Lanxess (Alonzo, 2007).

As heterogeneous catalyzed processes have advantages over homogeneous catalyzed, and are easier to operate, more commercial applications will certainly be introduced in the near future with an important impact on the biodiesel production.

3.3.2.2. *Biological catalyzed biodiesel production process*
Biological catalysis includes both enzymes and living organisms for biodiesel production. It can be implemented either in solution or supported (for example in biological films in packed beds), and are considered to be one of the most promising alternatives for future use.

Although extensive research has been devoted to this area, currently there are not a lot of examples of practical applications of biological catalysis to biodiesel production (Ming et al., 1998; Dossat et al., 1999; Shimada et al., 2002; Oliveira and Rosa, 2006; Akoh et al., 2007; Rathore and Madras, 2007; Ibrahim et al., 2008; Ranganathan et al., 2008; Shaw et al., 2008).

Enzymes can be used for the transesterification of oils, in particular lipases that are present in many living cells. As catalysts they are more efficient, selective, require less energy (reaction temperature is lower), and produce less side products or waste when compared with other types of catalysed processes. On the other hand, enzymes or living organisms can be genetically engineered to improve not only its performance but also its resilience and capacity to operate in harsher conditions.

Nevertheless their advantages, enzymes are still not extensively used at industrial practice. Some of the main problems include:
- the difficulty in determining what are the best enzymes or microorganisms to do the job, depending on the characteristics of the feedstocks and impurities that may exist;
- reaction conditions, in particular what are the optimal molar ratio of reactants, solvents to be used, temperature, and water content;
- how to operate the reactor;
- how the enzymes will be used, supported, in solution or using whole cells;
- how to recover and reuse the enzymes;
- how to avoid the deactivation of enzymes, or the living organisms death;
- adequacy of the enzymes or micro-organisms used and of the feed characteristics, in particular the FFA content, and other impurities and contaminants.

Presently the focus of research is focused on the utilization of lipases obtained from different biological sources to perform the transesterification. When compared with microorganisms they do not require the utilization of nutrients and simplify the downstream processing, avoiding the need to remove the alkali or acid catalyst.

Some studies can be found in literature that addresses some of the problems listed above. For example Shimada et al. (2002) concluded that the best way to avoid the inhibition or deactivation of enzymes by the oil or methanol is their stepwise addition to the mixture, in order to maintain the oil/methanol ratio at certain optimal levels. This way it is possible to maintain the enzyme activity for longer periods of time.

Different enzymes have different capacities, as shown by Kaieda et al. (2001) who reported that certain lipases obtained from various microorganisms can be used in high water and methanol contents. When methanol is used in an appropriate concentration, Al-Zuhair (2007) shown that faster reaction rates are possible when compared with the homogenous process.

Also, the addition of co-solvents appears in some cases to have a positive effect on the enzyme stability (Akoh et al., 2007), although there is some work to be done in order to identify the most adequate solvents and how they influence the ongoing reaction.

Enzymes as lipases can also be used as an alternative route for biodiesel production, commonly known as alcoholysis, a form of transesterification, or through an interchange reaction between ester groups, increasing the process flexibility.

Other studies have focused on the chemical reaction kinetics and its dependence on water and reactants concentration (Canakci and Van Gerpen, 1999, 2001, 2003; Van Gerpen et al., 2004; Rice et al., 1997).

Although being seen as a viable option when compared with existing processes, there is still a lot of work to be done before biological catalysis becomes an alternative and has a significant impact on biodiesel production.

3.3.2.3. Reaction in supercritical media

A non-catalytic route for biodiesel production is to carry out the reaction in supercritical methanol or ethanol. Under high temperature and pressure conditions the alcohol is in a supercritical gaseous state and the triglycerides are somewhat dissolved on it.

When compared to the conventional process, the reaction rates in supercritical media are much faster, achieving full conversion of triglycerides in a matter of minutes generating soap-free glycerol (Soetaert and Vandamme, 2009). Moreover, there is no need to recover the catalyst and the adverse influence of the water presence, so strong in the homogeneous process, is negligible and even advantageous, as reported by some authors (Al-Zuhair, 2007).

A continuous operation mode is possible in various forms, as for example in tubular reactors (Chen et al., 2009), and some expertise already exists in operating units with supercritical fluids, for example for the removal of caffeine from coffee beans and temperature sensitive aromas and flavours, although in most of these cases carbon dioxide is the solvent used.

The reasons for this behaviour are not yet fully understood, but are certainly related to the high solubility of triglycerides in supercritical alcohol and solvent effects (Madras et al., 2004). Some authors have observed a dependence on the type of alcohol and triglyceride used (Alonso, 2007).

Notwithstanding its clear advantages over other processes, significant hurdles remain for full scale implementation of supercritical production units. First of all, high temperatures and pressures are necessary to ensure that the alcohol is in supercritical state. Usually temperatures and pressures of more than 550 K and 100 atm are necessary, requiring the utilization of special equipment designed to support these conditions. This will lead to high equipment and operational costs, making the process economics not so attractive when compared to other options. Also, the excess of methanol used in the reaction is much larger when compared to the conventional process, normally oil/methanol molar ratio of 1:42 is used, which needs to be recovered and recycled back to the reactor further complicating the process design.

3.3.2.4. *Novel reactors with improved mixing*
Other alternatives are being proposed for biodiesel production. In this section some of them are presented and its merits evaluated.

New reactor designs and mixing systems such as ultrasonic mixing and cavitation reactors are examples of such processes. Although these may be more complex to operate than simple stirred reactors, they can dramatically increase the reaction rate and the time needed for a complete conversion. For example, ultrasounds applied to chemical reactions (also called sonochemistry) promote the acoustic cavitation of fluids (Quintana, 2002; Santos *et al.*, 2009; Colucci *et al.*, 2005; Kalva *et al.*, 2008; Deshmane *et al.*, 2009). The acoustic cavitation phenomenon provides the mechanical energy for agitation and mixing of reagents and the required activation energy for initiating the transesterification reaction, increasing the mass transfer between the liquid phases and the reaction rate.

Colucci *et al.* (2005) report that with ultrasonic agitation, for a 6:1 methanol/oil ratio and an average temperature of 40°C, the reaction rate constants are three to five times higher than those reported in the literature for mechanical agitation, and conversion to esters is greater than 99.4% after about 15 min. These results confirm the existence of mass transfer limitations in transesterification reaction systems. The small bubbles created by the ultrasounds increase the interfacial area and mass transfer between the oil and alcohol phases, and consequently the reaction rate.

Cavitation reactors are other possibility that is receiving increased attention (Gogate, 2008; Gogate and Pandit, 2000a, 2000b; Gogate *et al.*, 2001; Gogate *et al.*, 2006; Gogate and Kabadi, 2009; Kelkar *et al.*, 2008). In practice they are based on the formation of an emulsion of bubbles originated by the fluid flow, and its implosive collapse and compression, producing intense local heating and cooling rates, high pressures and liquid jet streams. The reaction times are reduced to less than 30 sec, reducing the size of equipment and simplifying the purification and recycling process steps.

Due to the high interfacial area and reduced path for the diffusion of reactant molecules, the ultrasonic mixing and cavitation reactors allows the reaction to proceed faster and in a way less sensitive to the water and soap presence. This way it is possible to reduce the conventional 1 to 4 h of the batch processing time to min or even seconds, and change the production from batch to continuous flow processing. Because these reactor types have such a profound impact in reducing the reaction time, equipment can be much smaller for the same production capacity, the methanol/oil molar ratios can be reduced, easing the purification of biodiesel and the recycling of catalyst and unused alcohol, and reducing process investment and operational costs.

The ultrasonic mixing and cavitation reactors have already been well studied and implemented at industrial scale for performing different tasks, with various work intensities and production capacities. For example, Hielscher (http://www.hielscher.com) supplies industrial ultrasonic processing equipment worldwide and CAVITRON (www.cavitron.de) supplies cavitation rotary reactors and mixers.

When compared to ultrasound mixing, cavitation reactors have the advantage of being simpler and do not need any external equipment such as the ultrasound generator. Thus, they are easier to operate and control in a continuous mode. In this case to promote the mixture of oil, with alcohol and catalyst, special reactor design obstacles or static mixer units may be created such as for example the orifices or the immersed surfaces.

3.3.2.5. *In situ transesterification*

In situ transesterification is another production possibility that simply or even eliminate the need to perform the pre-processing of feedstocks, in particular the oil extraction and refining steps. The transesterification reaction is performed directly in the macerated oil seeds, such as soybeans flakes) or animal fats containing lipids material (Siler-Marinkovic and Tomasevic, 1998; Alonzo, 2007, Carrapiso *et al.*, 2000; Georgogianni *et al.*, 2008; Zeng, *et al.*, 2009; Mondala *et al.*, 2009).

Although already proposed some years ago, *in situ* transesterification has not yet been used extensively. Reasons for this may be the large molar ratios between oil and alcohol that are necessary to obtain full oil conversion, and the dependence on the seeds characteristics and its oil content (Siler-Marinkovic and Tomasevic, 1998). This method can be simple but it is still not fully worked out for practical applications and is not economic efficient.

3.3.2.6. *Microwave assisted transesterification*

Microwave assisted transesterification is another possibility for biodiesel production (Cravotto *et al.*, 2008; Lertsathapornsuk *et al.*, 2008; Perin *et al.*, 2008; Refaat *et al.*, 2008). Aczan and Danisman (2008) have considered microwave heating to perform the transesterification of rapeseed oil and showed that increased yields and reduced reaction times are possible.

3.3.2.7. Thermal cracking or pyrolysis

Thermal cracking (also called pyrolysis) is another possibility that it is currently pursued. Instead of transesterification reactions, now one is more interested in thermal decomposition or cleavage of the triglycerides and other organic compounds presented in the feedstock in simpler molecules, namely alkans, alkenes, aromatics, carboxylic acids, among others (Bahadur et al., 1995; Babu, 2008; Boateng et al., 2008).

Actually, pyrolysis is not a method for biodiesel production. However, as this process produces chemicals that can be used in diesel-like fuel, and the biodiesel companies are starting to diversify their feedstock and product portfolio, thermal cracking is increasingly seen as an interesting option. Yet, currently transesterification still appears to be the best option for biodiesel production in the large quantities required, and with the quality characteristics mandatory for its utilization in existing engines.

The composition of the products obtained by the pyrolysis is highly dependent on the feedstocks used, and conditions in which it is performed (Bahadur et al., 1995; Babu, 2008; Boateng et al., 2008). Thermal cracking can be used for a wide range of feedstocks, including many with negligible lipids content. From the resulting mixture besides fuel for transportation, also known as bio-oil, other valuable chemicals can be obtained (Abdullah and Gerhauser, 2008). However, the process normally requires high temperatures, with associated high energy costs, and is highly dependent on the conditions in which it is performed, making it difficult to control.

3.3.2.8. Catalytic cracking

A variant of thermal cracking is catalytic cracking, extensively used in the petrochemical industry to produce a significant percentage of the fossil fuels currently used. This possibility is also been pursued for the production of biodiesel from a wide variety of feedstocks. The utilization of a catalyst permits the utilization of milder conditions of temperature and pressure, with a better control of the resulting final products (Twaiq et al., 2004; Chew and Bhatia, 2009).

Different combinations of reactors and catalysts can be used, as for example pillared clays, alumina metal supported catalysts, zeolites, among others. Also, the huge experience gathered in the petrochemical industry can be relevant in the development and implementation of cracking processes for biodiesel production.

3.3.4. *Biodiesel post-treatment and blending*

Although different routes can be considered for biodiesel production the most commonly used at industrial scale is the homogeneous alkalis-catalyzed process. After the transesterification reaction, the mixture is allowed to separate into an upper layer of methyl esters and a lower layer of glycerine diluted with methanol. The unreacted methanol is then air-stripped or vacuum distilled away and recycled back to the process. Depending on the process, water can be used to wash catalyst residues and sodium soaps from the methyl esters. Also small amounts of concentrated phosphoric acid (H_3PO_4) can be added to the raw methyl esters to break catalyst residues and sodium soaps.

The glycerol, a by-product from biodiesel production, can be commercially valorised, *e.g.* as a raw-material for pharmaceutical, cosmetic or food industry. However, biodiesel development has created a flood in the glycerine supply resulting in a price collapse. Thus, there is a need to find alternative routes for glycerol valorisation. Currently, many biodiesel companies are burning glycerol for heat production in the biodiesel production process, as it is not economically viable to further process or purify it in order to meet the requirements imposed for its use in other applications.

Biodiesel can be blended with petroleum diesel or used as 100% biodiesel (called B100) in vehicles. Predominantly biodiesel is available in the market as mixtures formulated with petroleum diesel on a volume basis (v/v) to yield blends from B2 (2% biodiesel mixed with 98% petroleum diesel) to B99.9. In each country specific legislation defines limits for the biodiesel blends to be commercialised.

In the United States the standard ASTM D6751 specifies quality requirements for 100% biodiesel, B100. Recently the ASTM D7467-08 standard specification was approved for diesel fuel oil-biodiesel blend B6 to B20. In Europe the minimum quality requirements for 100% biodiesel are specified in the European standard EN 14214 which was created in collaboration with both engine manufacturers and biodiesel industry.

The European Fuel Standard EN 590 describes the physical properties that all diesel fuel must meet if it is to be sold in the European Union (EU), Iceland, Norway and Switzerland, allowing the addition of a maximum 5% biodiesel. In some European countries all diesel sold routinely contains this 95/5 mix. However biodiesel blends higher than B5 can also be commercialised with variations depending on the country. For example in France it is B30. In Germany it is common to use neat biodiesel (B100) instead of petroleum diesel. In Portugal it is possible to commercialise B100 and blends up to B15 according to national legislation. It is expected in the future that both standards and blends used will change, as a result of the evolution in the diesel engines and changes in regulations.

4. CONCLUSIONS

Biodiesel is already a viable option to address the existing dependence of the transportation sector on fossil fuels, while reducing its negative potential environmental impact, namely GHG and other pollutant emissions such as CO_2 and particulate matter. A wide variety of feedstocks can be used in their production, ranging from edible and non-edible vegetable oils to lipidic waste materials, which are renewable resources.

Despite the current developments concerning biodiesel production, it can be considered a relatively new area and many improvements and developments are expected in the near future throughout the entire biodiesel production chain. Starting with feedstocks, vegetal oil crops have been preferentially used, situation that will lead to severe problems as more tropical forests and peat lands with high biodiversity and carbon stock are cleared for its production, or as edible oil crops also used in human consumption are converted to biodiesel. To answer some of these questions new feedstock options with higher biomass productivity and oil yields than current feedstocks are being considered, as for example microalgae or other not directly integrated in the human food chain and with less land requirements.

Concerning the production processes, the large majority of production units currently operated, or even under construction, are based on the alkali homogenous catalysed process, operating many times in batch mode and taking a long time to ensure full reactants conversion. Also, the conventional process for biodiesel production is very sensitive to existing contaminants in particular to water that has a strong negative effect on the chemical reaction. Thus, there is a lot of room for process improvement, for example for the development of fast, continuous and robust processes that can handle a wide variety of feedstocks with different characteristics. Many options currently investigated and even under implementation are described in this article, some of them trying to improve existing processes, while others borrowing expertise and knowledge from other areas such as the petrochemical industry.

In the medium to long term future it is expected that the biodiesel production process will be further integrated in an ever increasingly bio based economy. Biomass conversion processes will be increasingly used to produce fuels, electricity, and all the chemicals needed for actual and future generations. The biorefinery concept is gaining momentum as a facility where biomass will be converted in a wide variety of products, in a way akin to a petroleum refinery, currently responsible for most of the fuels and essential to chemicals production. The biodiesel production, either based on the conversion of triglycerides or the production of other chemicals, such as esters and others, will have a natural role to play in an economy based on natural renewable resources and the with a more extensive use of biotechnology.

REFERENCES

1. Abdullah, N. and Gerhauser, H. 2008. Bio-oil derived from empty fruit bunches, *Fuel* **87**: 2606–13.
2. Agarwal, A.K. 2007. Biofuels (alcohols and biodiesel) applications as fuels for internal combustion engines, *Progr. Energ. Combust.* **33**: 233–71.

3. Akoh, C.C., Chang, S.W., Lee, G.C. and Shaw, J.F. 2007. Enzymatic Approach to biodiesel production, *J. Agr. Food Chem.* **55**: 8995–9005.
4. Albuquerque, M.C.G., Santamaría-González, J., Mérida-Robles, J.M., Moreno-Tost, R., Rodríguez-Castellón, E., Jiménez-López, A., Azevedo, D.C.S., Cavalcante, C.L. and Maireles-Torres, P. 2008. MgM (M = Al and Ca) oxides as basic catalysts in transesterification processes, *Appl. Catal. A-Gen.* **347(2)**: 162–8.
5. Allen, C.A.W., Watts, K.C., Ackmanb, R.G. and Pegg, M.J. 1999. Predicting the viscosity of biodiesel fuels from their fatty acid ester composition, *Fuel* **78**: 1319–26.
6. Almeida, R.M., Noda, L.K., Gonçalves, N.S., Meneghetti, S.M.P. and Meneghetti, M.R. 2008. Transesterification reaction of vegetable oils, using superacid sulfated TiO2-base catalysts, *Appl. Catal. A-Gen.* **347(1)**: 100–5.
7. Alonzo, D.E.L. 2007. Heterogeneous catalysis and biodiesel forming reactions. PhD Thesis in Chemical Engineering, Graduate School of Clemson University.
8. Al-Zuhair, S. 2007. Production of biodiesel: possibilities and challenges, *Biofuels, Bioproducts Biorefining* **1(1)**: 57–66.
9. Antunes, W.M., Veloso, C.O. and Henriques, C.A. 2008. Transesterification of soybean oil with methanol catalyzed by basic solids, *Catal. Today* **133–135**: 548–54.
10. Aranda, D.A.G., Santos, R.T.P., Tapanes, N.C.O., Ramos, A.L.D. and Antunes, O.A.C. 2008. Acid-Catalyzed Homogeneous Esterification Reaction for Biodiesel Production from Palm Fatty Acids, *Catal. Lett.* **122**: 20–25.
11. Arzamendi, G., Arguiñarena, E., Campo, I., Zabala, S. and Gandía, L.M. 2008. Alkaline and alkaline-earth metals compounds as catalysts for the methanolysis of sunflower oil, *Catal. Today* **133(135)**: 305–13.
12. Arzamendi, G., Campo, I., Arguiñarena, E., Sánchez, M., Montes, M. and Gandía, L.M. 2007. Synthesis of biodiesel with heterogeneous NaOH/alumina catalysts: Comparison with homogeneous NaOH, *Chem. Eng. J.* **134(1–3)**: 123–30.
13. Ataya, F. 2008. Mass Transfer Limitations in the Transesterification of Canola Oil to Fatty Acid Methyl Ester, PhD Thesis in Chemical Engineering, University of Ottawa.
14. Ataya, F., Dube, M.A. and Ternan, M. 2008. Transesterification of Canola Oil to Fatty Acid Methyl Ester (FAME) in a Continuous Flow Liquid-Liquid Packed Bed Reactor, *Energ. Fuels* **22**: 3551–6.
15. Azcan, N. and Danisman, A. 2008. Microwave assisted transesterification of rapeseed oil, *Fuel* **87**:1781–8.
16. Babu, B.V. 2008. Biomass pyrolysis: a state-of-the-art review, *Biofuels, Bioprod. Biorefining* **2(5)**: 393–414.
17. Bahadur, N.P., Boocock, D.G.B. and Konar, S.K. 1995. Liquid hydrocarbons from catalytic pyrolysis of sewage sludge lipid and canola oil: evaluation of fuel properties, *Energ. Fuels* **9**: 248–56.
18. Baker, C.J., Saxton, K.E., Chamen, W.C.T., Reicosky, D.C., Ribeiro, F., Justice, S.E. and Hobbs, P.R. 2006. No tillage seeding in conservation agriculture. 2nd Edition, CABI November.
19. Balat, M. 2008. Modeling vegetable oil viscosity, Energy Sources, Part A: Recovery, *Utilization and Environmental Effects* **30(20)**: 1856–69.
20. Barakos, N., Pasias, S. and Papayannakos, N. 2008. Transesterification of triglycerides in high and low quality oil feeds over an HT2 hydrotalcite catalyst, *Bioresource Technol.* **99**: 5037–42.
21. Benjapornkulaphong, S., Ngamcharussrivichai, C. and Bunyakiat, K. 2009. Al2O3-supported Alkali and Alkali Earth Oxides for Transesterification of Palm Kernel Oil and Coconut Oil, *Chem. Eng. J.* **145**: 468–74.

22. Bickel, U. and Dros, J.M. 2003. The impacts of soybean cultivation on Brazilian ecosystems. Three case studies. World Wildlife Fund. Forest Conversion Initiative.
23. Boateng, A.A., Mullen, C.A., Goldberg, N., Hicks, K.B., Jung, H.J.G. and Lamb, J.F.S. 2008. Production of Bio-oil from Alfalfa Stems by Fluidized-Bed Fast Pyrolysis, *Ind. Eng. Chem. Res.* **47**: 4115–22.
24. Boocock, D.G.B. 2003. Single phase process for production of fatty acid methyl esters from mixtures of triglycerides and fatty acids, US Patent 6642399B2.
25. Boocock, D.G.B. 2004. Process for production of fatty acid methyl esters from fatty acid triglycerides, Biox Corporation, US Patent 6712867.
26. Boocock, D.G.B., Konar, S.K., Mao, V. and Sidi, H. 1996. Fast one-phase oil-rich processes for the preparation of vegetable oil methyl esters, *Biomass and Bioener.* **11(1)**: 43–50.
27. Boz, N. and Kara, M. 2009. Solid base catalyzed transesterification of canola oil, *Chem. Eng. Commun.* **196(1)**: 80–92.
28. Bozbas, K. 2008. Biodiesel as an alternative motor fuel: Production and policies in the European Union, *Renewable and Sustainable Energy Reviews* **12**: 542–52.
29. Brito, A., Borges, M.E., Arvelo, R., Garcia, F., Diaz, M.C. and Otero, N. 2007. Reuse of fried oil to obtain biodiesel: zeolites y as a catalyst, *Int. J. Chem. React. Eng.* **5**: Article A104.
30. Canakci, M. 2007. The potential of restaurant waste lipids as biodiesel feedstocks, *Bioresource Technol.* **98(1)**: 183–90.
31. Canakci, M. and Sanli, H. 2008. Biodiesel production from various feedstocks and their effects on the fuel properties, *J. Ind. Microbiol. Biotechnol.* **35**: 431–41.
32. Canakci, M. and Van Gerpen, J. 1999. Biodiesel production via acid catalysis, *Trans. Am. Soc. Agr. Eng.* **42(5)**: 1203–10.
33. Canakci, M. and Van Gerpen, J. 2001. Biodiesel production from oils and fats with high free fatty acids, *Trans. Am. Soc. Agr. Eng.* **44(6)**: 1429–36.
34. Canakci, M. and Van Gerpen, J. 2003. A pilot plant to produce biodiesel from high free fatty acid feedstocks, *Trans. Am. Soc. Agr. Eng.* **46(4)**: 945–54.
35. Cao, W., Han, H. and Zhang, J. 2005. Preparation of Biodiesel from soybean oil using supercritical methanol and co-solvent, *Fuel* **84(4)**: 347–51.
36. Carrapiso, A.I., Timón, M.L., Petrón, M.J., Tejeda, J.F. and García, C. 2000. In situ transesterification of fatty acids from Iberian pig subcutaneous adipose tissue, *Meat Sci.* **56**: 159–64.
37. Chen, W., Wang, C., Ying, W., Wang, W., Wu, Y. and Zhang, J. 2009. Continuous production of biodiesel via supercritical methanol transesterification in a tubular reactor. part 1: thermophysical and transitive properties of supercritical methanol, *Energ. Fuels* **23**: 526–32.
38. Chew, T.L. and Bhatia, S. 2009. Effect of catalyst additives on the production of biofuels from palm oil cracking in a transport riser reactor, *Bioresource Technol.* **100**: 2540–5.
39. Chisti, Y. 2007. Biodiesel from microalgae, *Biotechnol. Adv.* **25(3)**: 294–306.
40. Chung, K.H., Chang, D.R. and Park, B.G. 2008. Removal of free fatty acid in waste frying oil by esterification with methanol on zeolite catalysts, *Bioresource Technol.* **99**: 7438–43.
41. Colucci, J.A., Borrero, E.E. and Alape, F. 2005. Biodiesel from an alkaline transesterification reaction of soybean oil using ultrasonic mixing, *JAOCS*, **82(7)**: 525–30.
42. Cravotto, G., Boffa, L., Mantegna, S., Perego, P., Avogadro, M. and Cintas, P. 2008. Improved extraction of vegetable oils under high-intensity ultrasound and/or microwaves, *Ultrason. Sonochem.* **15(5)**: 898–902.

43. Delucchi, M.A. 2003. A lifecycle emissions model (LEM): lifecycle emissions from transportation fuels, motor vehicles, transportation modes, electricity use, heating and cooking fuels, Main Report UCD-ITS-RR-03-17.
44. Demirbas, A. 2008. Comparison of transesterification methods for production of biodiesel from vegetable oils and fats, *Energ. Convers. Manage.* **49(1)**: 125–30.
45. Demirbas, A. 2009. Production of biodiesel fuels from linseed oil using methanol and ethanol in non-catalytic SCF conditions, *Biomass Bioenerg.* **33**: 113–8.
46. Deshmane, V.G., Gogate, P.R. and Pandit, A.B. 2009. Ultrasound-assisted synthesis of biodiesel from palm fatty acid distillate, *Ind. Eng. Chem. Res.* (in Press).
47. Dewulf, J. and Van Langenhove, H. 2006. Renewables-Based Technology: Sustainability Assessment, John Wiley & Sons Ltd., West Sussex PO19 8SQ, England.
48. Di Serio, M., Tesser, R., Pengmei, L. and Santacesaria, E. 2008. Heterogeneous Catalysts for Biodiesel Production, *Energ. Fuels* **22**: 207–17.
49. Doll, K.M., Sharma, B.K., Suarez, P.A.Z. and Erhan, S.Z. 2008. Comparing biofuels obtained from pyrolysis, of soybean oil or soapstock, with traditional soybean biodiesel: density, kinematic viscosity, and surface tensions, *Energ. Fuels* **22**: 2061–6.
50. Dossat, V., Combes, D. and Marty, A. 1999. Continuous enzymatic transesterification of high oleic sunflower oil in a packed bed reactor: influence of the glycerol production, *Enzyme Microb. Tech.* **25**: 194–200.
51. Eijsink, V.G.H., Vaaje-Kolstad, G., Varum, K.M. and Horn, S.J. 2008. Towards new enzymes for biofuels: lessons from chitinase research, *Trends Biotechnol.* **26(5)**: 228–35.
52. FAO, 2008. The state of food and agriculture, biofuels: prospects, risks and opportunities, Rome, Italy.
53. FAO, 2009. Oilseeds, oils and meals, food and agriculture organization of the United Nations. URL: http://www.fao.org/docrep/011/ai474e/ai474e07.htm#33.
54. Fjerbaek, L., Christensen, K.V. and Norddahl, B. 2009. A review of the current state of biodiesel production using enzymatic transesterification, *Biotechnol. Bioeng.* **102(5)**: 1298–1315.
55. Geissler, M. 2008. Biodiesel patterns reflect quality. LC-GC Europe, Applications Book, pp. 44–45.
56. Georgogianni, K.G., Kontominas, M.G., Pomonis, P.J., Avlonitis, D. and Gergis, V. 2008. Alkaline conventional and *in situ* transesterification of cottonseed oil for the production of biodiesel, *Energ. Fuels* **22(3)**: 2110–5.
57. Gilbert, R. and Perl, A. 2008. Transport revolutions: Moving people and freight without oil. Earthscan, UK.
58. Girard, P. and Fallot, A. 2006. Review of existing and emerging technologies for the production of biofuels in developing countries, *Energ. Sust. Dev.* **10(2)**: 92–108.
59. Gogate, P.R. 2008. Cavitational reactors for process intensification of chemical processing applications: A critical review, *Chem. Eng. Process.* **47**: 515–27.
60. Gogate, P.R. and Kabadi, A.M. 2009. A review of applications of cavitation in biochemical engineering/biotechnology, *Biochem. Eng. J.* **44**: 60–72.
61. Gogate, P.R. and Pandit, A.B. 2000a. Engineering design method for cavitational reactors: I-Sonochemical reactors, *AIChE J.* **46(2)**: 372–9.
62. Gogate, P.R. and Pandit, A.B. 2000b. Engineering design method for cavitational reactors: II-Hydrodynamic cavitation, *AIChE J.* **46(8)**: 1641–9.
63. Gogate, P.R., Shirgaonkar, I.Z., Sivakumar, M., Senthilkumar, P., Vichare, N.P. and Pandit, A.B. 2001. Cavitation reactors: efficiency assessment using a model reaction, *AIChE J.* **47(11)**: 2526–38.

64. Gogate, P.R., Tayal, R.K. and Pandit, A.B. 2006. Cavitation: A technology on the horizon, *Current Sci.* **91(1)**: 35–46.
65. Grima, M.E., Belarbi, E.H., Fernaìndez, F.G.A., Medina, A.R. and Chisti, Y. 2003. Recovery of microalgal biomass and metabolites: Process options and economics, *Biotechnol. Adv.* **20(7–8)**: 491–515.
66. Guan, G., Kusakabe, K., Sakurai, N. and Moriyama, K. 2009a. Transesterification of vegetable oil to biodiesel fuel using acid catalysts in the presence of dimethyl ether, *Fuel* **88**: 81–86.
67. Guan, G., Sakurai, N. and Kusakabe, K. 2009b. Synthesis of biodiesel from sunflower oil at room temperature in the presence of various cosolvents, *Chem. Eng. J.* **146**: 302–6.
68. Han, H., Cao, W. and Zhang, J. 2005. Preparation of biodiesel from soybean oil using supercritical methanol and CO_2 as co-solvent, *Process Biochem.* **40(9)**: 3148–51.
69. Han, M., Yi, W., Wu, Q., Liu, Y., Hong, Y. and Wang, D. 2009. Preparation of biodiesel from waste oils catalyzed by a Brønsted acidic ionic liquid, *Bioresource Technol.* **100**: 2308–10.
70. Hodaifa, G., Martiìnez, M.E. and Saìnchez, S. 2008. Use of industrial wastewater from olive-oil extraction for biomass production of *Scenedesmus obliquus*, *Bioresource Technol.* **99(5)**: 1111–7.
71. Ibrahim, N.A., Guo, Z. and Xu, X. 2008. Enzymatic interesterification of palm stearin and coconut oil by a dual lipase system, *J. Am. Oil Chem. Soc.* **85**: 37–45.
72. Ioannidou, O., Zabaniotou, A., Antonakou, E.V., Papazisi, K.M., Lappas, A.A. and Athanassiou, C. 2009. Investigating the potential for energy, fuel, materials and chemicals production from corn residues (cobs and stalks) by non-catalytic and catalytic pyrolysis in two reactor configurations, *Renew. Sust. Energ. Rev.* **13**: 750–62.
73. Issariyakul, T., Kulkarni, M.G., Dalai, A.K. and Bakhshi, N.N. 2007. Production of biodiesel from waste fryer grease using mixed methanol/ethanol system, *Fuel Process. Technol.* **88(5)**: 429–36.
74. Kaieda, M., Samukawa, T., Kondo, A. and Fukuda, H. 2001. Effect of methanol and water contents on production of biodiesel fuel from plant oil catalyzed by various lipases in a solvent-free system, *J. Biosci. Bioeng.* **91(1)**: 12–15.
75. Kalva, A., Sivasankar, T. and Moholkar, V.S. 2008. Physical mechanism of ultrasound-assisted synthesis of biodiesel, *Ind. Eng. Chem. Res.* **48(1)**: 534–44.
76. Kelkar, M.A., Gogate, P.R. and Pandit, A.B. 2008. Intensification of esterification of acids for synthesis of biodiesel using acoustic and hydrodynamic cavitation, *Ultrason. Sonochem.* **15**: 188–94.
77. Kiss, A.A., Omota, F., Dimian, A.C. and Rothenberg, G. 2006. The heterogeneous advantage: biodiesel by catalytic reactive distillation, *Topics Catal.* **40(1–4)**: 141–50.
78. Kuramochi, H., Maeda, K., Osako, M., Nakamura, K. and Sakai, S. 2008. Superfast transesterification of triolein using dimethyl ether and a method for high-yield transesterification, *Ind. Eng. Chem. Res.* **47(24)**: 10076–9.
79. Kurki, A., Hill, A. and Morris, M. 2006. Biodiesel: the sustainability dimensions, ATTRA-National Sustainable Agriculture Information Service pp. 1–12.
80. Laherrere, J. 2005. Forecasting production from discovery, ASPO Lisbon May pp. 19–20. URL: http://www.peak-oil-crisis.com Laherrere_PeakOilReport May2005.pdf.
81. Lertsathapornsuk, V., Pairintra, R., Aryusuk, K. and Krisnangkura, K. 2008. Microwave assisted in continuous biodiesel production from waste frying palm oil and its performance in a 100 kW diesel generator, *Fuel Process. Technol.* **89**: 1330–6.

82. Liu, Y., Lotero, E., Goodwin, J.G. Jr. and Lu, C. 2007. Transesterification of triacetin using solid Brønsted bases, *J. Catal.* **246**: 428–33.
83. Madras, G., Kolluru, C. and Kumar, R. 2004. Synthesis of Biodiesel in supercritical fluids, *Fuel* **83(14–15)**: 2029–33.
84. Mao, V.W.L. 1995. Biodiesel Kinetics: One-phase Base-catalyzed Methanolysis of Soybean oil Using Tetrahydrofuran as Co-solvent. Master Thesis, Dept. Chemical Engineering and Applied Chemistry, University of Toronto, Canada.
85. Ming, L.O., Ghazal, H.M. and Letb, C.C. 1998. Effect of enzymatic transesterification on the fluidity of palm stearin-palm kernel olein mixtures, *Food Chem.* **63(2)**: 155–9.
86. Mondal, P., Basu, M. and Balasubramanian, N. 2008. Direct use of vegetable oil and animal fat as alternative fuel in internal combustion engine, *Biofuels Bioprod. Biorefining* **2**: 155–74.
87. Mondala, A., Liang, K., Toghiani, H., Hernandez, R. and French, T. 2009. Biodiesel production by *in situ* transesterification of municipal primary and secondary sludges, *Bioresource Technol.* **100**: 1203–10.
88. Montero, J.M., Gai, P., Wilson, K. and Lee, A.F. 2009. Structure-sensitive biodiesel synthesis over MgO nanocrystals, *Green Chem.* **11**: 265–8.
89. Murakami, M., Yamada, F., Nishide, T., Muranaka, T., Yamaguchi, N. and Takimoto, Y. 1998. The biological CO_2 fixation using Chlorella sp. with high capability in fixing CO_2, *Stud. Surf. Sci. Catal.* **114**: 315–20.
90. Noureddini, H. and Zhu, D. 1997. Kinetics of transesterification of soybean oil, *J. Am. Oil Chem. Soc.* **74(11)**: 1457–63.
91. Noureddini, H., Harkey, D. and Medikonduru, V. 1998. A continuous process for the conversion of vegetable oils into methyl esters of fatty acids, *J. Am. Oil Chem. Soc.* **75(12)**: 1775–83.
92. Oliveira, A.C. and Rosa, M.F. 2006. Enzymatic transesterification of sunflower oil in an aqueous-oil biphasic system, *JAOCS* **83(1)**: 21–25.
93. Otera, J. 2003. Esterification: Methods, Reactions and Applications. Wiley-VCH Verlag, Weinheim.
94. Perin, G., Álvaro, G., Westphal, E., Viana, L.H., Jacob, R.G., Lenardão, E.J. and D'Oca, M.G.M. 2008. Transesterification of castor oil assisted by microwave irradiation, *Fuel* **87**: 2838–41.
95. Quintana, E.E.B. 2002. Optimization Studies for the Alkaline Transesterification Biodiesel Reaction Using Ultrasound Mixing. Master Thesis in Chemical Engineering, University of Puerto Rico.
96. Ranganathan, S.V., Narasimhan, S.L. and Muthukumar, K. 2008. An overview of enzymatic production of biodiesel, *Bioresource Technol.* **99**: 3975–81.
97. Rathore, V. and Madras, G. 2007. Synthesis of biodiesel from edible and non-edible oils in supercritical alcohols and enzymatic synthesis in supercritical carbon dioxide, *Fuel* **86(17–18)**: 2650–9.
98. Refaat, A.A., El Sheltawy, S.T. and Sadek, K.U. 2008. Optimum reaction time, performance and exhaust emissions of biodiesel produced by microwave irradiation, *Int. J. Environ. Sci. Tech.* **5(3)**: 315–22.
99. Rice, B., Frohlich, A., Leonard, R. and Korbitz, W. 1997. Bio-diesel production based on waste cooking oil: promotion of the establishment of an industry in Ireland. Final Report, ALTENER CONTRACT No. XVII/4.1030/AL/77/95/IRL.
100. Richmond, A. 2004. Handbook of Microalgal Culture: Biotechnology and Applied Phycology. Blackwell Science Ltd.
101. Santos, F.F.P., Rodrigues, S. and Fernandes, F.A.N. 2009. Optimization of the production of biodiesel from soybean oil by ultrasound assisted methanolysis, *Fuel Process. Technol.* **90**: 312–6.

102. Sawangkeaw, R., Bunyakiat, K. and Ngamprasertsith, S. 2007. Effect of co-solvents on production of biodiesel via transesterification in supercritical methanol, *Green Chem.* **9(6)**: 679–85.
103. Schenk, P.M., Thomas-Hall, S.R., Stephens, E., Marx, U.C., Mussgnug, J.H., Posten, C., Kruse, O. and Hankamer, B. 2008. Second generation biofuels: high-efficiency microalgae for biodiesel production, *Bioenerg. Res.* **1**: 20–43.
104. Shaw, J.F., Chang, S.W., Lin, S.C., Wu, T.T., Ju, H.Y., Akoh, C.C., Chang, R.H. and Shieh, C.J. 2008. Continuous enzymatic synthesis of biodiesel with novozym 435, *Energ. Fuels* **22**: 840–4.
105. Shi, H. and Bao, Z. 2008. Direct preparation of biodiesel from rapeseed oil leached by two-phase solvent extraction, *Bioresource Technol.* **99**: 9025–8.
106. Shimada, Y., Watanabe, Y., Sugihara, A. and Tominaga, Y. 2002. Enzymatic alcoholysis for biodiesel fuel production and application of the reaction to oil processing, *J. Mol. Catal. B- Enzym.* **17**: 133–42.
107. Siler-Marinkovic, S. and Tomasevic, A. 1998. Transesterification of sunflower oil *in situ*, *Fuel* **77(12)**: 1389–91.
108. Soetaert, W. and Vandamme, E.J. 2009. Biofuels. John Wiley & Sons Ltd.
109. Toman, M., Curtright, A.E., Ortiz, D.S., Darmstadter, J. and Shannon, B. 2008. Unconventional fossil-based fuels economic and environmental trade-offs, sponsored by the national commission on energy policy. RAND Corporation.
110. Twaiq, F.A.A., Mohamad, A.R. and Bhatia, S. 2004. Performance of composite catalysts in palm oil cracking for the production of liquid fuels and chemicals, *Fuel Process. Technol.* **85**: 1283–1300.
111. Van Gerpen, J. 2005. Biodiesel processing and production, *Fuel Process. Technol.* **86**: 1097–1107.
112. Van Gerpen, J.V., Shanks, B., Pruszko, R., Clements, D. and Knothe, G. 2004. Biodiesel production technology, national renewable energy laboratory, NREL/SR-510-36244, Colorado, USA.
113. Watanabe, Y., Pinsirodom, P., Nagao, T., Yamauchi, A., Kobayashi, T., Nishida, Y., Takagi, Y. and Shimada, Y. 2007. Conversion of acid oil by-produced in vegetable oil refining to biodiesel fuel by immobilized Candida antarctica lipase, *J. Mol. Catal. B-Enzym.* **44(3–4)**: 99–105.
114. Watanabe, Y., Shimada, Y., Sugihara, A. and Tominaga, Y. 2001. Enzymatic conversion of waste edible oil to biodiesel, fuel in a fixed-bed bioreactor, *JAOCS* **78(7)**: 703–7.
115. Watanabe, Y., Shimada, Y., Sugihara, A., Noda, H., Fukuda, H. and Tominaga, Y. 2000. Continuous production of biodiesel fuel from vegetable oil using immobilized candida antarctica lipase, *JAOCS* **77(4)**: 355–60.
116. Yin, J.Z., Xiao, M. and Song, J.B. 2008. Biodiesel from soybean oil in supercritical methanol with co-solvent, *Energ. Convers. Manage.* **49**: 908–12.
117. Zabaniotou, A., Ioannidou, O. and Skoulou, V. 2008. Rapeseed residues utilization for energy and 2nd generation biofuels, *Fuel* **87**: 1492–1502.
118. Zeng, J., Wang, X., Zhao, B., Sun, J. and Wang, Y. 2009. Rapid *in situ* transesterification of sunflower oil, *Ind. Eng. Chem. Res.* **48**: 850–6.
119. Zhang, Y., Dubé, M.A., McLean, D.D. and Kates, M. 2003. Biodiesel production from waste cooking oil: 1-Process design and technological assessment, *Bioresource Technol.* **89**: 1–16.

Index

A
Acoustic drying 170
Air drying 166, 168
Al/SiC 228,
Average value 141
Axial/longitudinal dispersion 179, 182

B
Biodiesel 313, 314

C
Catalyst 315, 326
Cavitation 244, 245
Cellular material 141
Chemical reaction 293, 300
Chemical Vapor Deposition (CVD) 208, 251
Composite channel 271, 288
Composites 107, 126
Computed tomography 107, 119
Conjugated problem 287
Channel flow 286
Continuum flow 2
Coordinate system 279, 286
Correlation 179, 271, 273
Critical point 293, 295
Crystallite 208, 213, 220
Crytallinity degree 224

D
Diffusion 158, 161
Dry cleaning 293, 300
Drying kinetics 161, 162

E
Effective thermal conductivity 108, 131, 132
Electrochemistry 32, 44
Energy 24, 25
Energy demand 156,
Energy recovery system 271
Engineering system 271
Equilibrium grain boundaries 49, 95
Esterification 315, 321
Extraction 293, 296

F
Fatty acid alkyl esters 315
Finite Element method 105, 110, 140, 149
Fluid chromatography 293, 302
Forchheimer drag 2, 18
Free fatty acids 315, 321
Fuel Cell 21, 24

G
Gas flow 190, 198
Gas sensors 240, 264
Gold 263
Grain boundaries 49, 50

H
Heat transfer coefficient 154, 159
Homogenised 148,
Hydrogen 23, 24

Hydroxyapatite 220, 262

I

Impregnation 300, 304
Industrial (Commercial) application 329
Inertial effects 3
Inter-crystalline diffusion 86
Interface 273, 275
Isosteric heat 161

K

Knudsen number 1, 2

L

Lattice monte carlo method 105, 107
Lattice-Boltzmann 16
Lightweight construction 146
Liquid flow 186, 199

M

Mechanical alloying 209, 210
Modeling 21, 26, 271, 273
Molecular self-assembly 209, 208
Mössbauer spectroscopy 64, 69

N

Nano particles formation 311
Nanoamorphous 247, 258
Nanochemistry 232
Nanocomposite 212, 220
Nanocrystalline materials 69, 80
Nanodiffusion 255, 258
Nanoparticle 208, 232, 244, 251
Nanostructured Materials 95, 208, 248
Non equilibrium grain boundaries 94, 95
Numerical methods 117, 119, 281

O

Optimization 153, 161

P

Packed beds 181, 184
Peclet number 179, 180
Perforated hollow sphere structure 131
Periodic model structure 149
Permeability 1, 2
Phase change material 108, 121
Porous materials 36
Porous media 1, 2, 179, 181, 271, 272
Production chain 313, 315
Production processes 313, 317

R

Radial/transverse dispersion 179, 184, 193
Reynolds number 2, 3, 180, 181

S

Sandwich panel 149
Schmidt number 179, 180
SiC 228
Silica 236
Slip-flow 10, 14
SnO/ZnO 248
Sol-Gel 208, 232
Solid Oxide 21, 24
Sonochemistry 207, 244
Supercritical carbon dioxide 300, 304
Supercritical fluid 293, 294
Supercritical water 298, 303

T

Temperature dependencies 107, 115
Tetraethoxisilane (TEOS) 236, 237
Thermal analysis 107, 108
Tissue engineering 261
Transesterification 315, 321
Transient analysis 107
Transport processes 21, 30

Transport resistances 161
Triglycerides 315, 318
Turbulence 271, 273

U

Unit cell 131, 132

W

Water activity 153, 154

Z

ZnO 240